Workbook 3

Hydraulic Fluids and Contamination Control

Dr. Medhat Kamel Bahr Khalil, Ph.D, CFPHS, CFPAI.
Director of Professional Education and Research Development,
Applied Technology Center, Milwaukee School of Engineering,
Milwaukee, WI, USA.

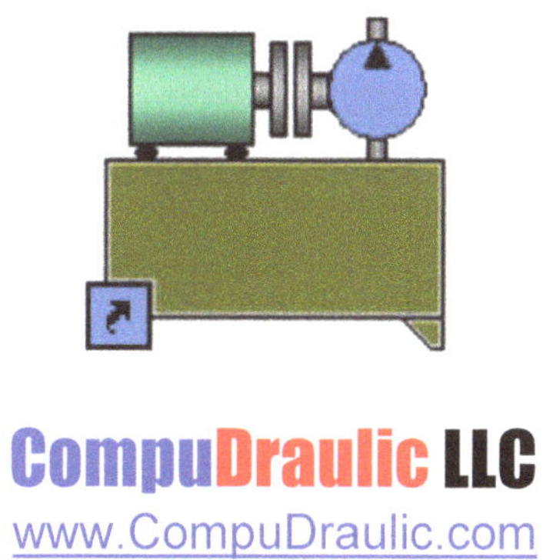

CompuDraulic LLC
www.CompuDraulic.com

CompuDraulic LLC

Workbook 3

Hydraulic Fluids and Contamination Control

ISBN: 978-0-9977816-4-9

Printed in the United States of America
First Published by October 2019
Revised by Sept. 2023

3850 Scenic Way, Franksville, WI, 53126 USA.
www.compudraulic.com

Disclaimer

It is always advisable to review the relevant standards and the recommendations from the system manufacturer. However, the content of this book provides guidelines based on the author's experience.

Any portion of information presented in this book could be not applicable for some applications due to various reasons. Since errors can occur in circuits, tables, and text, the publisher assumes no liability for the safe and/or satisfactory operation of any system designed based on the information in this book.

The publisher does not endorse or recommend any brand name product by including such brand name products in this book. Conversely the publisher does not disapprove any brand name product by not including such brand name in this book. The publisher obtained data from catalogs, literatures, and material from hydraulic components and systems manufacturers based on their permissions. The publisher welcomes additional data from other sources for future editions.

Workbook 3
Hydraulic Fluids and Contamination Control
Table of Contents

PREFACE

This Workbook is a complementary part to the textbook of the same title. This book is used as a workbook for students to take notes during the course delivery. It contains colored printout of the PowerPoint slides that are designed to present the course. Each chapter is followed by a number of review questions and assignments for homework.

Dr. Medhat Kamel Bahr Khalil

Chapter 1
Introduction

Objectives:

This chapter introduces the scope of hydraulic fluids conditioning and contamination control. The chapter also overviews various organizations who are involved in developing standards and set standard test methods for fluid power components and systems.

0

0

Brief Contents:

1

1

Hydraulic fluid contamination is defined as?

A. Air or water content in hydraulic fluid.
B. Overheated hydraulic fluid.
C. Particulate contaminants in hydraulic fluid.
D. Any of the above mentioned.

2

2

Hydraulic fluid contamination can be broadly defined as any internal or external reason that can change the properties or performance

3

3

1.1- Hydraulic Fluids Conditioning and Contamination Control

The following matter of facts justify the importance of controlling *contamination* in hydraulic fluids:

- There is no single substance on the earth that is 100% pure.
- Hydraulic fluids are not an exception.
- About 80% of hydraulic systems failures are due to contamination.
- Cost of contamination goes beyond system failure, it affects system productivity.
- Many warranty claims are rejected because of contamination-related reasons.
- Solid particles are not the only contaminants in hydraulic fluids.
- Filtration is not the only action needed for controlling hydraulic fluid contamination.

Video 407 (1.5 min)

4

4

Which of the following statements is FALSE?

A. Proper filtration is the only way to keep the oil clean.
B. Hydraulic fluid contamination is the main source of hydraulic system failures.
C. Hydraulic fluid contamination could be physical or energetic.
D. Overheating hydraulic fluid is one form of contamination.

5

5

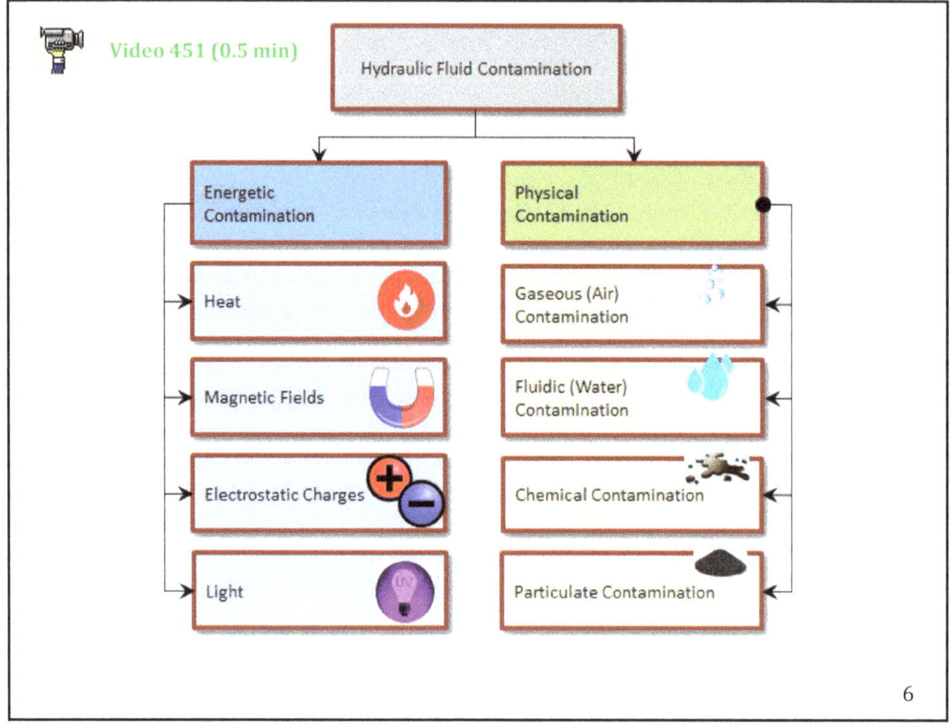

6

As a result of hydraulic fluid contamination, any of the following consequences can occure:

- Hydraulic fluid degradation, color change, depletion of additives, and growth of bacteria.
- Orifice blockage, loss of control, and improper actuator motion.
- Component wear, leakage, noise, and vibration.
- Reduction in components and system efficiency and loss of productivity.
- Increased operational cost due to high energy consumption, frequent oil and filter changes, costly flushing processes, and costly disposal products of used fluids and filters.
- Component failure, damage, pump cavitation, valve seizing, and sealing element failure.
- Possible unsafe operation of the machine.

7

7

Which item has the least effect on the overall cost of contamination?

A. Cost of contamination control.

B. Cost due to loss of system performance.

C. Cost due to inefficient components.

D. Cost due to system downtime.

8

8

1.2- Cost of Contamination

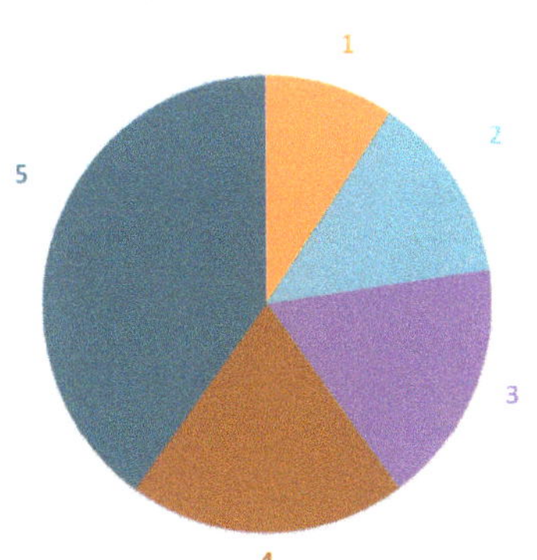

Fig. 1.2 - Cost Analysis for Contamination Control

1. **Cost of Contamination Control:** Standard filtration, condition monitoring, etc.
2. **Cost due to Loss of System Performance:** Slower actuators, less productivity, and general performance degradation.
3. **Cost due to Inefficient Components:** High energy consumption, high cooling demand, etc.
4. **Cost due to Equipment Repair:** Labor, components, filter and fluid cost, testing, etc.
5. **Cost due to System Downtime:** Lost production revenue, warranty claims, overhead costs, etc.

9

9

Which item has the highest effect on the overall cost of contamination?

A. Cost of contamination control.
B. Cost due to loss of system performance.
C. Cost due to inefficient components.
D. Cost due to system downtime.

10

10

Hydraulic Systems Volume 3: Hydraulic Fluids and Contamination Control

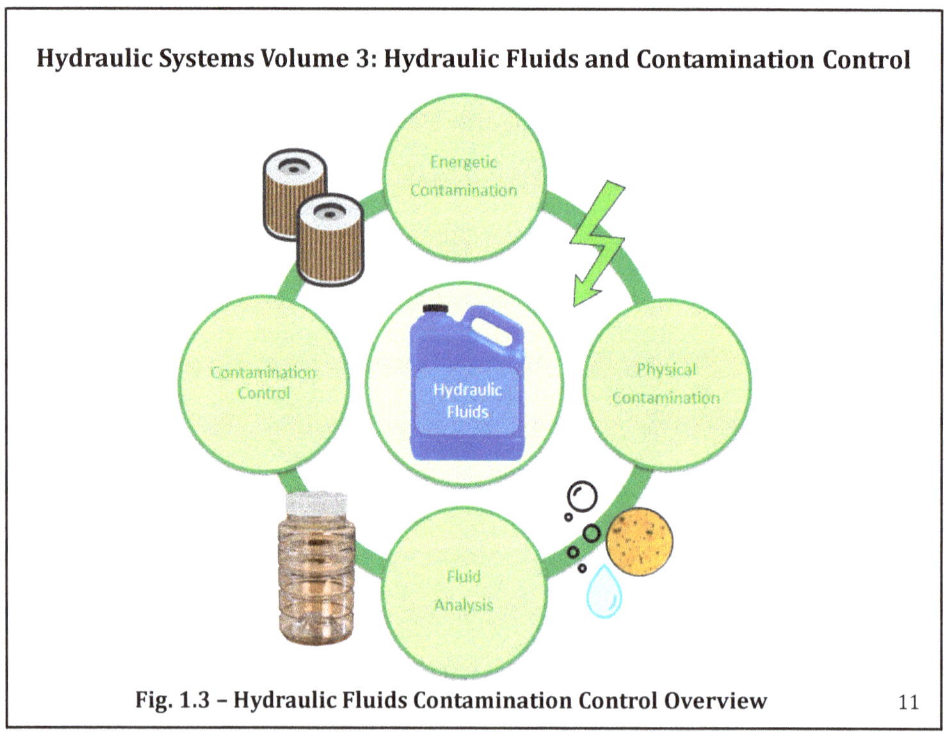

Fig. 1.3 – Hydraulic Fluids Contamination Control Overview

11

11

Hydraulic Systems Volume 4: Hydraulic Fluids Conditioning

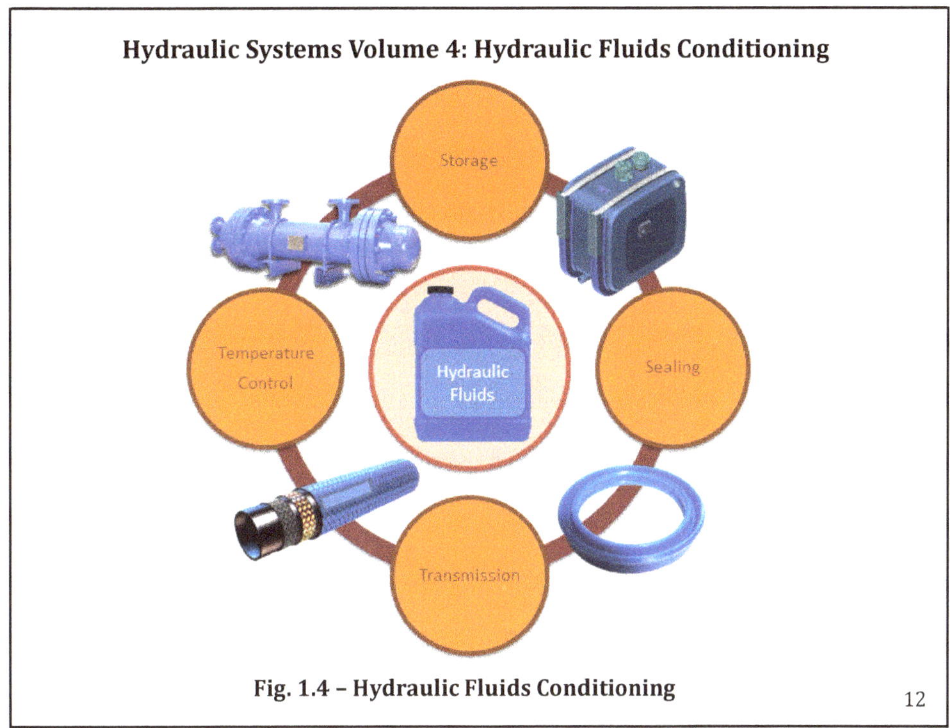

Fig. 1.4 – Hydraulic Fluids Conditioning

12

12

1.3- Sources for Standard Test Methods

1.3.1- International Organization for Standardization (ISO)
www.iso.org

1.3.2- American Society for Testing and Materials (ASTM)
www.astm.org

1.3.3- Society of Automotive Engineers (SAE)
www.sae.org

1.3.4- American National Standards Institute (ANSI)
www.ansi.org

1.3.5- German Institute for Standardization (DIN)
www.din.de/en

1.3.6- Fluid Power Systems and Components Vocabulary (ISO 5598)

13

13

Chapter 1 Reviews

1. Hydraulic fluid contamination is defined as?
   A. Air or water content in hydraulic fluid.
   B. Overheated hydraulic fluid.
   C. Particulate contaminants in hydraulic fluid.
   D. Any of the above mentioned.

2. Which of the following statements is FALSE?
   A. Proper filtration is the only way to keep the oil clean.
   B. Hydraulic fluid contamination is the main source of hydraulic system failures.
   C. Hydraulic fluid contamination could be physical or energetic.
   D. Overheating hydraulic fluid is one form of contamination.

3. Which item has the least effect on the overall cost of contamination?
   A. Cost of contamination control.
   B. Cost due to loss of system performance.
   C. Cost due to inefficient components.
   D. Cost due to system downtime.

4. Which item has the highest effect on the overall cost of contamination?
   A. Cost of contamination control.
   B. Cost due to loss of system performance.
   C. Cost due to inefficient components.
   D. Cost due to system downtime.

Chapter 1 Assignment

Student Name: -- Student ID: ------------------

Date: --- Score: -----------------------

Assignment 1: List the different various types of energetic and physical contaminations for hydraulic fluids that you know.

Assignment 2: List three standards that are involved in fluid power technology.

Chapter 2 Hydraulic Fluids

Objectives:

This chapter provides an overview of the commonly used hydraulic fluids including petroleum-based, water-based, chemical-based, fire-resistant, and environmental-friendly types of hydraulic fluids. The chapter discusses thoroughly 21 various properties and the relevant standard test methods of hydraulic fluids. Fluid properties are categorized as physical, thermal, and chemical properties. The chapter introduces the best practices for hydraulic fluid selection, replacement, and storage.

0

0

Brief Contents:

1

1

QUIZ

Which of the following statements describes liquid?

A. A substance that can't overcome shear forces.
B. A substance that takes the interior shape of a container with a flat surface.
C. A substance that has internal friction is represented by a property named "viscosity"
D. All the above.

2

2

2.1-Basic Definition

By definition, the term "*Fluid*" means a substance that:

- Continuously deform under shear stresses.
  - *Newtonian* Fluid (Constant rate of deformation).
  - *Non-Newtonian* Fluid (variable rate of deformation).
- Can't maintain a solid physical shape.
- Can be liquid or gas.

3

3

a ***Liquid*** means a substance that:

- Takes the shape of the container with a flat surface.
- Has surface tension phenomena.
- Is practically incompressible.
- Its viscosity is the most critical property.

a ***Gas*** means a substance that:

- Takes the full shape of the container, i.e. occupies the whole volume of the container.
- Has negligible surface tension.
- Is highly compressible.
- Gas compressibility is the most important property.

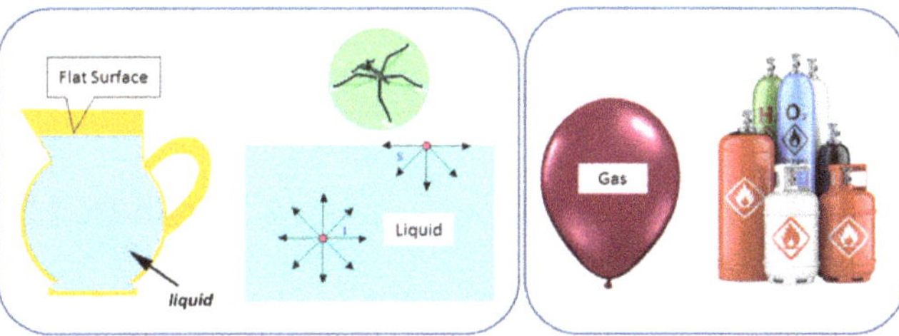

Fig. 2.1 – Characteristics of Liquids and Gases

4

4

Main duty of hydraulic fluid is to?

A. Resist foaming.

B. Transmit energy between the pump and the actuator

C. Resist fire.

D. Lubricate, cool, clean and seal inside the hydraulic devices.

5

5

Auxiliary duties of hydraulic fluid are to?

A. Resist foaming.
B. Transmit energy between the pump and the actuator
C. Resist fire.
D. Lubricate, cool, clean and seal inside the hydraulic devices.

6

6

2.2- Hydraulic Fluid Contribution

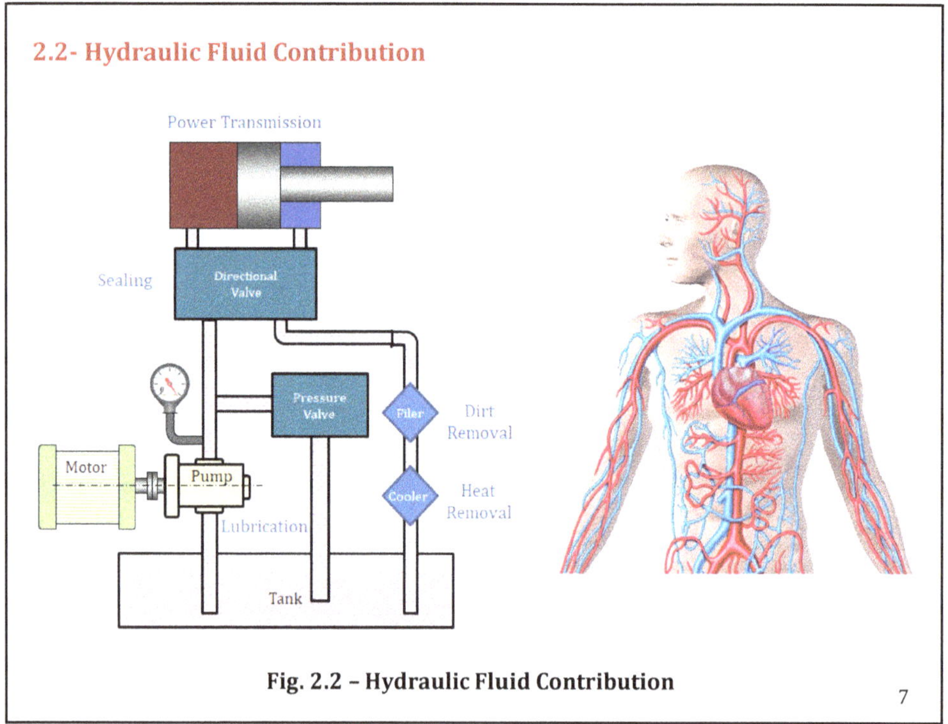

Fig. 2.2 – Hydraulic Fluid Contribution

7

7

2.3- Historical Background

- **Water**: used early at the beginning of the 18th century.
  - Example: 1872 retrofitting the mechanism of the London Bridge to operate with a water hydraulic system.
  - Water is: cheap, available, and fire-resistant.
  - Water is: a poor lubricant, corrodes the metal components, and contains significant contaminants.

Fig. 2.3 – Water Hydraulics used in the London Bridge

8

8

- **Petroleum-Based Fluids:** First use was in1920.
  - Good lubricants and
  - Preserve machinery.

- **Additives:** First use was in 1940.
  - Improve the properties of hydraulic fluids.

- **Fire-Resistant and Synthetic Fluids:** Developed in the 1940s
  - for applications that are associated with fire hazards.

- **Environmental-Friendly Fluids:** Were developed when the importance of keeping the environment clean and hydraulic systems were involved in some applications such as agricultural and off-shore.

9

9

2.4- Properties and Test Methods for Hydraulic Fluids

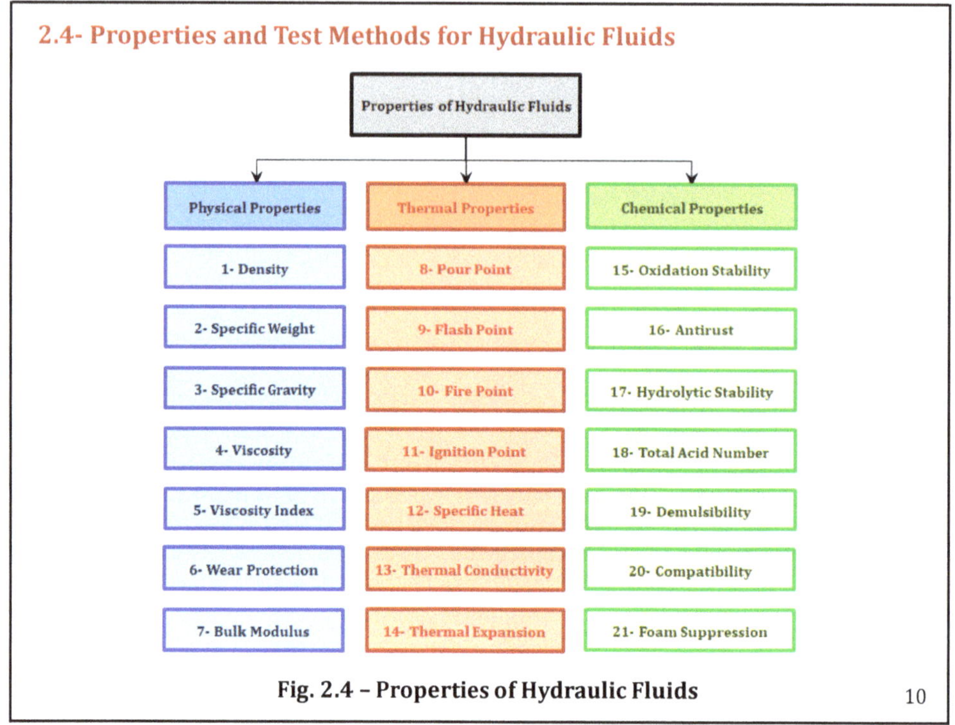

Fig. 2.4 – Properties of Hydraulic Fluids

10

10

2.4.1- Fluid Density

Density = Mass/Unit Volume.

$$\text{Density} = \rho \left[\frac{lb_m}{ft^3}\right] = \frac{\text{Mass}}{\text{Volume}} \quad 2.1.\mathbf{A}$$

$$\text{Density} = \rho \left[\frac{kg_m}{m^3}\right] = \frac{\text{Mass}}{\text{Volume}} \quad 2.1.\mathbf{B}$$

Density of Water = ρ_w = 62.4 [lb_m/ft^3] = 1000 [kg_m/m^3].

Hydraulic Fluid	Density [kg/m³] @16 °C (60 °F)
Mineral	870 - 900
Vegetable-Based	910 - 930
Water-Glycol	1060
Phosphate Esters	1150

Table 2.1 – Density of Hydraulic Fluids

11

11

2.4.2- Fluid Specific Weight

Specific Weight = Weight/Unit Volume.

$$\textbf{Specific Weight } \gamma \left[\frac{lb}{ft^3}\right] = \frac{\textbf{Weight}}{\textbf{Volume}} = \rho g \qquad 2.2.\textbf{A}$$

$$\textbf{Specific Weight } \gamma \left[\frac{kg}{m^3}\right] = \frac{\textbf{Weight}}{\textbf{Volume}} = \rho g \qquad 2.2.\textbf{B}$$

Where g = Gravitational Acceleration = 9.81 [m/s^2] = 32.2 [ft/s^2].

Specific Weight of water = γ_w = 62.4 [lb/ft^3] = 1000 [kg/m^3].

12

12

2.4.3- Fluid Specific Gravity

$$\textbf{Specific Gravity} = \textbf{SG} = \frac{\rho_f}{\rho_w} = \frac{\gamma_f}{\gamma_w} \qquad 2.3$$

- ❑ SG < 1 → fluid is less dense than water.
- ❑ SG >1 → fluid is denser than water.

Hydraulic Fluid	SG @ @16 °C (60 °F)
Mineral	0.87 – 0.9
Water-Based	1.0
Vegetable-Based	0.91 - 0.93
Water-Glycol	1.06
Phosphate Esters	1.150

Table 2.2 – Specific Gravity of Hydraulic Fluids

SG <1
Water
SG >1

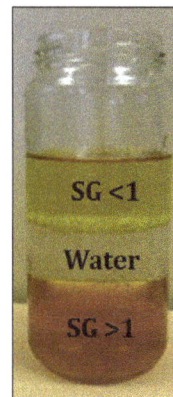

Fig. 2.5 – Specific Gravity Demonstration

Standard Test Method (ASTM D1298):

- This test use a calibrated hydrometer.
- Primarily used to determine % of water to fluid in water oil emulsions and water glycols.

13

13

2.4.4- Fluid Viscosity

Viscosity is the most important fluid property
Will be discussed more comprehensively than other properties.

2.4.4.1- Definition of Fluid Viscosity

- **Operational:** How fast a fluid can flow.
- **Chemical:** how strong the molecular bonds are in the chemical structure.
- **Physical:** Fluid resistivity to deformation under shear stresses.

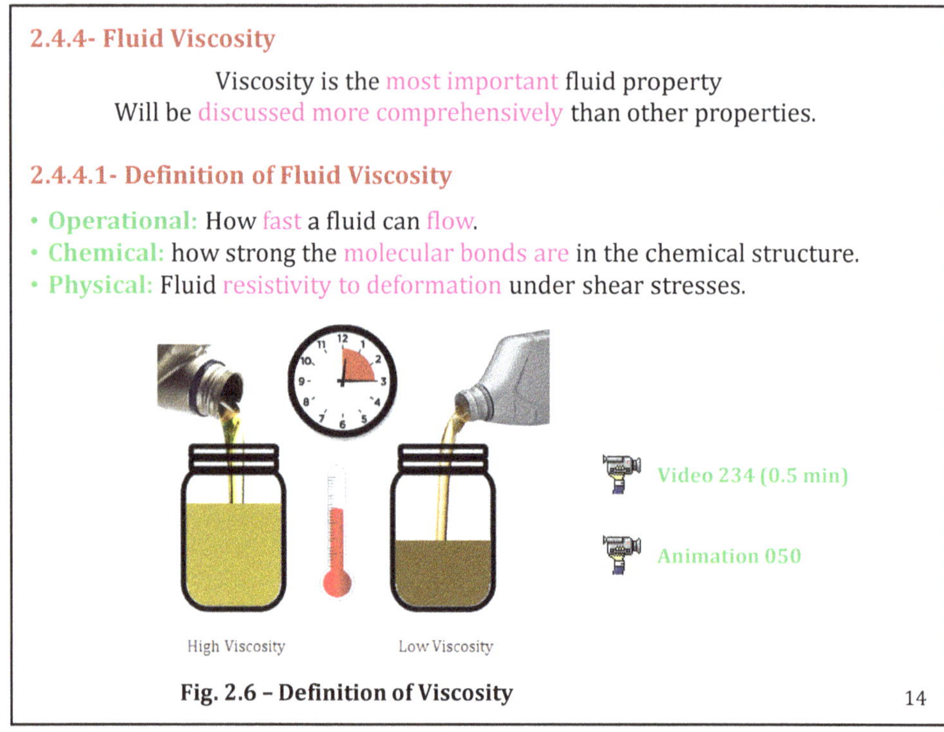

Fig. 2.6 – Definition of Viscosity

14

14

2.4.4.2- Effect of Fluid Viscosity on System Performance

Fluid Viscosity is a vital property for:

- Responsive power transmission.
- Component self-lubrication.
- System energy efficiency.
- System performance such as noise and reliability.

Fluid viscosity is a challenging property for:

- Designers to specify.
- Operators to maintain.

15

15

Which of the following viscosity ranges results in decreasing reliability, reducing productivity, and increasing power consumption of a hydraulic system?

A. Viscosity lower than the recommended range.

B. Viscosity higher than the recommended range.

C. Viscosity lower or higher than recommended range.

D. None of the above.

16

16

V701-LB-013-Viscosity and Leakage

Viscosity too Low	Viscosity too High
Insufficient lubrication	High fluid internal friction
Higher rate of wear	Higher pressure drops across lines and components
Higher noise level	Difficult machine starting
Leakage losses increase	Possibility of pump cavitation
Actuator speeds slowing down	Sluggish valve response
Volumetric eff. decrease	Mechanical eff. decreases
Higher power consumption	Higher power consumption
Inefficient system operation	Inefficient system operation
More heat generation	More heat generation
Less system reliability	Less system reliability

Table 2.3 – Symptoms when Fluid Viscosity Changes Beyond the Recommended Limits

Video 729

17

17

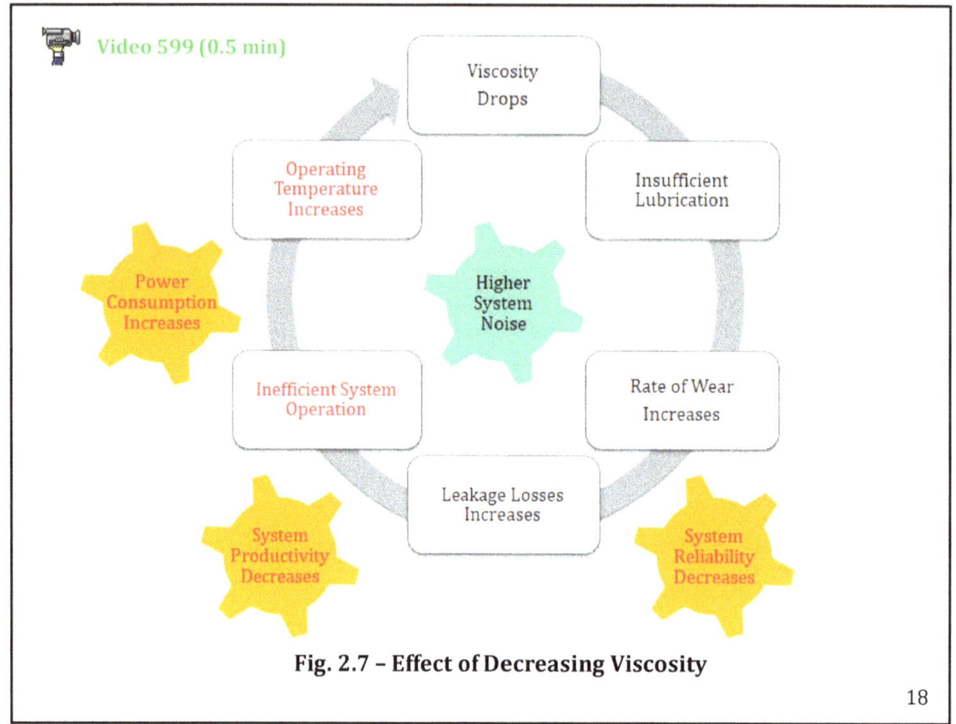

Fig. 2.7 – Effect of Decreasing Viscosity

18

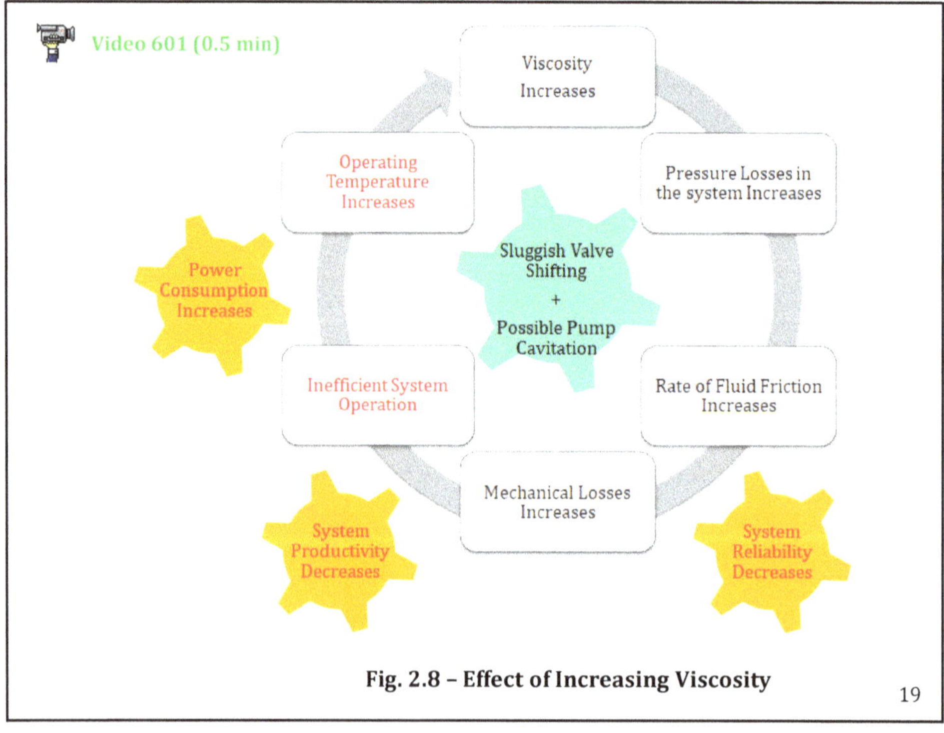

Fig. 2.8 – Effect of Increasing Viscosity

19

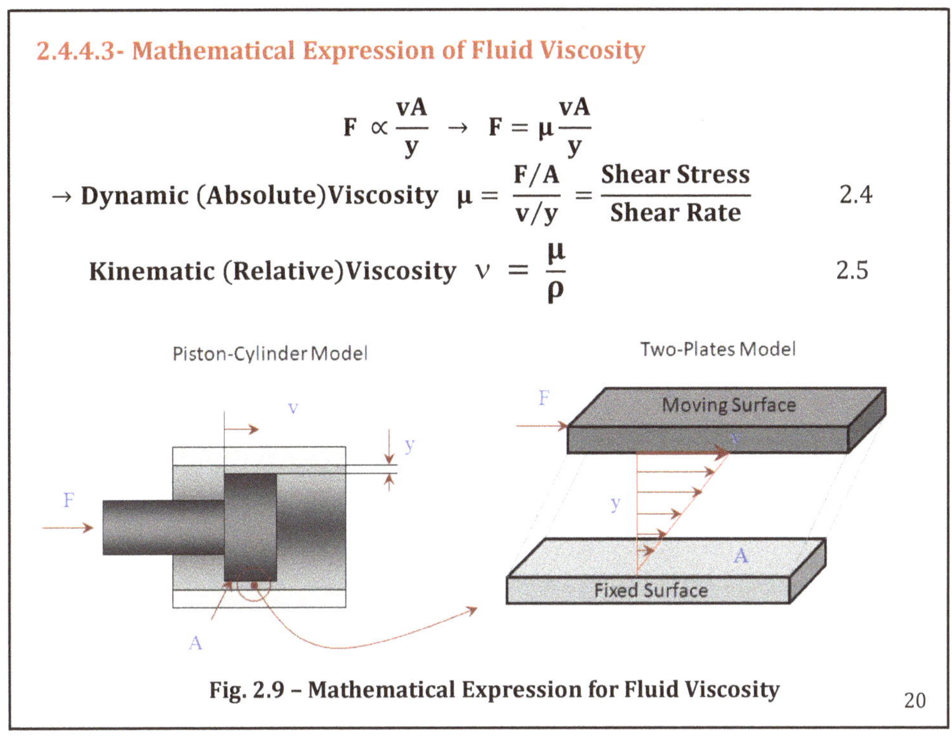

2.4.4.3- Mathematical Expression of Fluid Viscosity
$F \propto \frac{vA}{y} \rightarrow F = \mu \frac{vA}{y}$
$\rightarrow$ Dynamic (Absolute)Viscosity $\mu = \frac{F/A}{v/y} = \frac{\text{Shear Stress}}{\text{Shear Rate}}$ 2.4
Kinematic (Relative)Viscosity $\nu = \frac{\mu}{\rho}$ 2.5
Piston-Cylinder Model
Two-Plates Model
F
v
y
A
Moving Surface
Fixed Surface
20

Fig. 2.9 – Mathematical Expression for Fluid Viscosity

20

2.4.4.4- Newtonian versus Non-Newtonian Fluids

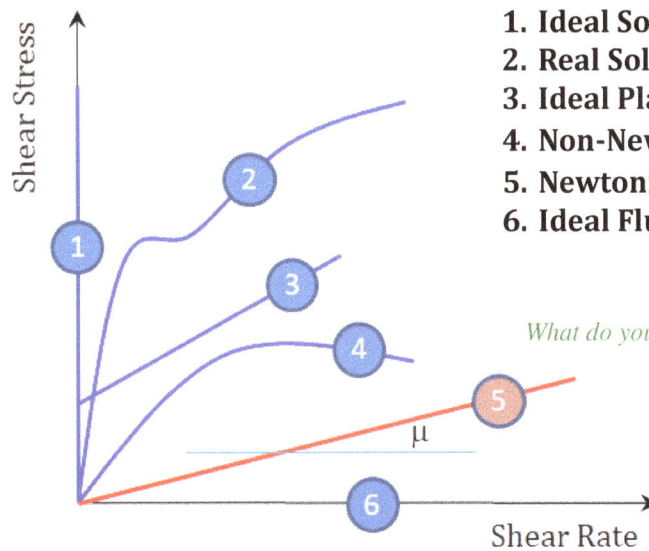

1. Ideal Solid.
2. Real Solid.
3. Ideal Plastic.
4. Non-Newtonian Fluid.
5. Newtonian Fluid.
6. Ideal Fluid.

What do you think about blood?

Shear Stress

Shear Rate

μ

Fig. 2.10 – Newtonian versus Non-Newtonian Fluids

21

21

2.4.4.5- Units of Fluid Viscosity

Metric Units of Dynamic (Absolute) Viscosity $\mu = \frac{\text{Shear Stress}}{\text{Shear Rate}}$ is $\left[\frac{N.s}{m^2}\right]$

Industrial Unit of Dynamic Viscosity is **Centipoise**, $1cP = 10^{-3}\left[\frac{N.s}{m^2}\right]$

Metric Units of Kinematic (Relative) Viscosity $= \frac{\mu}{\rho} =$ is $\left[\frac{m^2}{s}\right]$

Industrial Unit of Kinematic Viscosity is **Centistoke**, $1cSt = 10^{-6}\left[\frac{m^2}{s}\right] = \left[\frac{mm^2}{s}\right]$.

22

22

For balancing the units, Eq. 2.5 →

Kinematic (Relative) Viscosity $\left[\frac{m^2}{s}\right] = \frac{\mu\left[\frac{N.s}{m^2}\right]}{\rho\left(\frac{kg}{m^3}\right)}$

$$\rightarrow = 10^{-6}\,(cSt) = \frac{\mu\,(cP) x\, 10^{-3}}{\rho\left(\frac{kg}{m^3}\right)}$$

$$\rightarrow \nu\,(\mathbf{cSt}) = \frac{1000\ \mathbf{x}\ \boldsymbol{\mu}\,(\mathbf{cP})}{\boldsymbol{\rho}\left(\frac{\mathbf{kg}}{\mathbf{m^3}}\right)} = \frac{\mathbf{1000\ x}\ \boldsymbol{\mu}\,(\mathbf{cP})}{\mathbf{SG} \times \boldsymbol{\rho}_{\mathbf{W}}\left(\frac{\mathbf{kg}}{\mathbf{m^3}}\right)} \qquad 2.6$$

23

23

Knowing that the water density equal 1000 kg/m3, equation 2.6 can be used to solve for the dynamic viscosity as follows:

$$\mu\,(\mathbf{cP}) = \frac{\left[(\mathbf{cSt}) \times \mathbf{SG} \times \rho_W \left(\frac{\mathbf{kg}}{\mathbf{m}^3}\right)\right]}{1000}$$

$$= (\mathbf{cSt}) \times \mathbf{SG} \times \rho_W \left(\frac{g}{cc}\right) \qquad 2.7$$

Water has dynamic viscosity of 1 cP at 25 ^{0}C. and kinematic viscosity of 1 cSt.

24

24

2.4.4.6- Fluid Viscosity Standard Test Methods

Method 1: using Saybolt Viscometer:

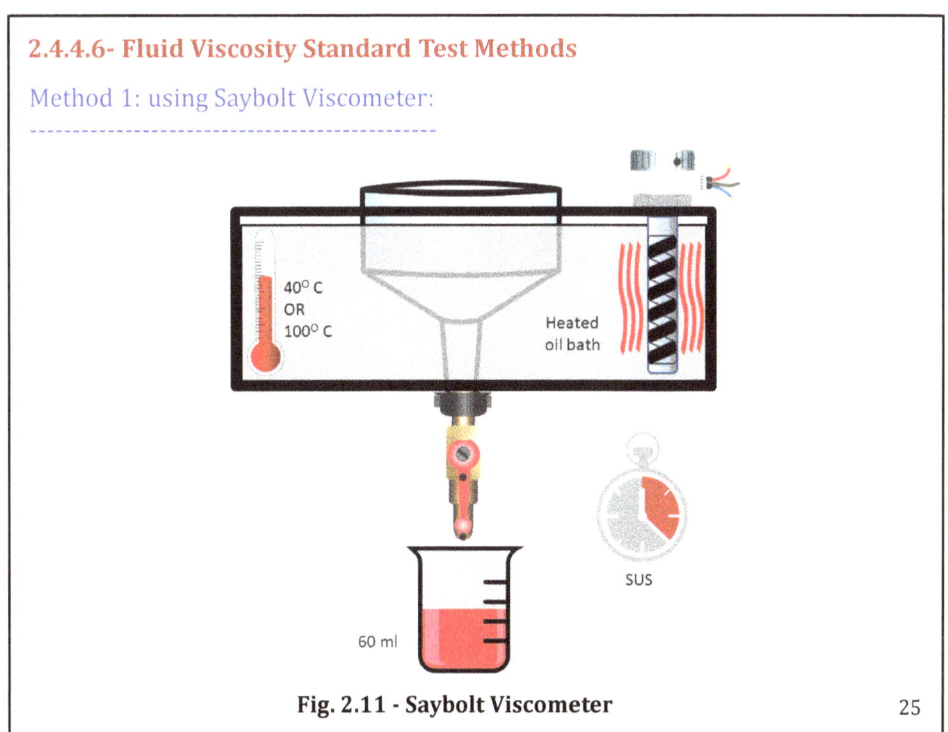

Fig. 2.11 - Saybolt Viscometer

25

25

Converting SUS to equivalent Kinematic Viscosity:

Empirical Formulas:

$$\nu\,(\mathbf{cSt}) = 0.226\,\mathbf{SUS} - \frac{195}{\mathbf{SUS}} \quad \textbf{for SUS} \le 100 \qquad 2.8$$

$$\nu\,(\mathbf{cSt}) = 0.220\,\mathbf{SUS} - \frac{135}{\mathbf{SUS}} \quad \textbf{for SUS} > 100 \qquad 2.9$$

Centistokes (cSt)	Saybolt Universal Second (SUS)	Centistokes (cSt)	Saybolt Universal Second (SUS)
5	~40	100	~465
7	~50	150	~700
9	~55	170	~770
10	~60	220	~1000
15	~75	320	~1500
22	~105	460	~2120
32	~150	680	~3000
46	~215	1000	~5000
68	~215	1500	~7000

Table 2.4 – Kinematic Viscosity to SUS Conversion Table

26

26

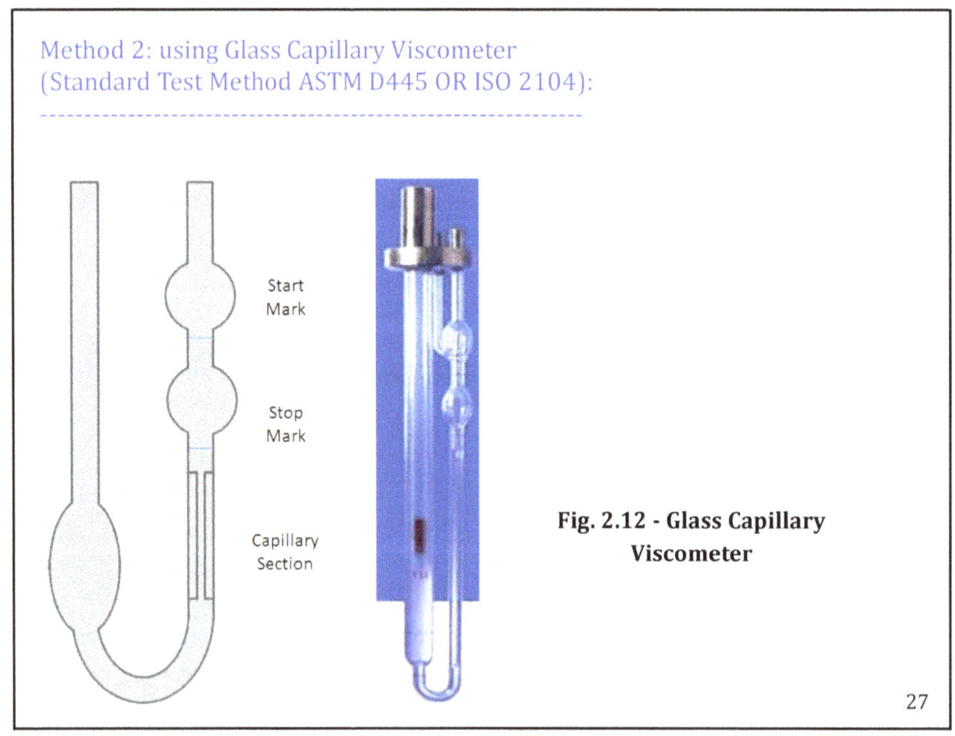

Fig. 2.12 - Glass Capillary Viscometer

27

27

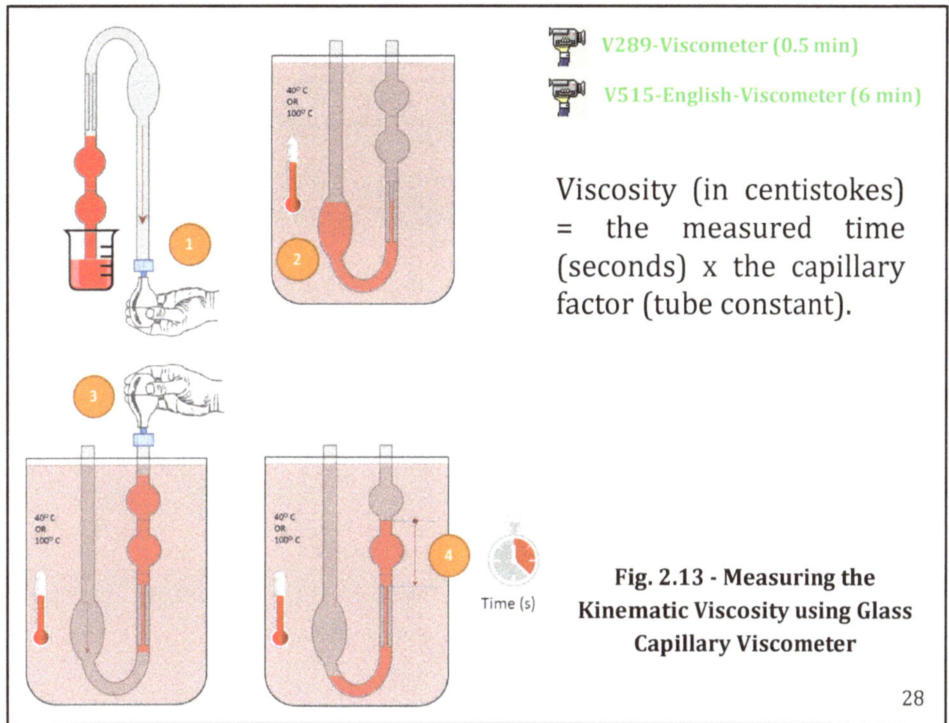

V289-Viscometer (0.5 min)
V515-English-Viscometer (6 min)
Viscosity (in centistokes) = the measured time (seconds) x the capillary factor (tube constant).
1
2
3
4
40° C OR 100° C
Time (s)
28

Fig. 2.13 - Measuring the Kinematic Viscosity using Glass Capillary Viscometer

28

Method 3: using Portable Viscometer (Standard Test Method ASTM D8092):

ASTM certified portable viscometer, model *MiniVisc*, for measuring kinematic viscosity. Such a viscometer has the following features:

- Portable and convenient for in-field viscosity measuring.
- Battery-operated.
- Easy to use and requires only few drops of oil sample.
- Controlled temperature at 40 ^{0}C.
- Results appear on a digital screen.
- Accuracy within plus or minus 3%.
- Easy to clean after measurement by wiping the cell without solvents.

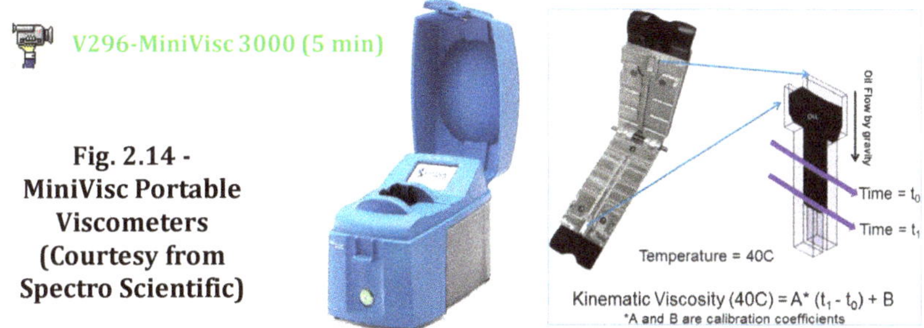

Fig. 2.14 - MiniVisc Portable Viscometers (Courtesy from Spectro Scientific)

29

29

2.4.4.7- Viscosity Standard Grades

ISO Viscosity Grade	Midpoint Kinematic Viscosity mm^2/s at 40°C (104°F)	Kinematic Viscosity Limit mm^2/s at 40°C (104°F) Minimum	Kinematic Viscosity Limit mm^2/s at 40°C (104°F) Maximum
ISO VG 2	2.2	1.98	2.42
ISO VG 3	3.2	2.88	3.52
ISO VG 5	4.6	4.14	5.06
ISO VG 7	6.8	6.12	7.46
ISO VG 10	10	9.00	11.0
ISO VG 15	15	13.5	16.5
ISO VG 22	22	19.8	24.2
ISO VG 32	32	29.8	35.2
ISO VG 46	46	41.4	50.6
ISO VG 68	68	61.2	74.8
ISO VG 100	100	90.0	110
ISO VG 150	150	135	165
ISO VG 220	220	198	242
ISO VG 320	320	288	352
ISO VG 460	460	414	506
ISO VG 680	680	612	748
ISO VG 1000	1000	900	1100
ISO VG 1500	1500	1350	1650
ISO VG 2200	2200	1980	2420
ISO VG 3200	3200	2880	3520

Table 2.5 – ISO Standard 3448 (ASTM D-2422) Viscosity Grades 30

30

SAE has various viscosity grades used in automotive applications such as: crank case engine oils, gear lubricants, and axle lubricants.

SAE Viscosity Grade	Low-Temp (°C) Cranking Viscosity (cP) Max	Low-Temp (°C) Pumping Viscosity (cP) Max (with no yield stress)	Kinematic Viscosity (cSt) at 100°C Min	Kinematic Viscosity (cSt) at 100°C Max	High Shear Viscosity (cP) at 150°C Min
0W	6200 @ -35	60,000 @ -40	3.8	-	-
5W	6600 @ -30	60,000 @ -35	3.8	-	-
10W	7000 @ -25	60,000 @ -30	4.1	-	-
15W	7000 @ -20	60,000 @ -25	5.6	-	-
20W	9500 @ -15	60,000 @ -20	5.6	-	-
25W	13000 @ -10	60,000 @ -15	9.3	-	-
16	-	-	6.1	<8.2	2.3
20	-	-	6.9	<9.3	2.6
30	-	-	9.3	<12.5	2.9
40	-	-	12.5	<16.3	3.5* / 3.7**
50	-	-	16.3	<21.9	3.7
60	-	-	21.9	<26.1	3.7

* For 0W-40, 5W-40 and 10W-40 Grades ** For 15W-40, 20W-40, 25W-40 and 40 Grades

Table 2.6 - SAE Standard J300 Viscosity Grades

5W-20 means the oil performs as:

- 5W-grade under cold conditions
- 20-grade fluid under normal working temperatures.

31

31

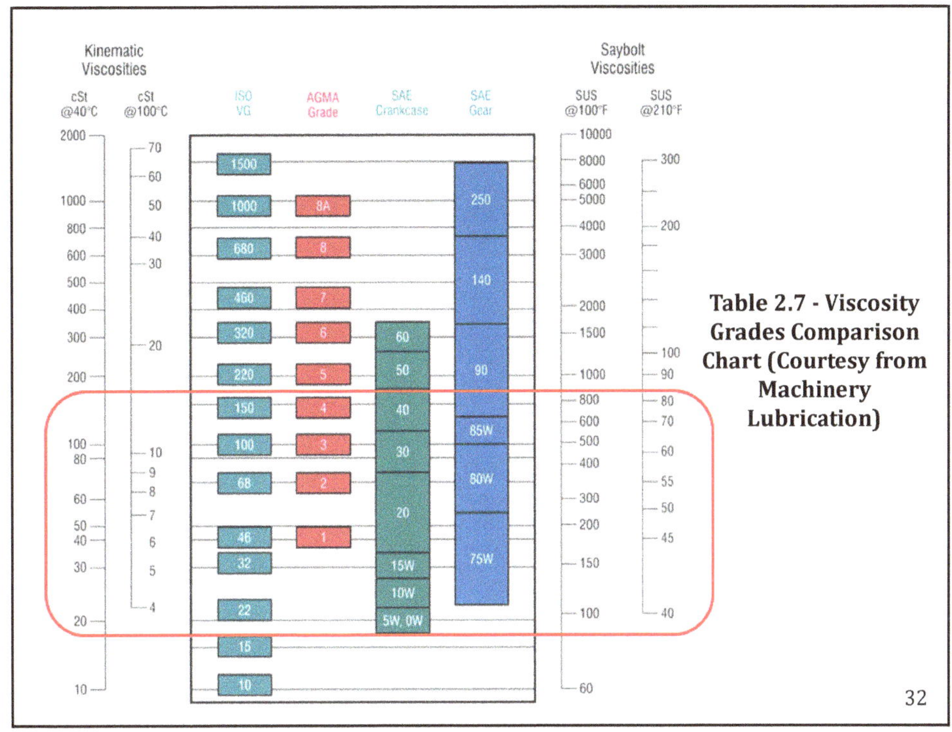

Table 2.7 - Viscosity Grades Comparison Chart (Courtesy from Machinery Lubrication)

32

32

Hydraulic fluids with higher viscosity index can?

A. Work better at high level of contamination.

B. Maintain viscosity for wider range of working temperature.

C. Maintain viscosity for wider range of working pressure.

D. All the above.

33

33

2.4.5- Viscosity Index
2.4.5.1-Definition of Viscosity Index

Video 602 (0.5 min)

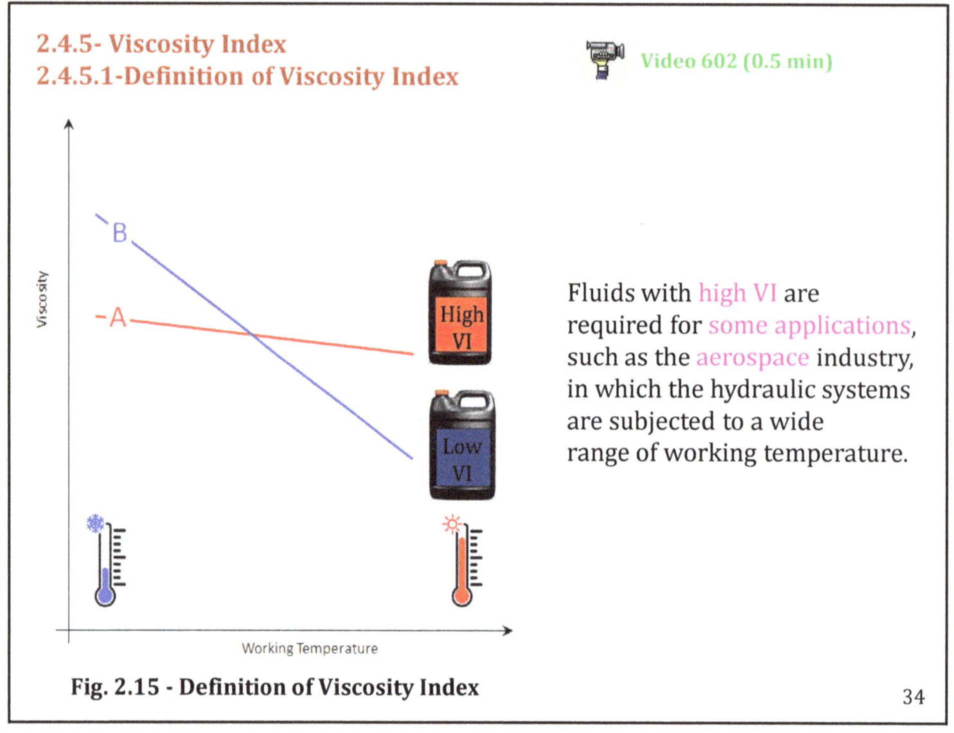

Fluids with high VI are required for some applications, such as the aerospace industry, in which the hydraulic systems are subjected to a wide range of working temperature.

Fig. 2.15 - Definition of Viscosity Index

34

34

2.4.5.2- Viscosity Index Standard Test Method (ASTM D2270-226)

ISO 2909 (ASTM D2270-226) → method and tables

Table 2.8 shows oils which were chosen as the basis for comparison with commercially available oils in the 1930.

Hydraulic Fluids should have at least 90.

Hydraulic Fluid	Viscosity Index	Kinematic Viscosity cSt (SUS) at 40 °C	Kinematic Viscosity cSt (SUS) at 100 °C
Paraffinic (Pennsylvania crude)	95	40 (210)	6.1 (46.5)
Naphthenic (Gulf Coast crude)	60	40 (210)	5.6 (45)
Multi-grade	150	31 (155)	6.2 (47)

Table 2.8 - Magnitude of Viscosity Index

35

35

Figure 2.15 shows the viscosity-temperature dependency for ISO-Graded fluids. The figure shows that the most common usable viscosity range in hydraulic systems are as follows:

- ISO VG 15 to 22: Recommended for arctic conditions.
- ISO VG 22 to 32: Recommended for winter conditions in central Europe and equivalent.
- ISO VG 32 to 46: Recommended for mild summer conditions.
- ISO VG 46 to 68: Recommended for tropical conditions or areas with high temperature.
- ISO VG68 to 100: Recommended for extremely high temperature conditions.

Such viscosities are defined at a refence temperature of 40 ^{0}C (104 ^{0}F).

36

36

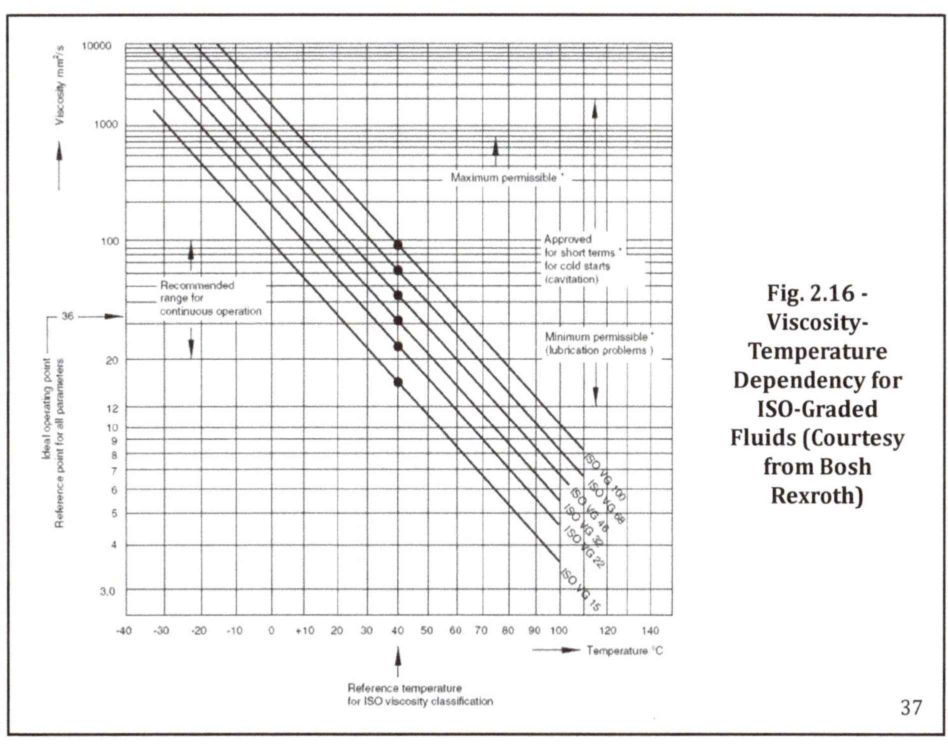

Fig. 2.16 - Viscosity-Temperature Dependency for ISO-Graded Fluids (Courtesy from Bosh Rexroth)

37

37

2.4.6- Wear Protection

2.4.6.1- Definition and Importance of Wear Protection

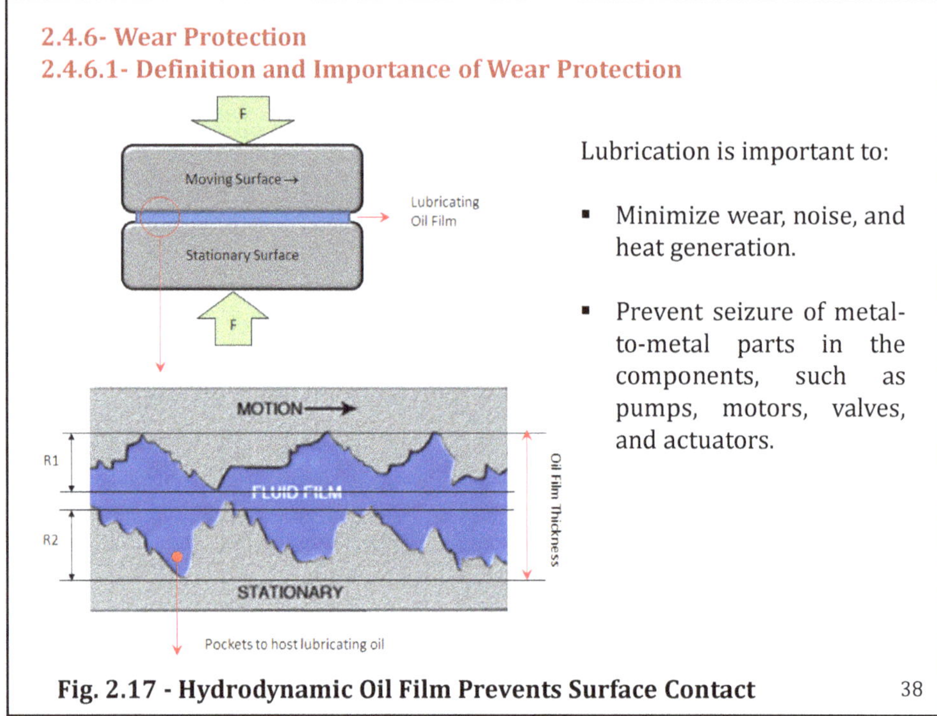

Lubrication is important to:

- Minimize wear, noise, and heat generation.
- Prevent seizure of metal-to-metal parts in the components, such as pumps, motors, valves, and actuators.

Fig. 2.17 - Hydrodynamic Oil Film Prevents Surface Contact 38

38

2.4.6.2- Factors Affecting Fluid Anti-Wear

A-Fluid Composition:

- **Type of the fluid (**molecular structure) of the fluid.
- **Fluid Viscosity:** High viscous fluids should be good lubricants, but they are difficult to get into fine clearances.
- **Additive package:** anti-wear additives improves friction removals

B-Working Conditions:

- **Working Pressure:** High working pressure is capable to sustain oil film thickness, particularly if the moving surfaces subject to high external load.
- **Working Load:** high working load on rubbing surfaces squeezes the lubrication film increasing the friction.
- **Working Temperature:** working temperature should be compromised with the fluid viscosity to get the optimized lubricity of the fluid.
- **Working Speed:** High working speed increases the hydrodynamic oil film thickness.

39

39

C-Design of Lubricated Surface:

Fig. 2.18- Design of Lubricated Surfaces in Hydraulic Components

Surface Coating:

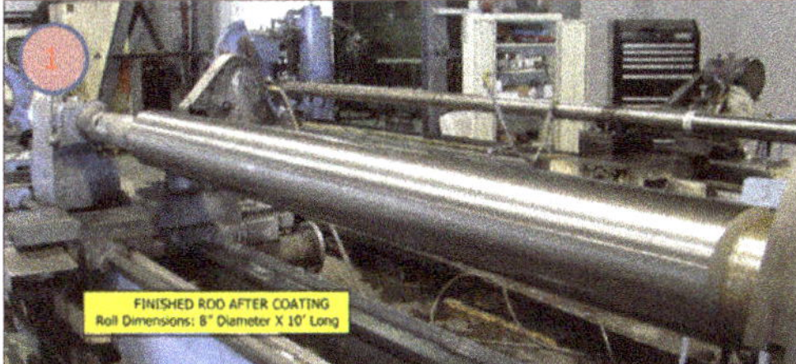

Hydraulic Rod Coated with HVOF Applied Tungsten Carbide As A Hard Chrome Replacement For Marine Application (Courtesy from Thermal Spray Solutions)

Hydrodynamic Lubrication

Design of Slipper Pads in Swash Plate Pumps

Advanced Surface Manufacturing

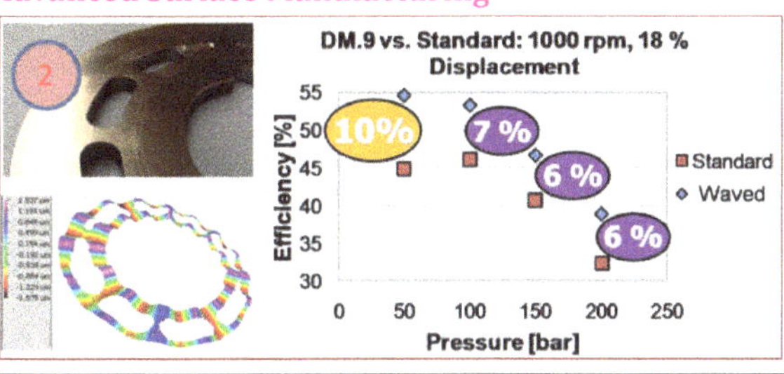

Micro- Wavy Port Plate Design Improves Swash Plate Pump Efficiency (Courtesy of CCEFP)

40

40

2.4.6.3- Anti-Wear Performance Test (ASTM D2882)

For many years, Vickers vane pump test (ASTM D2882) served as the benchmark for fluid wear performance.

Table 2.8 shows results of a laboratory test:

- Vane pump.
- 1800 rpm.
- 2000 psi (137 bar).

Test Specifications	Non-Anti-Wear Oil	Anti-Wear Fluid
Viscosity at 100 °F	150 SUS	155 SUS
Viscosity at Operating Temperature	60 SUS	60 SUS
Test Duration	1000 Hours	1000 Hours
Weight loss of Ring and 13 Vanes	**0.78 grams**	**0.04 grams**

Table 2.9 - Wear Rate Comparison of Two Fluids

41

41

2.4.7- Fluid Bulk Modulus
2.4.7.1- Definition of Fluid Bulk Modulus

Definition:

Bulk Modulus ~ Modulus of Elasticity ~ Spring Constant

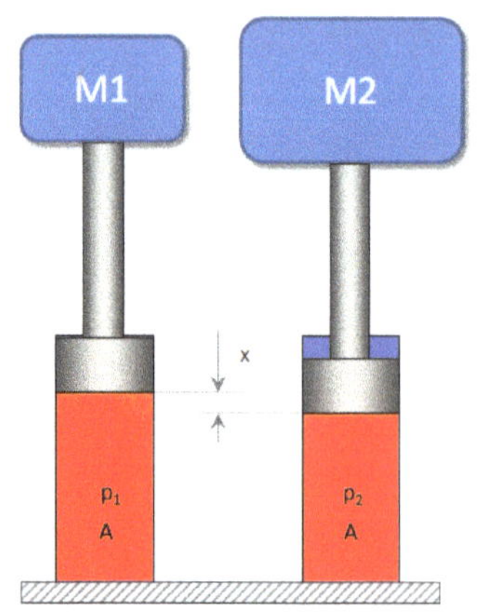

Fig. 2.19 - Hydraulic Fluid Bulk Modulus vs. Compressibility

Bulk modulus:

- Decreases with increasing temperature.
- Slightly increases with increasing pressure.
- Drastically decreases with entrained air.

Anim-013

2.4.7.2- Bulk Modulus Standard Test Method (ASTM D6793 – 02(2012)

This method measures the volumetric change of certain volume of oil under certain pressure and constant temperature.

42

42

2.4.7.3- Mathematical Expression for Fluid Bulk Modulus

V705-LB-017-Bulk Modulus

$$\beta(\text{bar or psi}) = -\frac{\Delta p}{\left(\frac{\Delta V}{V}\right)} \qquad 2.10$$

Fluid Type	Bulk Modulus at 20 °C and 10,000 psi
Water-Glycol	500,000 psi
Water-in Oil Emulsion	333,000 psi
Phosphate Ester	440,000 psi
ISO 32 Mineral Oil	260,000 psi

Table 2.10 - Bulk Modulus of Basic Hydraulic Fluids

COMPRESSIBILITY-VOLUME CHANGE (at 100 bar = 1450 psi)	
Fluid Type	**% ΔV Reduction**
Mineral Oil	0.7%
Vegetable-based Oil	0.5%
Water and Emulsified Water-Oil	0.4%
Water-Glycol and Synthetic Fluids (Polymers)	0.35%

Table 2.11 - Compressibility of Basic Hydraulic Fluids

43

43

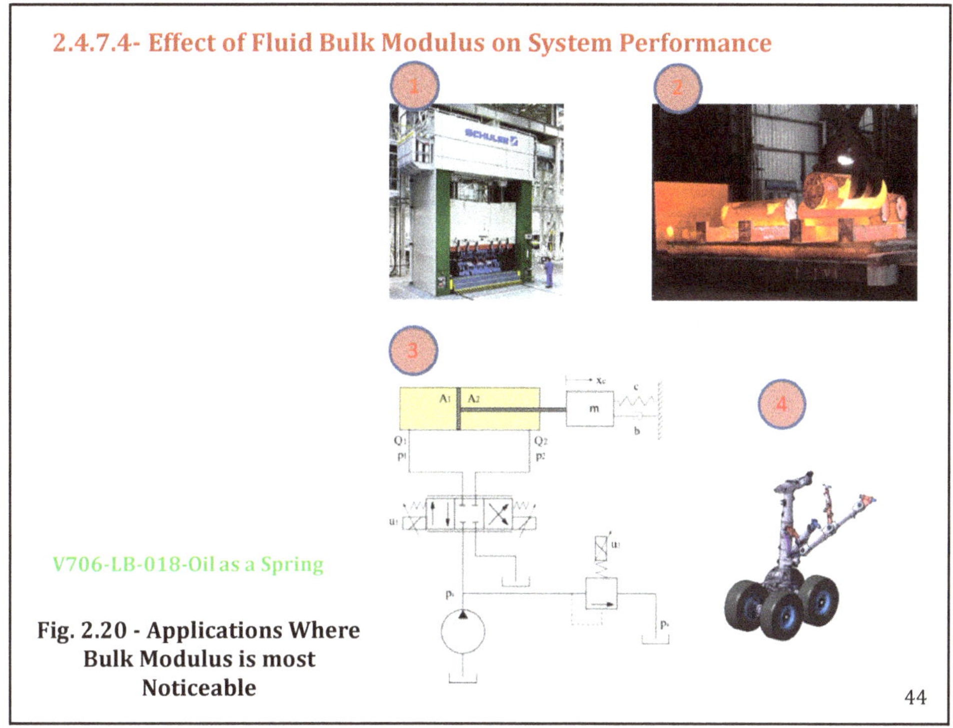

Fig. 2.20 - Applications Where Bulk Modulus is most Noticeable

44

The critical temperature, Pour Point, of a hydraulic fluid is defined as?

A. The lowest temperature at which the fluid can be poured + 5.4 F (3 °C).

B. The lowest temperature at which a liquid gives off vapor sufficient to ignite momentarily or flash when a flame is applied.

C. The lowest temperature at which the fluid continues to burn even if the flame is removed.

D. The lowest temperature at which a liquid ignites without an external flame or spark.

45

45

2.4.8- Pour Point

2.4.8.1- Pour Point Definition and Standard Test Method (ASTM D-97)

Definition: is the lowest temperature at which a test sample of a hydraulic fluid can be poured plus 5.4 °F (3°C).

ASTM D-97 Standard Test: the hydraulic fluid sample is placed in the test jar and cooled in (3°C) increments until the sample does not flow. The pour point is reported as 5.4 °F (3°C) above this temperature.

V297-Pour Point Test (1.5 min)

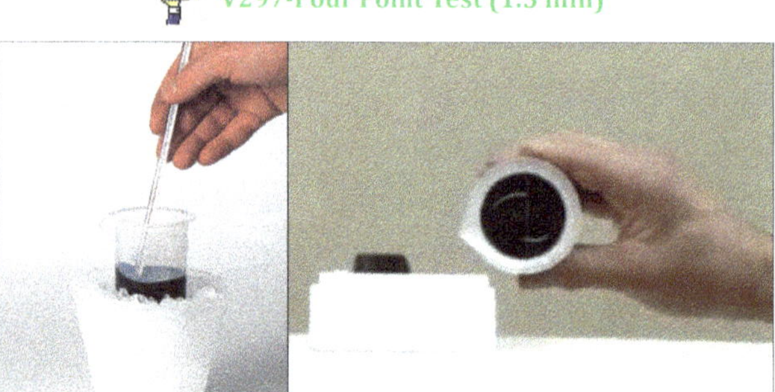

Fig. 2.21 - ASTM D-97 Standard Test for Pour Point 46

46

2.4.8.2-Pour Point Ratings

- **Minimum working temperature** = pour point + (10-20) °C.
- **Pour point is recommended** to be as low as possible for machines that work in cold weather.
- **Higher Pour Point** → V702-LB-014-Pour Point
  - Difficult machine starting.
  - Possible pump cavitation.
- **Pour Point vs. Viscosity:**
- Viscosity ↓ → Pour Point ↓

Fig. 2.22 - Hydraulic Systems in Cold Weather Require Hydraulic Fluid with Low Pour Point 47

47

The critical temperature, Flash Point, of a hydraulic fluid is defined as?

A. The lowest temperature at which the fluid can be poured + 5.4 F (3 °C).

B. The lowest temperature at which a liquid gives off vapor sufficient to ignite momentarily or flash when a flame is applied.

C. The lowest temperature at which the fluid continues to burn even if the flame is removed.

D. The lowest temperature at which a liquid ignites without an external flame or spark.

48

48

2.4.9- Flash Point

2.4.9.1- Flash Point Definition and Standard Test Method (ASTM D-92)

Definition: is the lowest temperature at which a hydraulic fluid produce vapor sufficient to ignite momentarily when a flame is applied under specific test conditions. The fluid stops burning if the flame is removed.

Closed Cup Flash Test

 V298-Flash Point Test (4.0 min)

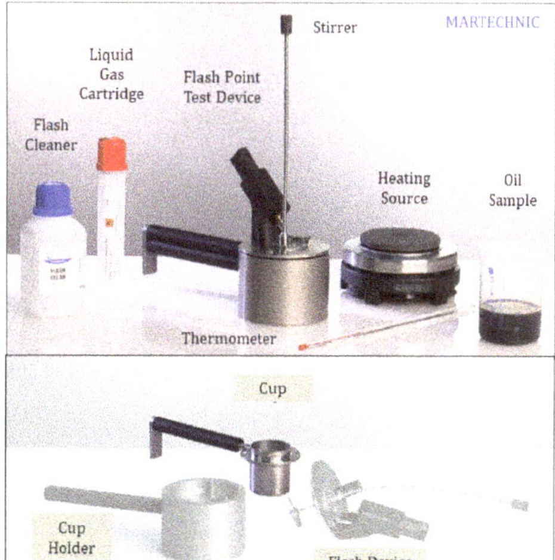

Fig. 2.23 - ASTM D-92 Standard Test for Flash Point

49

49

2.4.9.2- Flash Point Values

- **Flash Point:**.
  - Varies from 300°F (149 °C) for the lightest oils
  - To 510°F (265 °C) for the heaviest oils.
- **Flash point is recommended** to be as high as possible for a fluid to resist ignition.
- **For safety reasons:** flash point should be at least 68 °F (20 °C) **above the maximum** fluid working temperature.
- **Lower Flash Point** →
  - Work limitation at high temperature and some applications..
  - Risk of fire or explosion.
- **Flash Point vs. Viscosity:**
- Viscosity ↑ → Flash Point ↑

50

50

The critical temperature, <u>Fire Point</u>, of a hydraulic fluid is defined as?

A. The lowest temperature at which the fluid can be poured + 5.4 F (3 °C).

B. The lowest temperature at which a liquid gives off vapor sufficient to ignite momentarily or flash when a flame is applied.

C. The lowest temperature at which the fluid continues to burn even if the flame is removed.

D. The lowest temperature at which a liquid ignites without an external flame or spark.

51

51

The critical temperature, Ignition Point, of a hydraulic fluid is defined as?

A. The lowest temperature at which the fluid can be poured + 5.4 F (3 °C).

B. The lowest temperature at which a liquid gives off vapor sufficient to ignite momentarily or flash when a flame is applied.

C. The lowest temperature at which the fluid continues to burn even if the flame is removed.

D. The lowest temperature at which a liquid ignites without an external flame or spark.

52

52

2.4.10-Fire Point

Hydraulic fluid Fire Point is the temperature at which the fluid flashes when a flame is applied and continues to burn even if the flame is removed.

2.4.11- Ignition Point

Hydraulic fluid Ignition (spontaneous – Auto-Ignition) Point is the temperature at which the fluid ignites without application of an external flame or spark.

Critical Temperatures for Hydraulic Fluids			
Hydraulic Fluid	Flash Point °C	Fire Point °C	Ignition Point °C
Mineral Oil (ISO VG 15)	150	180	245
Fire Resistant (Phosphoric Ester Chlorinated)	310	330	610

Table 2.12 - Critical Temperatures for Mineral-based vs. Fire-Resistant Hydraulic Fluids

ISO Grade	VG22	VG32	VG46	VG68	VG100	VG150	VG220	Test
Flash Point °C	180	195	205	210	230	235	240	D-92
Pour Point °C	-30	-30	-27	-24	-21	-18	-15	D97

Table 2.13 - Critical Temperatures for ISO-Graded Hydraulic Fluids

53

53

2.4.12- Specific Heat

Definition: Specific Heat (c) is the amount of heat **Q** required to change the temperature **T** of a mass unit **m** of a substance by one degree.

Mathematical Expression:

$$dQ = m\,c\,dT \qquad 2.11A$$

$$c = dQ/mdT \qquad 2.11B$$

Where:

- **dQ** = heat supplied (J, kJ, Btu)
- **m** = unit mass (g, kg, lb).
- **c** = specific heat (J/g K), (kJ/kg ^{0}C), (kJ/kg K), (Btu/lb ^{0}F).
- **dT** = temperature change (K, ^{0}C, ^{0}F).

Importance: This property is required for calculating the change in working temperature based on the wasted energy in the system.

54

2.4.13- Thermal Conductivity

Definition: Thermal conduction is the transfer of heat from hotter to cooler parts of a body resulting in equalizing of temperature.

Mathematical Expression (Fourier Law):

- The heat transferred **Q**
- is directly proportional to:
  - **t** = the time elapsed.
  - **A** = the contact area between the two surfaces.
  - **ΔT** = the temperature gradient between the two surfaces.
- Inversely proportional to:
  - **x** = the distance between the two surfaces.
- The constant of proportionally K_{THC} is the Coefficient of Thermal Conductivity.

$$Q \propto \frac{t\,A\,\Delta T}{x} \rightarrow Q = K_{THC}\frac{t\,A\,\Delta T}{x} \rightarrow K_{THC} = \frac{Q}{t\,A\,\frac{\Delta T}{x}} \qquad 2.12$$

Importance: Required for :

- Heat exchangers design and selection.
- Calculation the heat conducted by the oil reservoir and pipelines.

55

2.4.14- Thermal Expansion

Definition:
If a confined volume of liquid is exposed to an increase in temperature, the liquid is thermally expanded intensifying the pressure.

Mathematical Expression:
Coefficient of Thermal Expansion K_{THE} is the ratio between the volumetric strain $\Delta V/V_0$ of the fluid and the change in the temperature ΔT.

$$\Delta V = K_{THE} V_0 \Delta T \rightarrow K_{THE}\left[\frac{1}{^0F} \text{ OR } \frac{1}{^0C}\right] = \frac{\Delta V/V_0}{\Delta T} \qquad 2.12$$

Importance:
- Calculation of the pressure intensification due to fluid thermal expansion.
- Oil reservoir sizing.

56

56

Hydraulic Fluid	Mineral Oil	Phosphate Ester	Water Glycol
Specific Heat c BTU/LB.^{0}F	0.45	0.32	0.8
Coefficient of Thermal Conductivity K_{THC} BTU/(HR)(FT^2)(^{0}F/FT)	0.08	0.067	0.025
Coefficient of Thermal Expansion K_{THE} 1/^{0}C	0.0005	0.00041	0.00034

Table 2.14 - Thermal Properties of Hydraulic Fluids

57

57

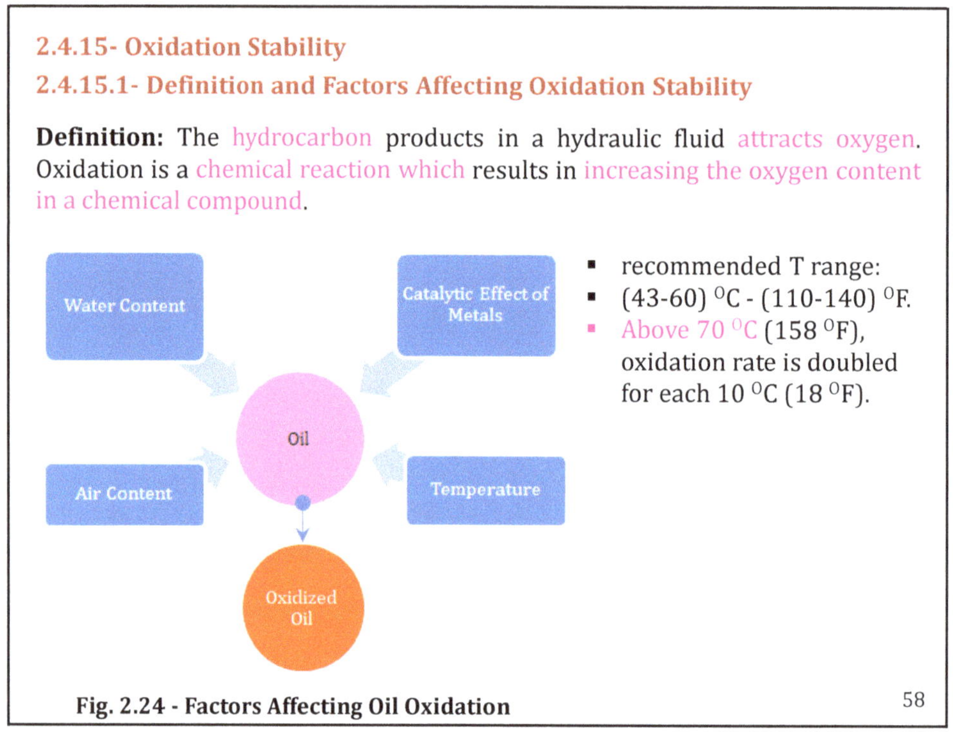

2.4.15- Oxidation Stability
2.4.15.1- Definition and Factors Affecting Oxidation Stability
Definition: The hydrocarbon products in a hydraulic fluid attracts oxygen. Oxidation is a chemical reaction which results in increasing the oxygen content in a chemical compound.
Water Content
Catalytic Effect of Metals
Oil
Air Content
Temperature
Oxidized Oil
▪ recommended T range:
▪ (43-60) °C - (110-140) °F.
▪ Above 70 °C (158 °F), oxidation rate is doubled for each 10 °C (18 °F).
58

Fig. 2.24 - Factors Affecting Oil Oxidation

58

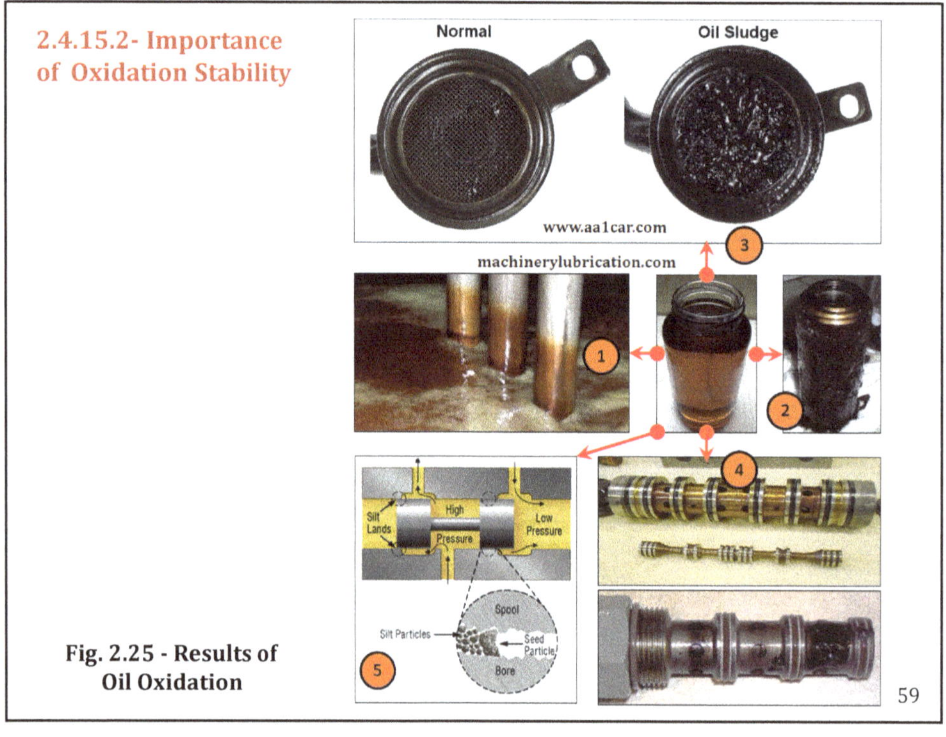

Fig. 2.25 - Results of Oil Oxidation

59

2.4.15.3- Oxidation Stability Standard Test Method (ASTM D-4636)

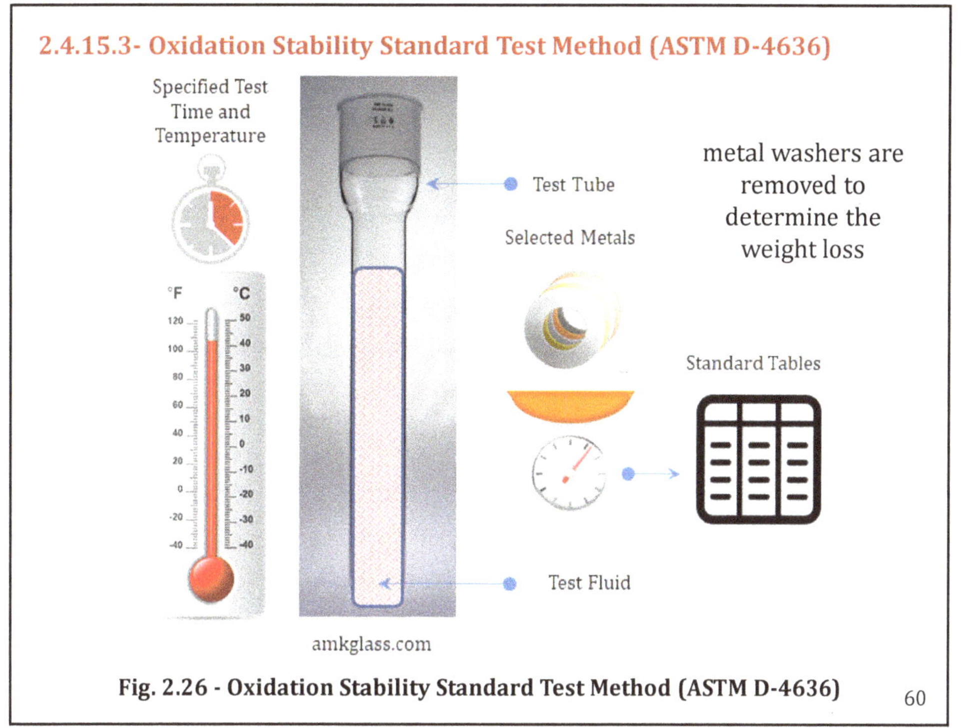

Fig. 2.26 - Oxidation Stability Standard Test Method (ASTM D-4636) 60

60

2.4.15.4- Corrosion Standard Test Method (ASTM D-130)

Hydraulic fluids can be corrosive to copper, zinc, and lead to unstable sulfur compounds present in additives and base oils.

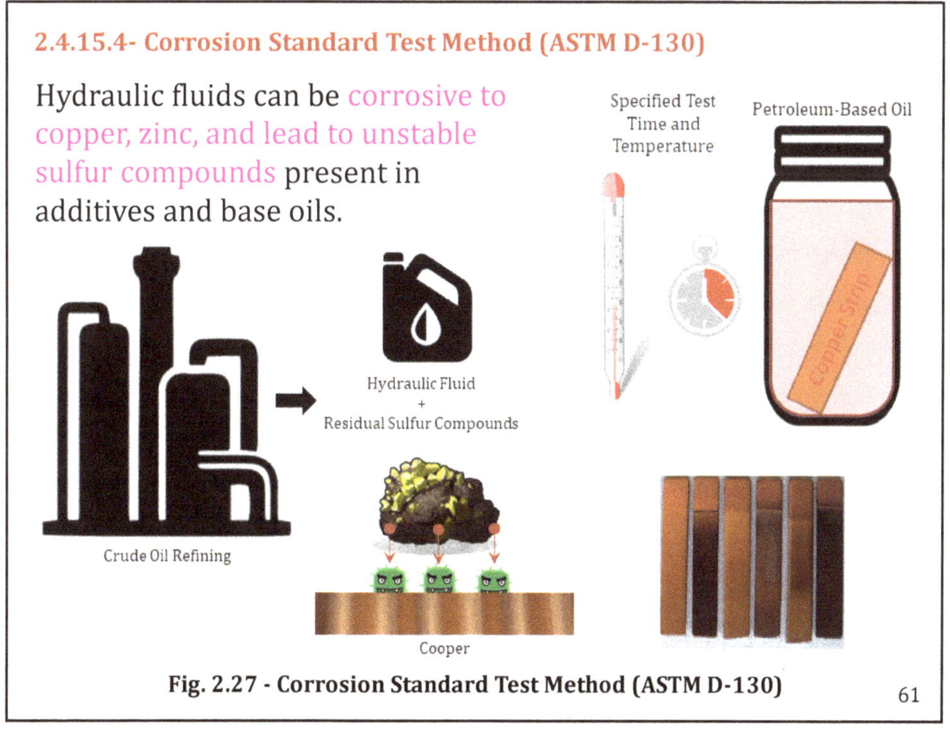

Fig. 2.27 - Corrosion Standard Test Method (ASTM D-130) 61

61

2.4.16- Antirust Property

2.4.16.1-Definition and Importance of Antirust Property

Definition: the ability of the hydraulic fluid to resist rusting due to water content in the fluid.

Importance:

- Mineral oils have better resistance against rusting.
- Rust particles are very abrasive and harmful to hydraulic components.

62

62

2.4.16.2- Antirust Standard Test Method (ASTM D665)

- A polished rounded steel bar is used as a specimen.
- The mixture should be continuously stirred over the test period.
- At the end of test period, the specimens are inspected for rust.

60° C (140° F)
Steel Rod
Stirrer
4 Hours
Oil-Water Mixture
10%-distilled water

Fig. 2.28 - Antirust Standard Test Method (ASTM D-665)

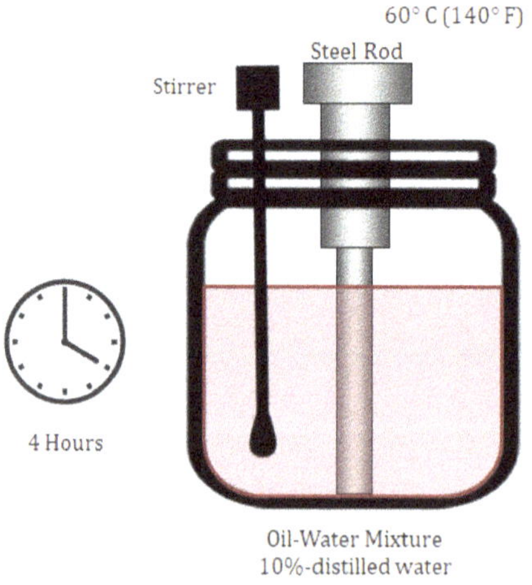

V300-ASTM D665 - Antirust (2.5 min)

63

63

2.4.17- Hydrolytic Stability

Definition: the ability of a hydraulic fluid to resist chemical decomposition due to the effect of water content in the fluid.

Importance:

- Additives and base oil molecular structure degradation
- → Affects fluid viscosity.
- → Forms acids.
  → Damage the hydraulic components and the seals.

- Additives are more susceptible to hydrolysis than base oil.

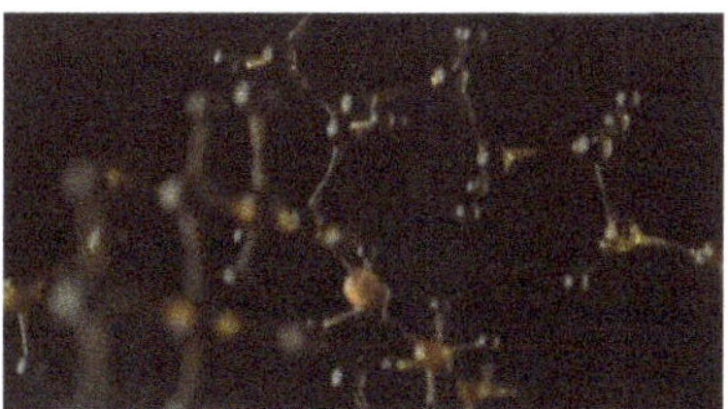

64

64

2.4.17.2- Hydrolytic Stability Standard Test Method (ASTM D2619-09)

- The mixture should be continuously rotated over the test period.
- The test fluid's total acidity number (TAN) is then determined.
- The results determine the fluid's hydrolytic stability.

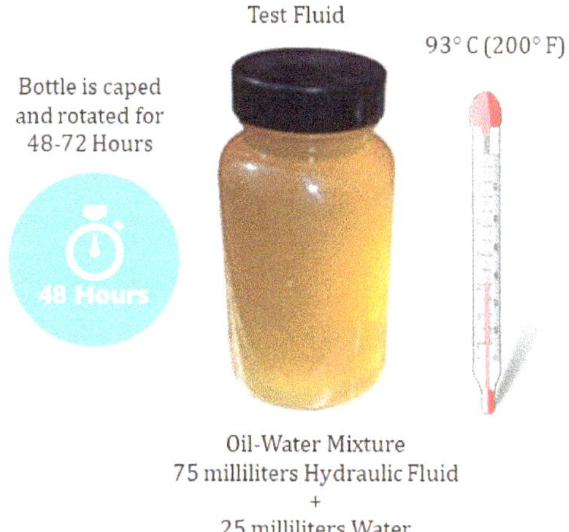

Fig. 2.29 - Hydrolytic Stability Standard Test (ASTM D2619-09)

65

65

2.4.18- Total Acid Number (TAN)

2.4.18.1- Definition and Importance of Total Acid Number

Definition: a measure of acidic concentration in a hydraulic fluid.

Neutralization Number = which is the number of milligrams of potassium hydroxide (KOH) required to neutralize one gram of oil.

Importance:

High concentration of acids → corrosion of machine parts and clogged oil filters due to the formation of varnish and sludge.

66

66

2.4.18.2- Total Acid Number Standard Test Method (ASTM D-664)

- This traditional test method requires (solvents, thorough technique, a well-trained chemist, and expensive equipment)
- Portable Infra-Red Spectrometer uses advanced data processing and a built-in oil application library to deliver immediate quantitative results.

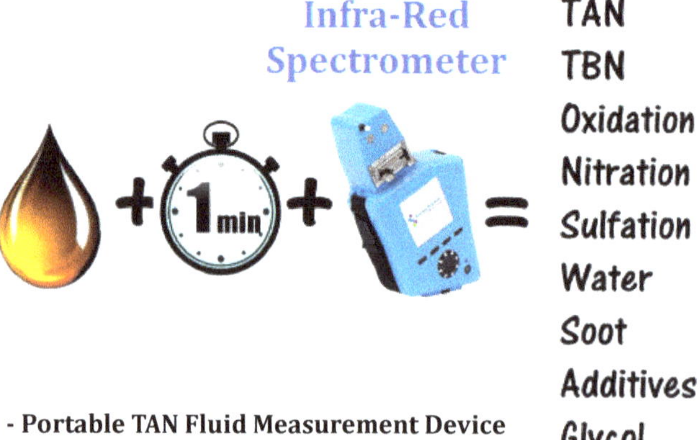

Fig. 2.30 - Portable TAN Fluid Measurement Device (Courtesy from Spectro Scientific)

67

67

2.4.19- Demulsibility (Water Separation)

2.4.19.1- Definition and Importance of Demulsibility

Definition: Demulsibility is the ability of a mineral or synthetic hydraulic fluid to resist solubility of water in the fluid or to separate from water.

Note: Demulsibility in water-based fluids is prohibited.

Importance:

Water contamination in mineral and synthetic hydraulic fluids:

→ Highly affects the fluid properties, particularly the lubricity.

→ Humid weather affects systems such as in marine applications.

68

68

2.4.19.2- Demulsibility Standard Tests Method (ASTM D-1401)

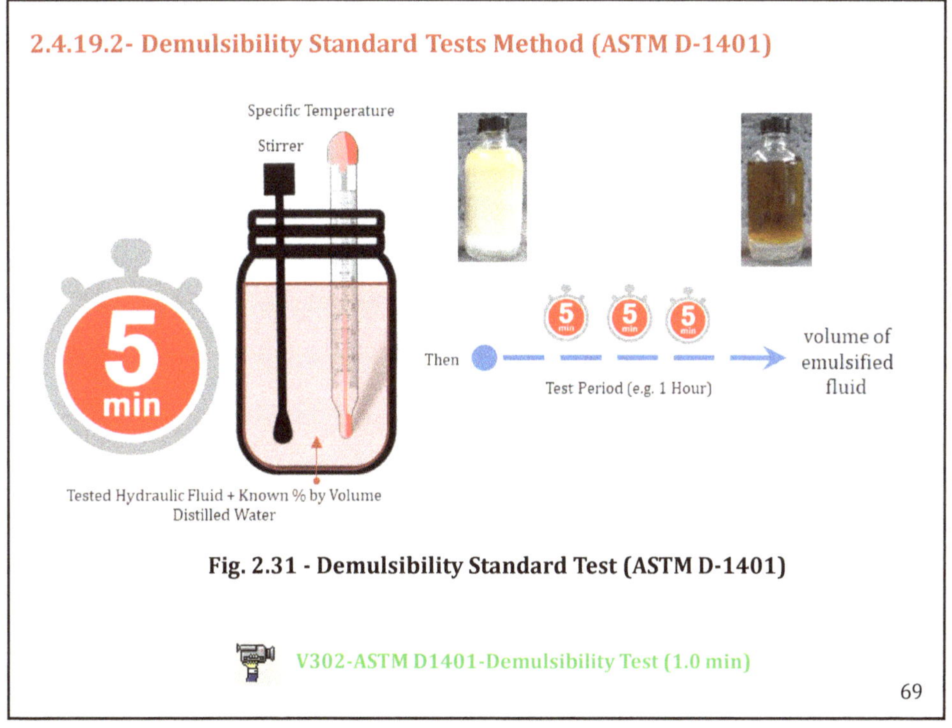

Fig. 2.31 - Demulsibility Standard Test (ASTM D-1401)

V302-ASTM D1401-Demulsibility Test (1.0 min)

69

69

2.4.20- Fluid Compatibility with Seals

2.4.20.1-Definition and Importance of Fluid Compatibility with Seals

Definition: Fluid *Compatibility with Seals* means the fluid does not chemically affect the seal material.

Importance:

- Some hydraulic fluids are not working friendly with the conventional rubber seals.
- Chemical reaction → seal deterioration → clogging control orifices → leakage → relevant consequences.

70

	Fluid Types					
Seal materials	Petroleum oil	Water-in-Oil Emulsion	Water Glycol	Phosphate Ester*	Chlorinated hydrocarbon	Synthetic with petroleum fractions
Buna-N (Acrylonitrile)	Excellent	Excellent	Very Good	Poor	Poor	Poor
Neoprene (Chloroprene)	Good	Good	Good	Poor	Poor	Poor
Butyl	Poor	Poor	Good	Fair to good	Poor	Poor
Silicone	Fair	Fair	Fair to poor	Fair to good	Poor to fair	Fair
Ethelene-Propylene	Poor	Poor	Good to excellent	Excellent	Fair	Poor
Viton® (Fluorocarbon)	Excellent	Excellent	Excellent	Good to Excellent	Good to Excellent	Good to Excellent
Metals	Conventional	Conventional	**	Conventional	Conventional	Conventional
Pipe Sealants	Conventional, Loctite® or Teflon® tape	Conventional, Loctite® or Teflon® tape	Loctite® or Teflon® tape	Loctite® or Teflon® tape	Loctite® or Teflon® tape	Loctite® or Teflon® tape

Table 2.15 - Compatibility of Common Hydraulic Fluids with Common Seal Materials (www.schoolcraftpublishing.com)

71

2.4.20.2- Fluid Compatibility Standard Test Methods (ASTM D6546-15 OR ISO 6072)

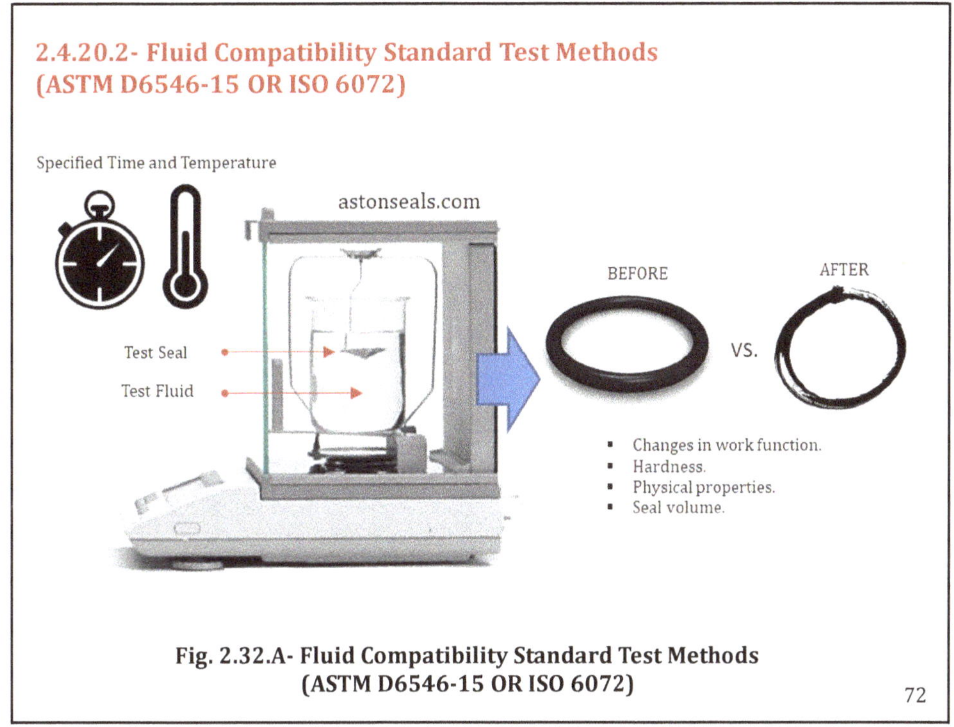

Fig. 2.32.A- Fluid Compatibility Standard Test Methods (ASTM D6546-15 OR ISO 6072)

72

72

ASTM D4289-15 is another standard test method for determining compatibility of Elastomer Seals for industrial hydraulic fluid applications. This test uses a rounded elastomeric disc as a specimen.

Measure Coupon Shore A Hardness | Measure Coupon Volume | Age Coupon in Oil at 158°F | Measure Change in Coupon Volume | Measure Change in Coupon Hardness

Fig. 2.32.B- Fluid Compatibility Standard Test Methods (ASTM D4289-15)

73

73

2.4.21-Foam Suppression

2.4.21.1-Definition and Importance of Foam Suppression

Definition: The ability of the fluid to get rid of air.

Measured by: Air Release Time + Volume of Air

Importance:

Foaming and Aeration can have detrimental effects on hydraulic system performance and durability. It can cause:

- Power Loss.
- Cavitation damage.
- Noisy operation.
- Poor lubrication.
- Oxidation.
- Erratic machine motion and control.
- Drastic reduction if fluid's bulk Modulus.

74

74

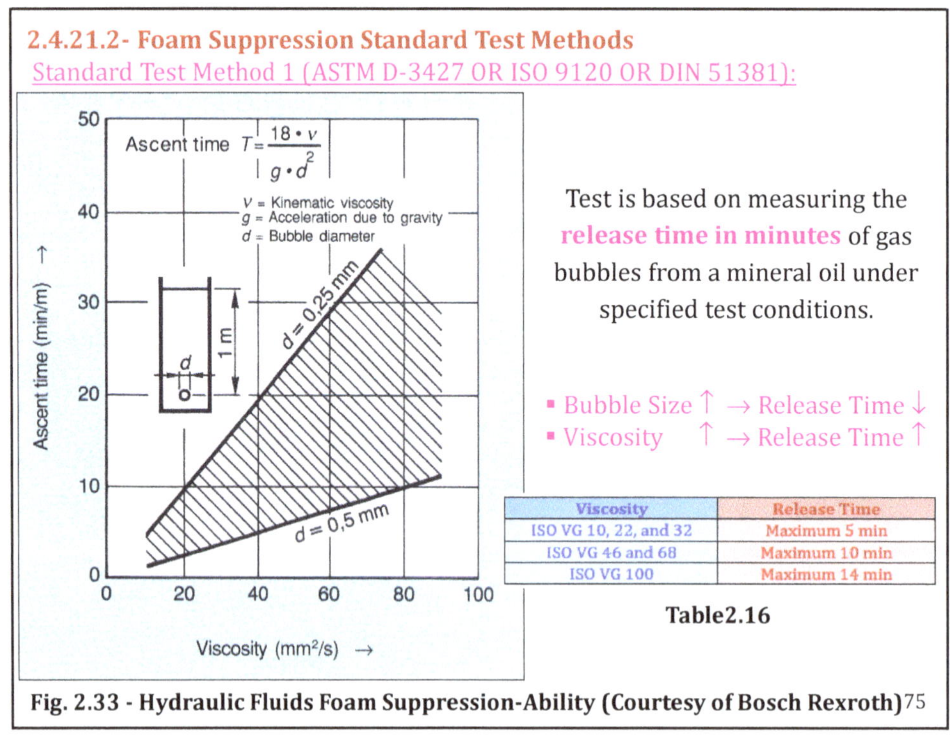
2.4.21.2- Foam Suppression Standard Test Methods
Standard Test Method 1 (ASTM D-3427 OR ISO 9120 OR DIN 51381):
Ascent time $T = \frac{18 \cdot v}{g \cdot d^2}$
v = Kinematic viscosity
g = Acceleration due to gravity
d = Bubble diameter
d = 0,25 mm
d = 0,5 mm
1 m
d
Ascent time (min/m) →
Viscosity (mm²/s) →
0
10
20
30
40
50
0
20
40
60
80
100
Test is based on measuring the release time in minutes of gas bubbles from a mineral oil under specified test conditions.
▪ Bubble Size ↑ → Release Time ↓
▪ Viscosity ↑ → Release Time ↑

Viscosity	Release Time
ISO VG 10, 22, and 32	Maximum 5 min
ISO VG 46 and 68	Maximum 10 min
ISO VG 100	Maximum 14 min

Table2.16

Fig. 2.33 - Hydraulic Fluids Foam Suppression-Ability (Courtesy of Bosch Rexroth) 75

75

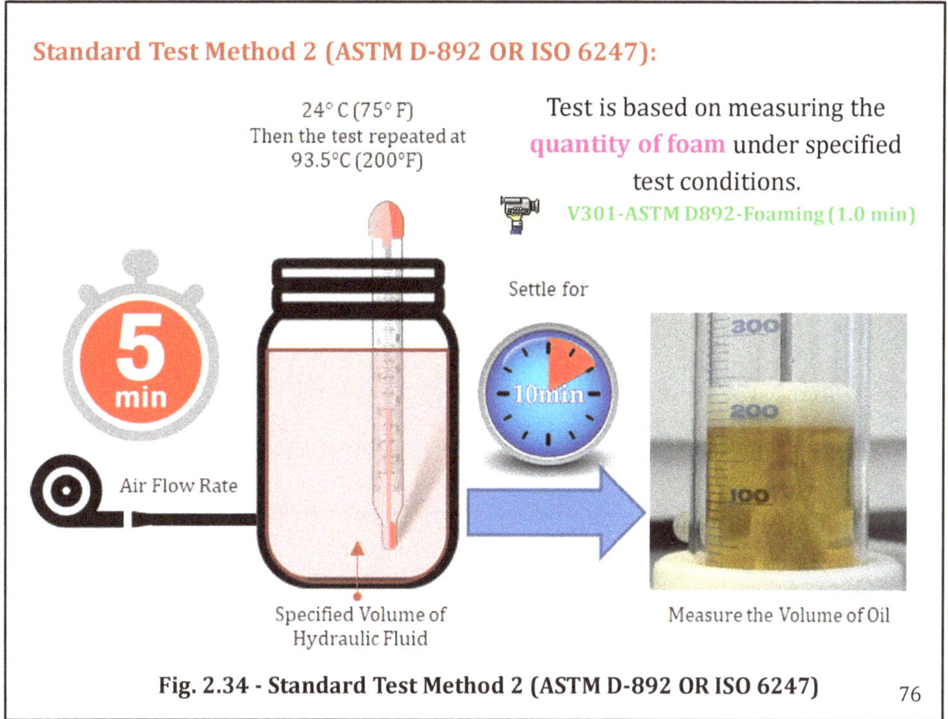

Fig. 2.34 - Standard Test Method 2 (ASTM D-892 OR ISO 6247) 76

76

Fig. 2.35 - A Typical Instrument for Measuring Foam Characteristics (Courtesy of Koehler Instrument) 77

77

Results of a hydraulic fluid high pour point are?

A. Chemical decomposition because of water contents in the oil.

B. Orifice blockage, spool sticking, filter clogging and possible pump cavitation.

C. Difficult machine starting and possible pump cavitation.

D. Deterioration of rubber seals and gaskets.

78

78

Results of a hydraulic fluid high oxidation rate are?

A. Chemical decomposition because of oil contamination by water.

B. Orifice blockage, spool sticking, filter clogging and possible pump cavitation.

C. Difficult machine starting and possible pump cavitation.

D. Deterioration of rubber seals and gaskets.

79

79

Results of a hydraulic fluid incompatibility with seals are?

A. Chemical decomposition because of oil contamination by water.

B. Orifice blockage, spool sticking, filter clogging and possible pump cavitation.

C. Difficult machine starting and possible pump cavitation.

D. Deterioration of rubber seals and gaskets.

80

80

Results of a hydraulic fluid low hydrolytic stability are?

A. Chemical decomposition because of oil contamination by water.

B. Orifice blockage, spool sticking, filter clogging and possible pump cavitation.

C. Difficult machine starting and possible pump cavitation.

D. Deterioration of rubber seals and gaskets.

81

81

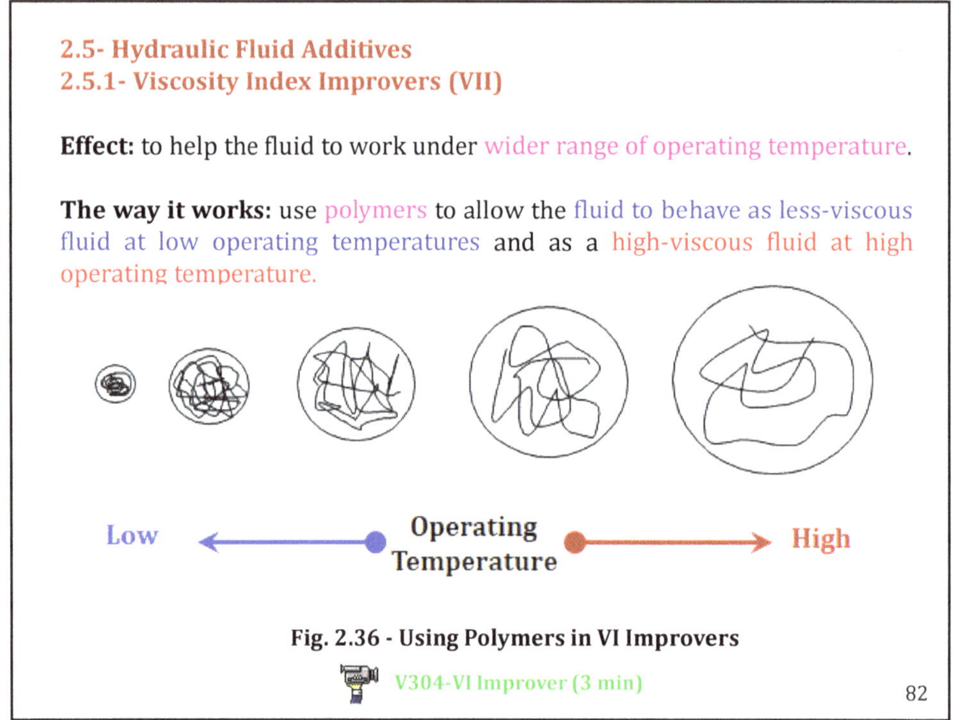

2.5- Hydraulic Fluid Additives
2.5.1- Viscosity Index Improvers (VII)
Effect: to help the fluid to work under wider range of operating temperature.
The way it works: use polymers to allow the fluid to behave as less-viscous fluid at low operating temperatures and as a high-viscous fluid at high operating temperature.
Low
Operating Temperature
High
V304-VI Improver (3 min)
82

Fig. 2.36 - Using Polymers in VI Improvers

82

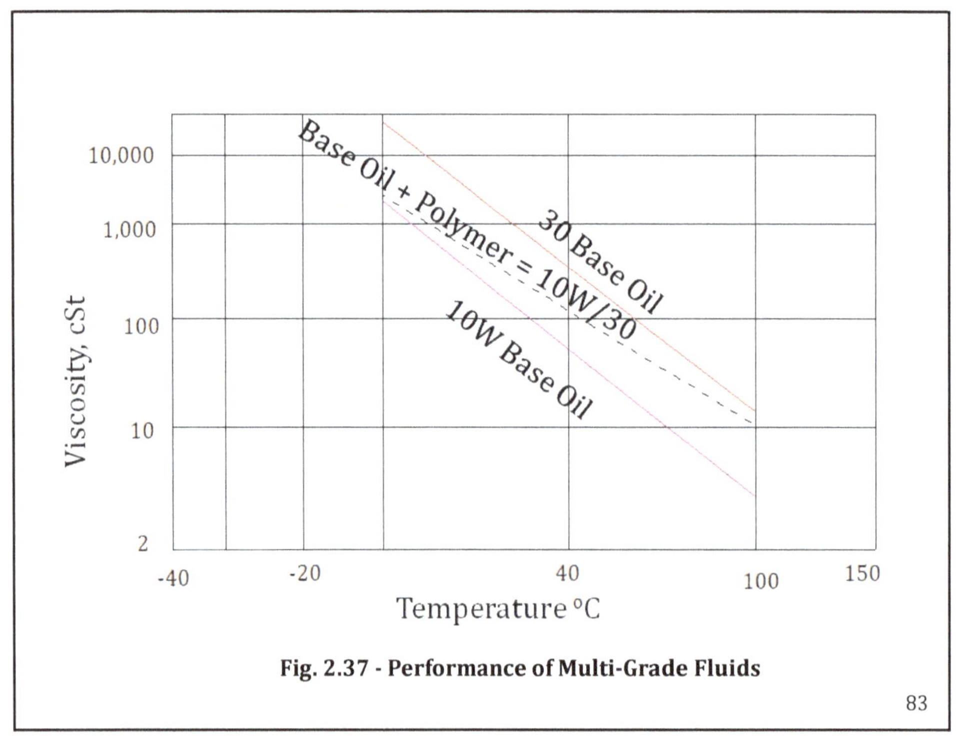

Fig. 2.37 - Performance of Multi-Grade Fluids

83

2.5.2- Rust and Oxidation (R & O) Inhibitors

Rust Inhibitors (Anti-Rust additives):
Effect: prevent corrosion of iron and carbon steel surfaces.

The way it works: Includes some compounds such as Calcium Phenate, Barium Sulfonate, and organic acids. These compounds function by plating out on iron and providing water repellant coating on the surface.

Oxidation Inhibitors (Anti-Oxidant additives):
Effect: extend the fluid life by preventing reaction between the fluid and oxygen at high temperature.

The way it works: Interrupt the oxidation chain reaction. That is why these additives may become depleted over the time.

84

2.5.3- Anti-Wear (AW) Additives
Effect: used to improve the frictional characteristics between metal surfaces; hence reduce wear and all the relevant consequences.

The way it works:

- **Classical Anti-Wear Additives:** work by forming a lubricating oil film with a thickness sufficient to separate the moving surfaces apart from each other. This type is good for moderate load and temperature conditions.

- **Extreme Pressure (EP) Anti-Wear Additives**: react with the moving surfaces to form a chemical insoluble film as a coating layer typically from Zinc and/or Sulfur. This type works better at high contact pressure.

85

2.5.4- Pour Point Depressant

Effect: improves low-temperature performance, help fluids to flow more comfortably at low temperatures, avoiding pump cavitation, and permitting cold weather operation of hydraulic systems.

The way it works: use chemicals that prevent the formation of crystals in fluids where crystals reduce the fluid's ability to freely flow.

2.5.5- Demulsifiers

Effect: help separate water from petroleum-based oil.

The way it works: increase the surface tension of petroleum-based hydraulic fluids so that they can easily separate from water. Water is then settled out and collected from the bottom of the reservoir.

2.5.6- Foam Suppressors

Effect: help separating foam from the hydraulic fluid.

The way it works: reduce the surface tension of a fluid to accelerate breaking down the bubbles, hence prevent foam forming.

86

86

Which of the following hydraulic fluids has better foam suppression ability?

A. Hydraulic fluids with higher oxidation resistance.

B. Fire resistant hydraulic fluids.

C. Hydraulic fluid with higher bulk modulus.

D. Hydraulic fluid with lower viscosity and blended with de-foaming additives.

87

87

How many broad categories of hydraulic fluids are available?

A. 1.
B. 2
C. 3.
D. 4.

88

88

2.6- Classification of Hydraulic Fluids

Video 467 (1.5 min)

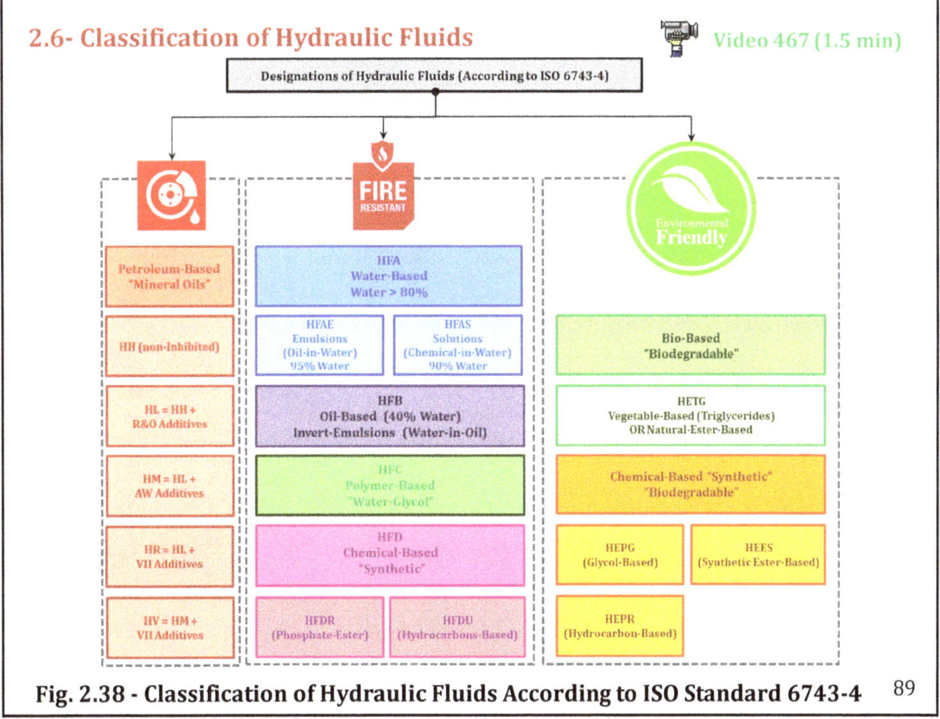

Fig. 2.38 - Classification of Hydraulic Fluids According to ISO Standard 6743-4

89

89

2.7-Petroleum-Based Hydraulic Fluids (Mineral Oils)
2.7.1- Main Features of Mineral Oils

Definition: Mineral Oil is a product of refining crude oil combined with appropriate additives.

Advantages:

- ☺ Cost and Availability.
- ☺ Viscosities
- ☺ Power Transmission.
- ☺ Heat Dissipation.
- ☺ Anti-Rusting.
- ☺ Additive Packages.
- ☺ Chemically Stability at140°F (60°C).
- ☺ Compatibility with common seals.

Disadvantages:

- ☹ Flammability.
- ☹ Environmental issues.
- ☹ Requirements for Ecologically Treated before Recycling.

90

90

Applications: Approximately 80% of the hydraulic fluids used to drive hydraulic systems in various applications.

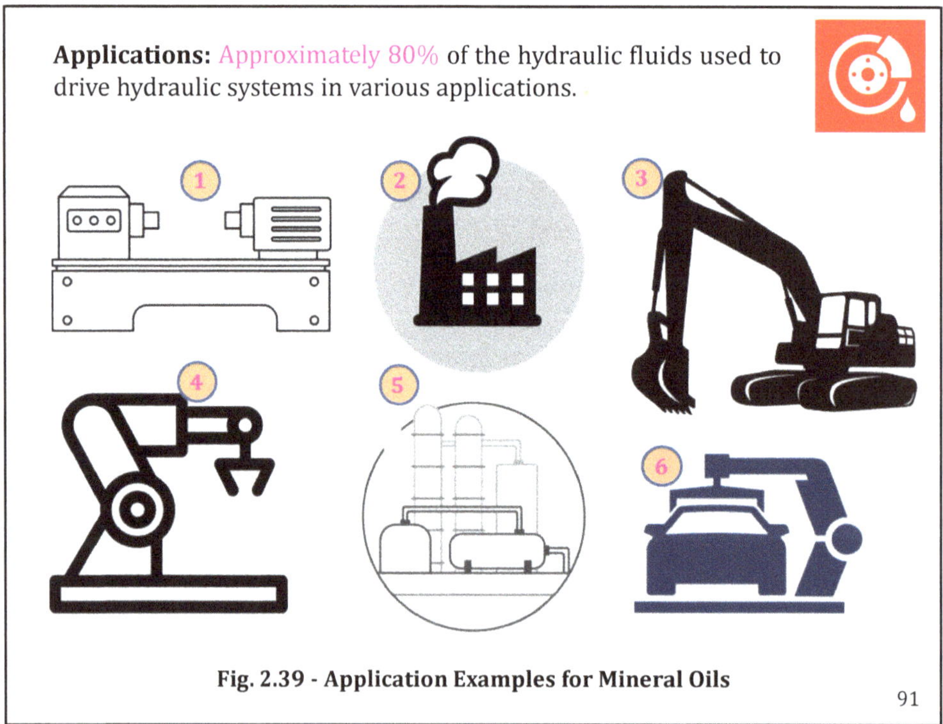

Fig. 2.39 - Application Examples for Mineral Oils

91

91

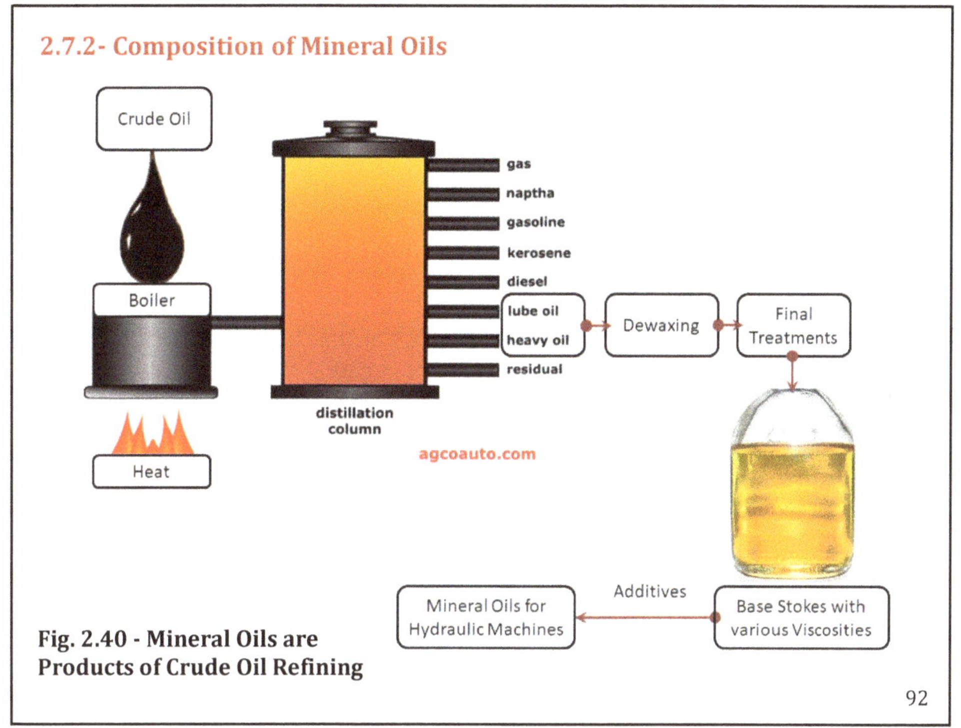

Fig. 2.40 - Mineral Oils are Products of Crude Oil Refining

92

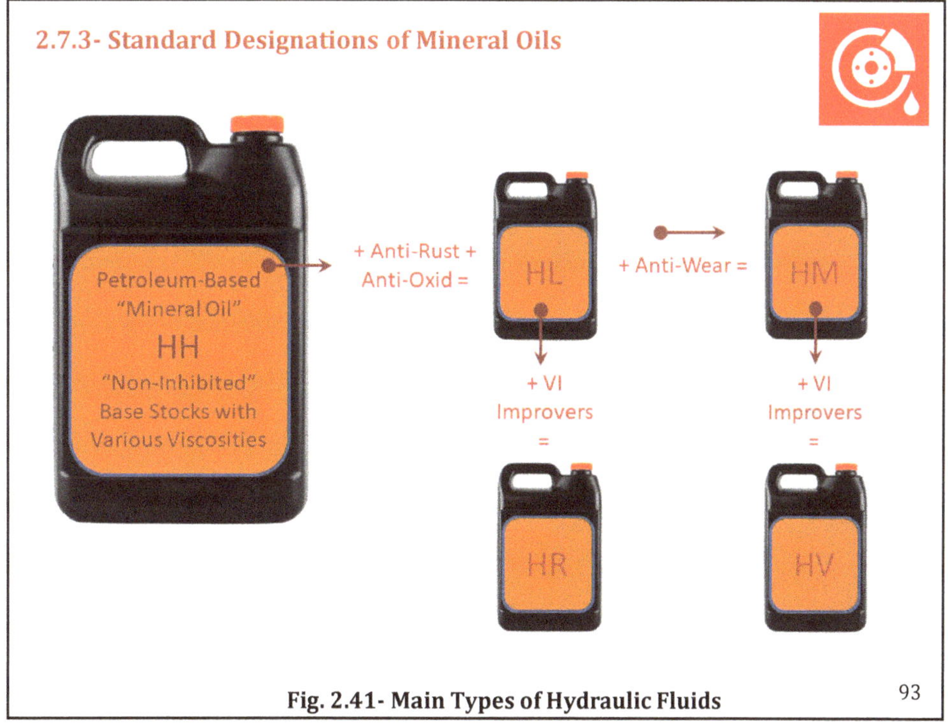

Fig. 2.41- Main Types of Hydraulic Fluids

93

DIN Code 51524	ISO Code 6743-4	Composition
H	HH	Non-inhibited refined mineral oil
HL	HL	Refined mineral oil with anti-rust and anti-oxidation properties
HLP	HM	Oils of type HL with improved anti-wear properties
HVLP	HR	Oils of type HL with improved viscosity-temperature properties
HVLP	HV	Oils of type HM with improved viscosity-temperature properties
	HPHM	Oils of type HM with hydrolytic stability, filterability, wear protection, and anti-foaming properties
	HPHV	Oils of type HV with hydrolytic stability, filterability, wear protection, and anti-foaming properties

Table 2.17 - Standard Designation of Mineral Oils (According to ISO 6743-4 and DIN 51524)

94

94

2.7.3.1- HH Mineral Oil

- Use for low pressure hydraulic systems.
- Able to perform the primary function of power transmission.
- Unable to withstand high temperatures.
- Have limited lubricating capabilities.

Fig. 2.42 - Applications for HH Mineral Oil

95

95

2.7.3.2- HL Mineral Oil

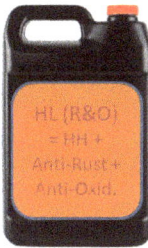

Pumps that contain yellow and silver metals requires zinc-free oil

Machine Tool Applications where P ≤ 138 bar (2000 psi)

- known as "R&O" oils because they contain anti-rust and anti-oxidation.
- Zinc-free, not aggressive for yellow metals (Brass and bronze)

Fig. 2.43- Applications for HL Mineral Oil

96

96

2.7.3.3- HM Mineral Oil

- HM fluid may contain zinc or some other type of anti-wear additives.
- Recent anti-wear additives utilize sulfur and phosphorus compounds to achieve satisfactory anti-wear performance.

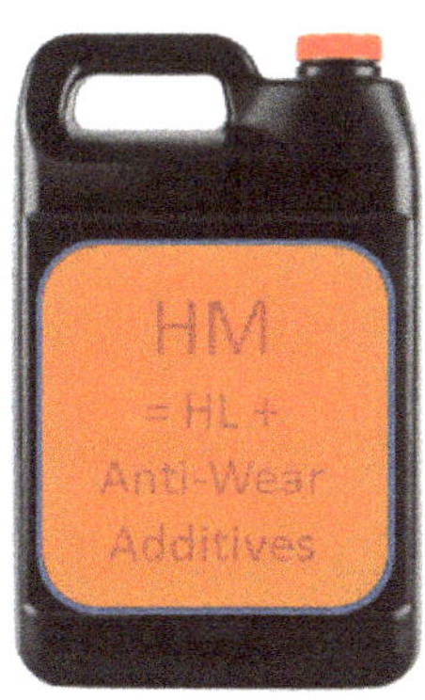

High Pressure Applications
P >138 bar (2000 psi)

Fig. 2.44 - Applications for HM Mineral Oil

97

97

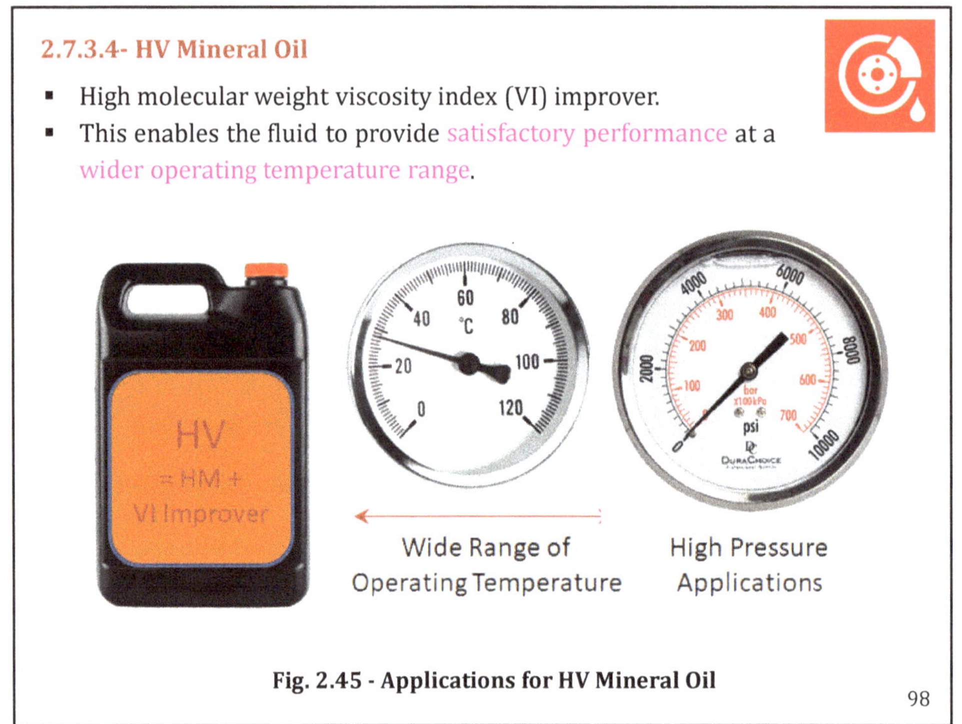

2.7.3.4- HV Mineral Oil
▪ High molecular weight viscosity index (VI) improver.
▪ This enables the fluid to provide satisfactory performance at a wider operating temperature range.
HV
= HM +
VI Improver
Wide Range of
Operating Temperature
High Pressure
Applications
98

Fig. 2.45 - Applications for HV Mineral Oil

98

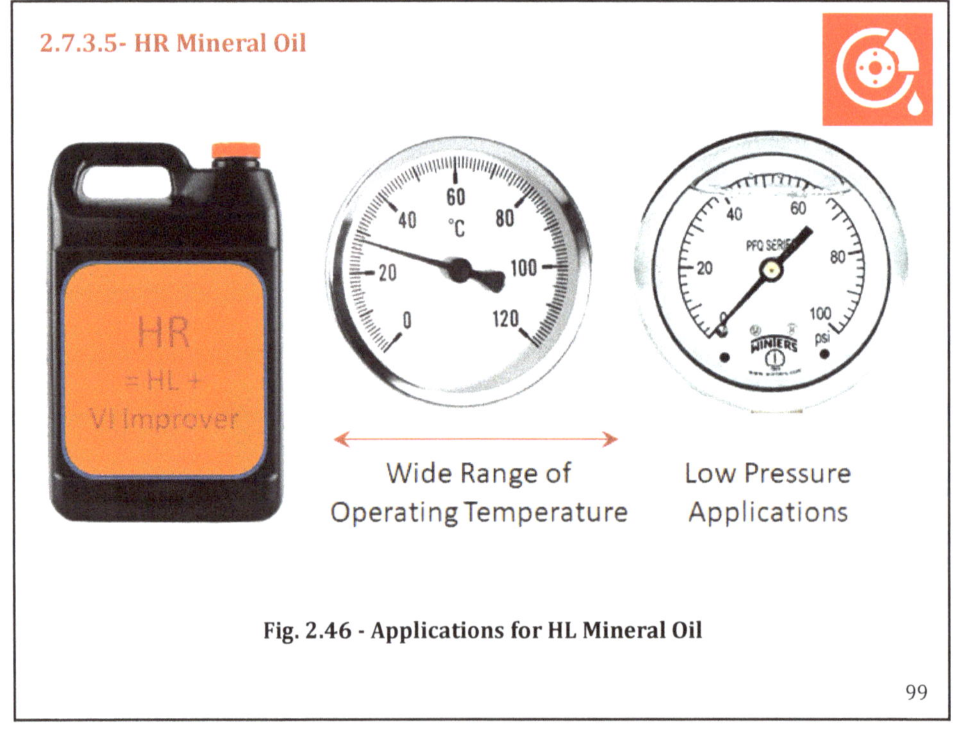

Fig. 2.46 - Applications for HL Mineral Oil

99

2.7.3.6- Special Mineral Oil

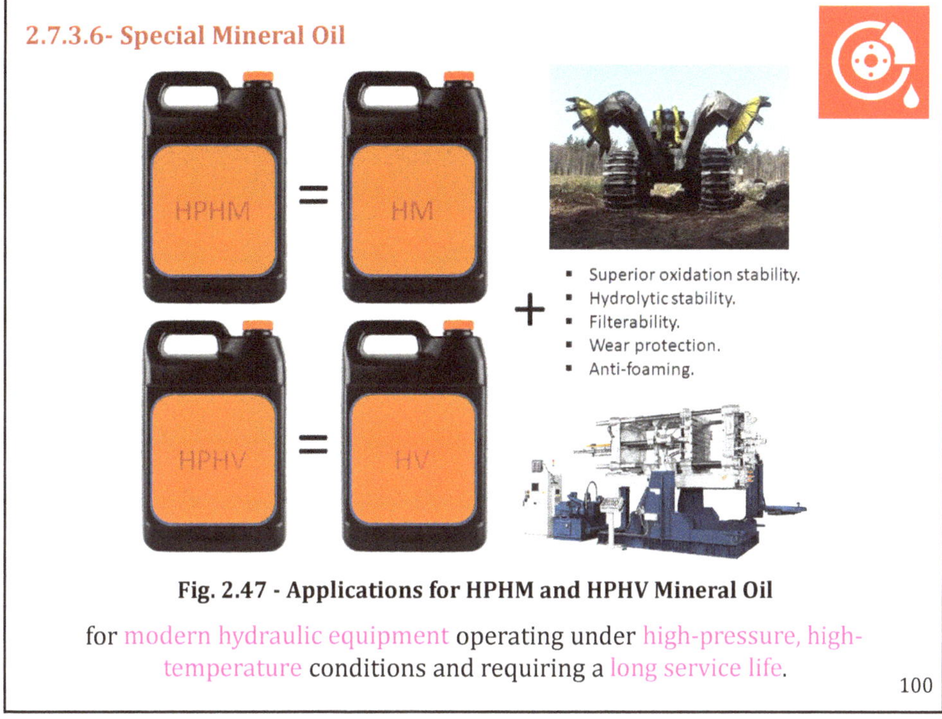

Fig. 2.47 - Applications for HPHM and HPHV Mineral Oil

for modern hydraulic equipment operating under high-pressure, high-temperature conditions and requiring a long service life.

100

100

2.7.3.7- Military-Graded Mineral Oil

Applications:

- Used for various hydraulic and brake systems of aircrafts.
- Other defense systems.
- Used as a superior hydraulic fluid for general hydraulic machinery.
- Pump manufacturer should be consulted when running military fluids.

Characteristics:

- Manufactured from highly refined "super clean" base oil -.
- Superior temperature-viscosity properties.
- Wide temperature range of -54°C to 135°C due to its superior temperature-viscosity properties.

101

101

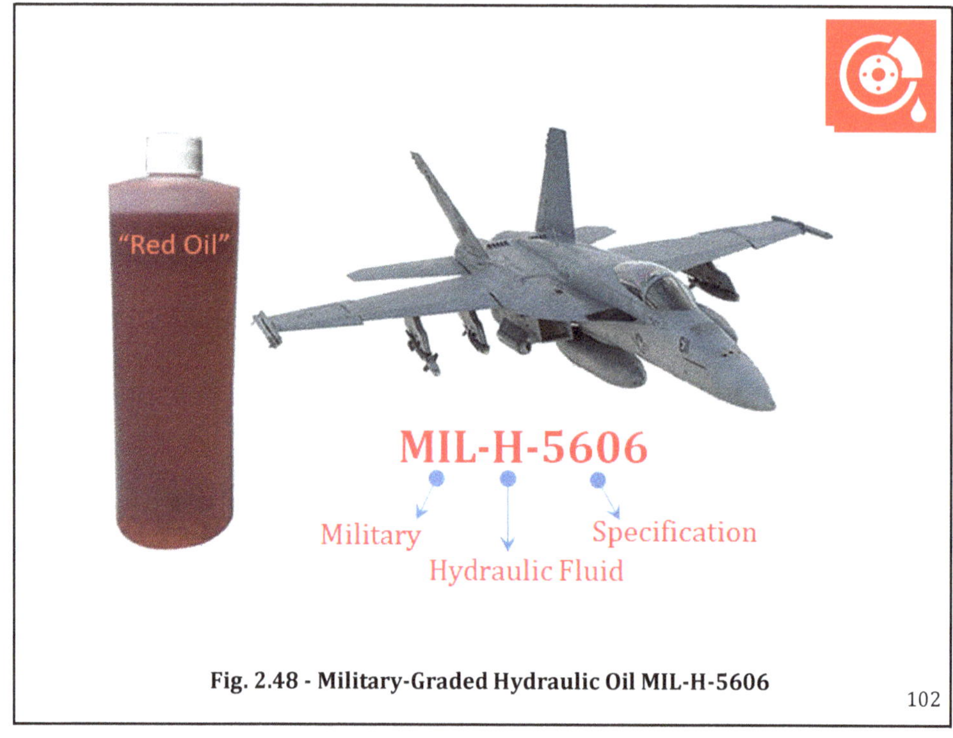

Fig. 2.48 - Military-Graded Hydraulic Oil MIL-H-5606

102

102

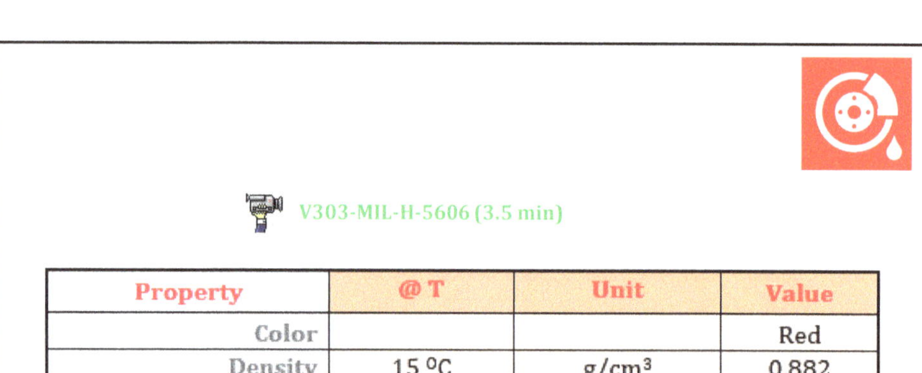

Property	@ T	Unit	Value
Color			Red
Density	15 °C	g/cm³	0.882
Kinematic Viscosity	-40 °C	mm²/s	491
Kinematic Viscosity	40 °C	mm²/s	13.8
Kinematic Viscosity	100 °C	mm²/s	5.1
Viscosity Index			Min 300
Flash Point		°C	96
Pour Point		°C	-77

Table 2.18 - Example of Properties for a Military-Graded Hydraulic Oil

103

103

2.8- Fire-Resistant Hydraulic Fluids

2.8.1- Definition of Fire-Resistant Hydraulic Fluids

Fire-Resistant Hydraulic Fluids are: the ones that:

- Resist ignition.
- Resist spreading of fire if they ignite.
- Will not continue to burn after removing the ignition source.

Using such fluids results in:

- Safer work conditions for the operators.
- Reduced insurance costs.

104

104

2.8.2- Composition and Standard Designations of Fire-Resistant Hydraulic Fluids

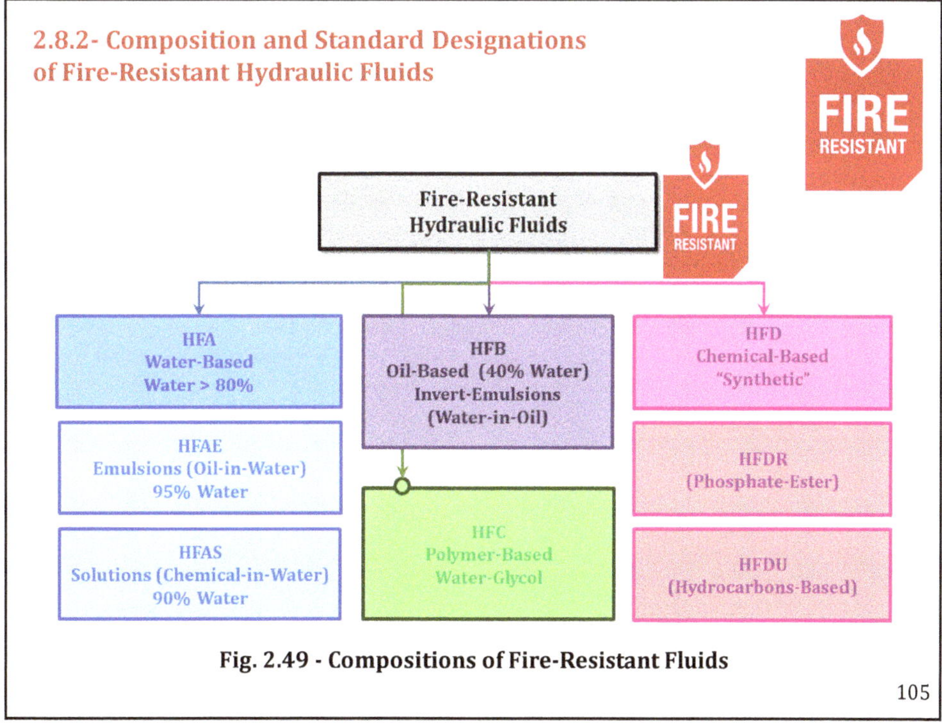

Fig. 2.49 - Compositions of Fire-Resistant Fluids

105

105

DIN Code 51502	ISO Code 6743-4	Description	Water %	Other Contents
-	HFA	Water-Based	> 80%	mineral oil + additives
HS-B	HFB	Oil-Based (Water-in-Oil) "Invert-Emulsion"	~ 40%	60% (mineral oil + add.)
HS-C	HFC	Polymer-Based (Water-Glycol)	~ 40%	60% Glycol
-	HFD	Synthetic	~ 0	free of water & mineral oil

Table 2.19 - Standard Designation for Fire-Resistant Hydraulic Fluids (According to ISO 6743-4 and DIN 51502)

106

106

2.8.3- General Properties of Fire-Resistant Hydraulic Fluids

Properties of the Four Groups of Non-Flammable Fluids

Properties	HFA	HFB	HFC	HFD
Kinematic Viscosity (cSt) at 40 °C (105 °F)	32-68	80-100	20-70	22-100
Specific Gravity at 40 °C (105 °F)	0.85-0.88	0.91-0.93	1.05-1.1	1.02-1.2
Working Temperature Range	3-55 °C (37-131 °F)	-25 to 60 °C (-13 to 140 °F)	-25 to 60 °C (-13 to 140 °F)	-20 to 150 °C (-4 °F to 302 °F)
Water Content (weight %)	80-89	35-55	35-55	Nil
Stability	Emulsions: poor Solution: very good	Acceptable	Good	Very good
Heat Transfer	Excellent	Very Good	Good	Poor
Lubrication	Acceptable	Good	Good	Excellent
Corrosion Resistance	Poor to acceptable	Good	Good	Excellent
Auto Ignition Temperature	Not possible	443 °C (830 °F)	443 °C (830 °F)	398 °C (750 °F)
Heat of Combustion	29.1 kJ/g	16.3 kJ/g	5.3 kJ/g	21.1 kJ/g
Environmental Risk	Least Cost to Waste	Special Waste	Special Waste	Special Waste
Seal Material	NBR, FKM		NBR	FKM, EPDM[1]

(1) Only for pure (mineral oil free) phosphate ester (HFD-R)

Table 2.20 - General Properties of Fire-Resistant Hydraulic Fluids (Courtesy of Parker)

107

107

2.8.4- HFA (Water-Based) Fire-Resistant Hydraulic Fluids

2.8.4.1- Composition and Standard Designations of HFA Fluids

FIRE RESISTANT

- Water-Based (water > 80%).
- Typically, sold as concentrates and diluted prior to use in service.
- Where possible, use distilled or de-ionized water to avoid introducing harmful contaminants that may cause emulsion problems.

HFA
Water-Based
Water > 80%
+
Concentrates
(oil or chemicals)

Water > 80%
Continuous Phase

Concentrates
(diluted prior to use)

Fig. 2.50 - Composition of HFA Fire-Resistant Hydraulic Fluids

108

108

DIN Code 51502	ISO Code 6743-4	Description	% Water Content	Other Contents
-	HFA	Water-Based fire-resistant HF	> 80%	mineral oil + additives
HS-A	HFAE	Oil-in-Water "Emulsion"	95%	5% (mineral oil + adds.)
-	HFAS	Chemicals-in-W. "Solutions:	90%	10% Synthetic

Table 2.21 - Standard Designation of HFA Fire-Resistant Hydraulic Fluids (According to ISO 6743-4 and DIN 51502)

109

109

2.8.4.2- Main Features of HFA Fire-Resistant Hydraulic Fluids

Advantages:

- ☺ Because of the high-water content, they offer better cooling than other FR hydraulic fluids.
- ☺ Less cost of disposal or recycling.
- ☺ Less environmental pollution in case of leakage.
- ☺ Higher bulk modulus because of the water content.

110

110

Disadvantages:

- ☹ Require distilled, de-ionized or demineralized water.
- ☹ Require anti-wear and anti-rust additives.
- ☹ Require anti-freezing additives and plumbing insulation.
- ☹ Require a positive head reservoir.
- ☹ Have low viscosity & low ability to lubricate or to seal tight clearances.
- ☹ Reduced life expectancy of hydraulic components, especially pumps.
- ☹ Work under reduced pressures (1000 psi).
- ☹ Work under reduced operating speeds (1000 rpm).
- ☹ Work under reduced operating temperature
- ☹ Due to water evaporation, max working T is limited to 60 ^{o}C (140 ^{o}F).
- ☹ Greater Potential for cavitation development (low water vapor pressure).
- ☹ Water loss (evaporation) must be continuously monitored and made up.
- ☹ Water loss increases the fluid viscosity.
- ☹ Water loss lowers fluid fire resistivity.
- ☹ Water loss increases concentration of additives in the fluid.
- ☹ Not compatible with some seal materials and paints.
- ☹ Heat, water, and Oxygen are good conditions for bacteria formation.

111

111

Applications:

- Because of low viscosity, they are used for low speed pumps of large displacements in applications such as deep drawing presses (1).
- Because of high-water content, they are used as machine tool coolant (2).

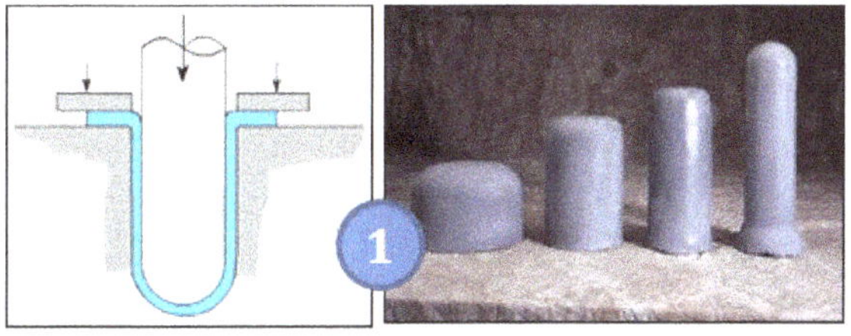

Fig. 2.51 - Application Examples for HFA Fire-Resistant Hydraulic Fluids

112

112

2.8.5- HFB (Oil-Based) Fire-Resistant Hydraulic Fluids

2.8.5.1- Composition and Standard Designations for HFB Fluids

- **HFB:** Water-in-Oil (Inverted Emulsions).
- Improves lubricating quality as compared to the emulsions.

60% Mineral Oil as Continuous Phase

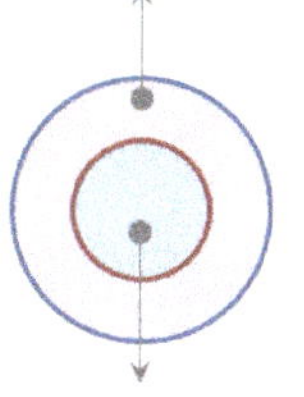

\+

- Emulsifying agents.
- Anti-wear additives.
- Rust inhibitors.

40% Water

Fig. 2.52 - Composition of HFB Fire-Resistant Hydraulic Fluids

113

113

2.8.5.2- Main Features of HFB Fire-Resistant Hydraulic Fluids

Maintenance of the Water Content:
- Must remain between 35 to 45 percent and monitored on daily basis.
- Water content < 35% → loss of fire-resistant properties
- Water content > 45% → loss anti-wear characteristics Therefore, it is

Operating Temperature:
- Should not exceed 50 ^{O}C (122 ^{O}F) to reduce water evaporation.
- Freezing temperatures can cause the emulsion to break.

Compatibility with Seals:
- Compatible with standard hydraulic seal material.

Foaming and Aeration:
- Can be greater problems with water-in-oil emulsions than mineral oils.
- Therefore, every effort must be made to eliminate foaming and aeration.

114

114

Solvent Effect:
- Invert emulsions may have a solvent effect on paints.
- Therefore, unpainted reservoirs are required for such fluids.

Viscosity:
- These fluids exhibit a temporary reduction in viscosity when subjected to the high shear rates, which exist in most hydraulic pumps.
- As a result, inverted emulsions are manufactured to a viscosity level somewhat higher than that of petroleum oils used in similar hydraulic applications.

Filtration Requirements:
- Inverted emulations have higher specific gravity
- → they hold particulate contamination in suspension much more than mineral oil.
- → Standard filtration for mineral oils may not be satisfactory
- → Use depth-type absorbent filters made of inorganic media or metal.
- → Do not use surface-type filters made of paper or wood.

115

115

2.8.6- HFC (Polymer-Based) Fire-Resistant Hydraulic Fluids

FIRE RESISTANT

Main Features:

- ☺ Provide better fire resistance than inverted emulsions.
- ☺ The viscosity of water-glycol fluids is comparable to mineral oils
- ☹ Highly alkaline, so they attack magnesium, cadmium and zinc.
- ☹ Dissolve most paints and varnishes.
- ☹ Require careful selection of components and filter media.

HFC Polymer-Based (Water-Glycol)

60% Glycol

40% Water

\+

- Anti-wear additives.
- Viscosity improvers.

Fig. 2.53 - Composition of HFC Fire-Resistant Hydraulic Fluids 116

116

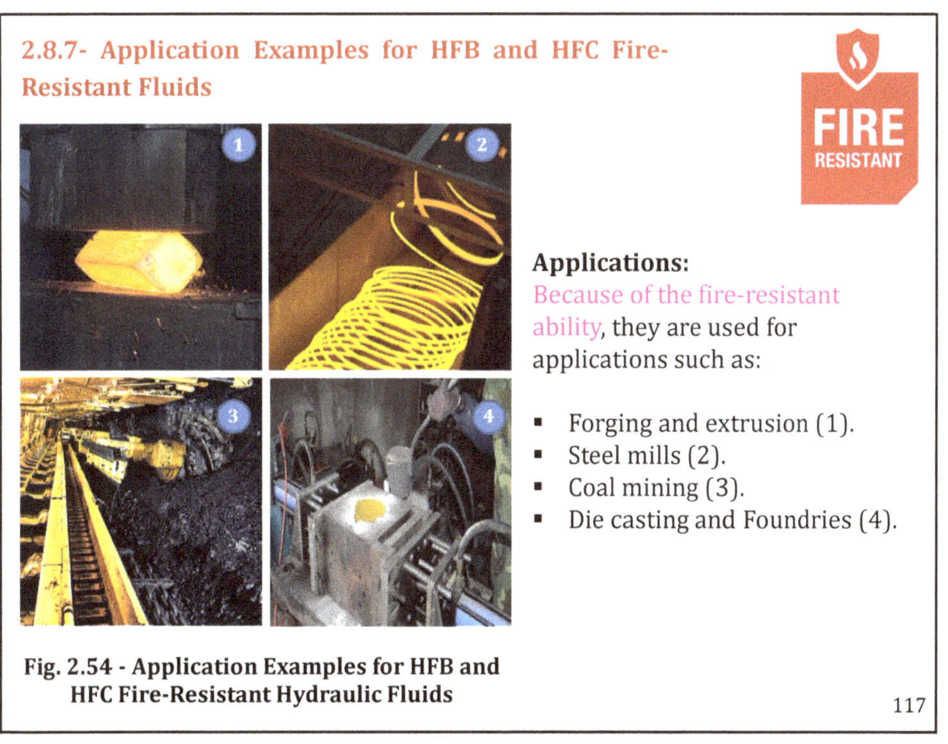

2.8.7- Application Examples for HFB and HFC Fire-Resistant Fluids
FIRE
RESISTANT
1
2
3
4
Applications:
Because of the fire-resistant ability, they are used for applications such as:
Forging and extrusion (1).
Steel mills (2).
Coal mining (3).
Die casting and Foundries (4).

Fig. 2.54 - Application Examples for HFB and HFC Fire-Resistant Hydraulic Fluids 117

117

2.8.8- HFD (Synthetic-Based) Fire-Resistant Hydraulic Fluids

Definition: Synthetic means custom-made hydraulic fluids containing additives that produce high levels of certain properties to meet operating conditions desired by end users.

2.8.8.1- Composition and Standard Designations for HFD Hydraulic Fluids

DIN Code 51502	ISO Code 6743-4	Description	% Water Content	% Other
-	HFD	Synthetic	-	free of water & mineral oil
HS-D	HFDR	Phosphate Esters	-	free of water & mineral oil
HS-D	HFDU	Hydrocarbon-Based	-	free of water & mineral oil

Table 2.22 - Standard Designation for HFD Fire-Resistant Hydraulic Fluids (According to ISO 6743-4 and DIN 51502)

118

118

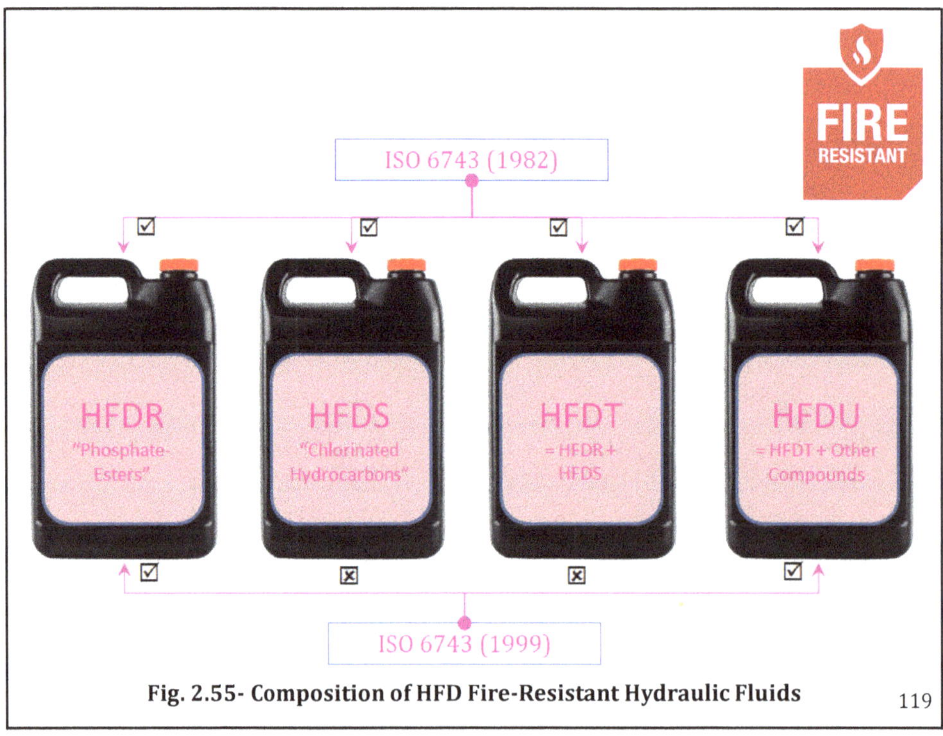

Fig. 2.55- Composition of HFD Fire-Resistant Hydraulic Fluids

119

119

2.8.8.2- Main Features of HFD Fire-Resistant Hydraulic Fluids

Advantages:

☺ Have superior fire resistivity and very high flash point.

☺ Have viscosities and lubricity similar to mineral oils.

☺ Have higher viscosity indexes and allow for high working temperature.

☺ Have superior oxidation stability.

☺ Does not affect pump life expectancy.

☺ Phosphate esters are compatible with all common metals except aluminum.

Disadvantages:

☹ Dissolve most paints, insulations and varnishes.

☹ Require special seals.

☹ Expensive as compared to other fire-resistant fluids.

120

120

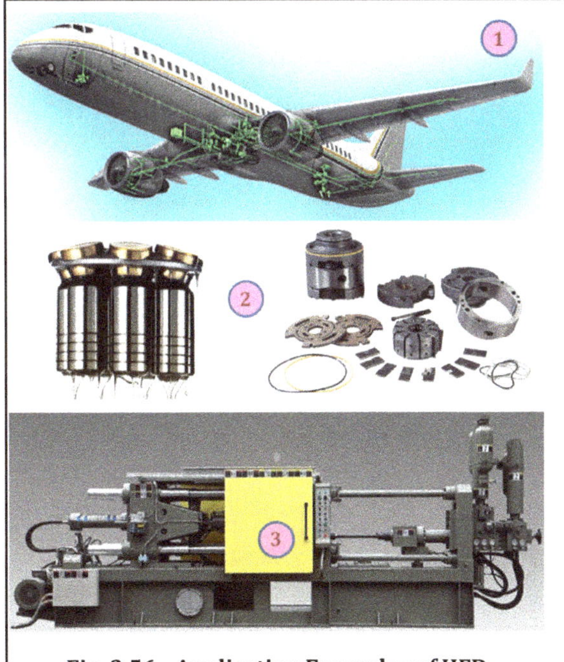

Applications: HFD synthetics have been used in a variety of fluid power applications including:

1. Aerospace
2. Superior hydraulic fluid for long-life components
3. High working temperature applications such Die-Casting machines.

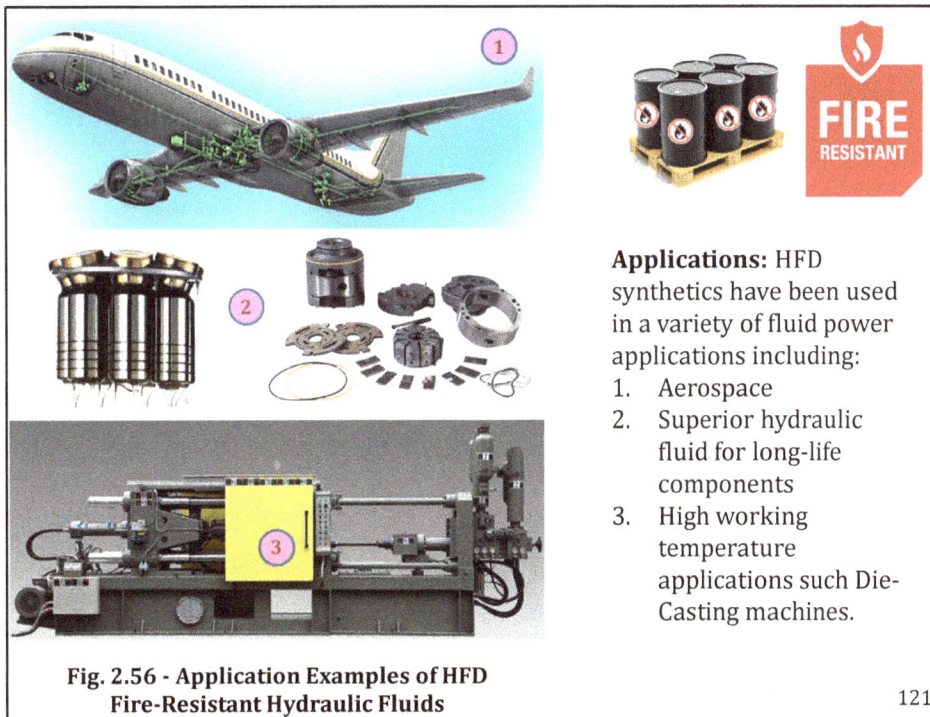

Fig. 2.56 - Application Examples of HFD Fire-Resistant Hydraulic Fluids

121

121

A water-in-oil emulsion

a. Is the same as an oil-in-water emulsion.

c. Is better for lubrication than oil-in-water emulsions.

d. Is a contamination issue that must be resolved.

e. Is the most effective fluid for removing heat.

122

122

2.8.9- Testing of Fire-Resistant Hydraulic Fluids

- V306-Flammability Test Fluid (2 min)
- V294-Flammability Test (1.5 min)

FIRE RESISTANT

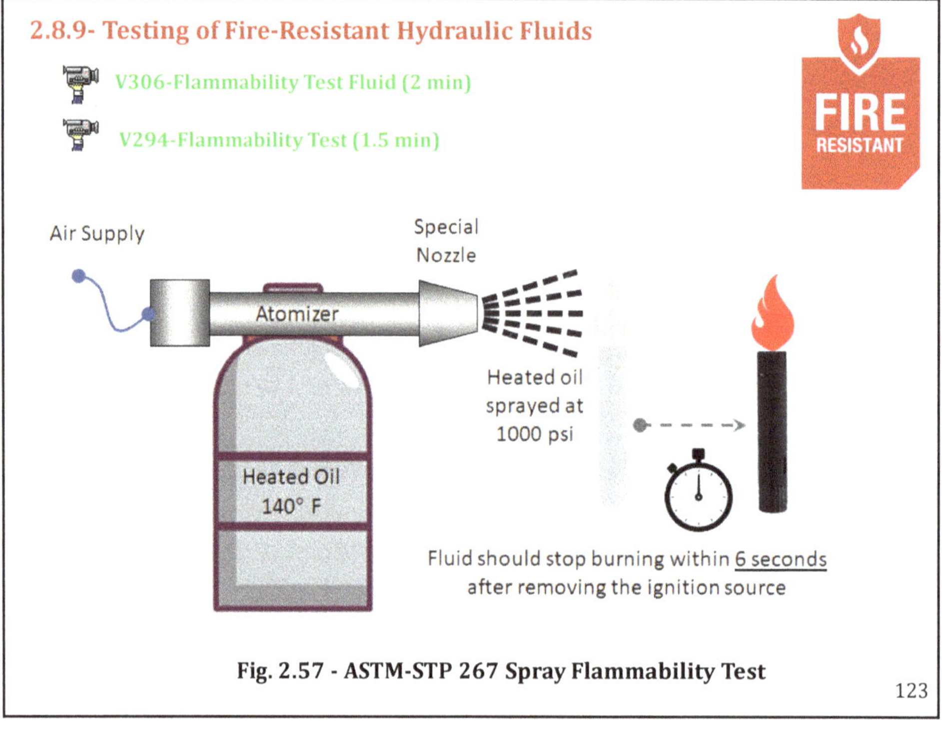

Fig. 2.57 - ASTM-STP 267 Spray Flammability Test

123

123

 V292-Fire Resistant Fluid (1 min)

Type of fluid	HFA Oil in water emulsion	HFB Water in oil emulsion	HFC Watery polymer solution	HFD Water free synthetic
Operating temperature*	5 – 55 °C [40 – 130 °F]	5 – 60 °C [40 – 140 °F]	-20 – 60 °C [-4 – 140 °F]	10 – 70 °C [50 – 160 °F]
Water content*	> 80%	> 40%	> 35%	–
Typical roller bearing life**	< 5%	30 – 35%	10 – 20%	50 – 100%

* The temperature range and the water content are based on the specific fluid properties.

** Mineral based fluid is 100%.

Table 2.23 - Operating Parameters for Typical Fire-Resistant Hydraulic Fluids

124

124

2.9- Environmental-Friendly Hydraulic Fluids

2.9.1- Definition of Environmental-Friendly Hydraulic Fluids

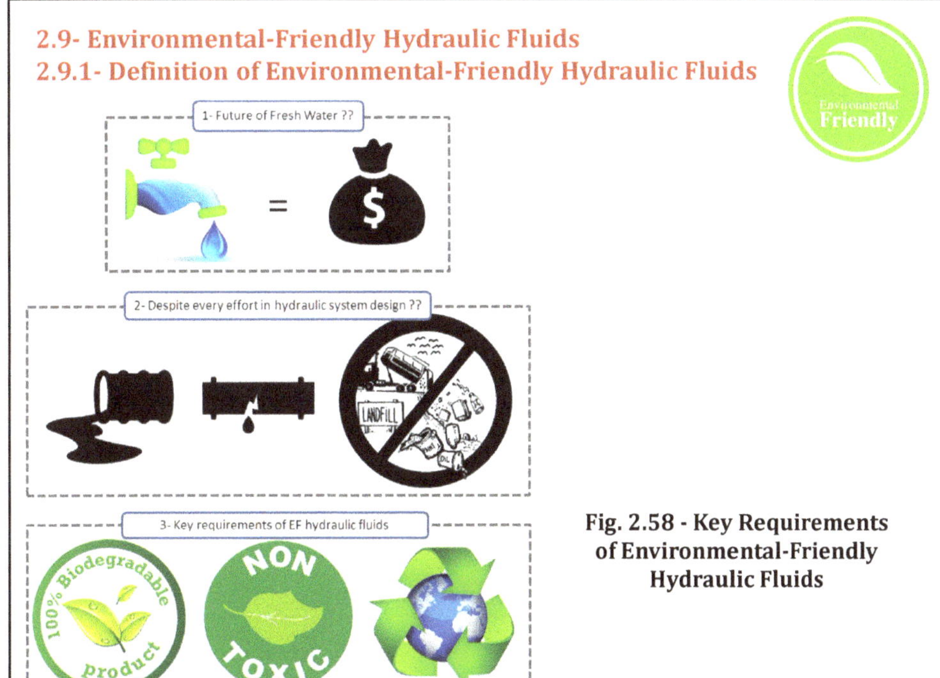

Fig. 2.58 - Key Requirements of Environmental-Friendly Hydraulic Fluids

125

125

▪ If a large volume of a biodegradable fluid is spilled, it must be treated like a mineral oil.
▪ Used biodegradable fluid must also disposed of according to federal, state, and local regulations.
Friendly
V288-Biodegradable Fluid (2.5 min)

Fig. 2.59 - Definition of Biodegradable Hydraulic Fluids

126

126

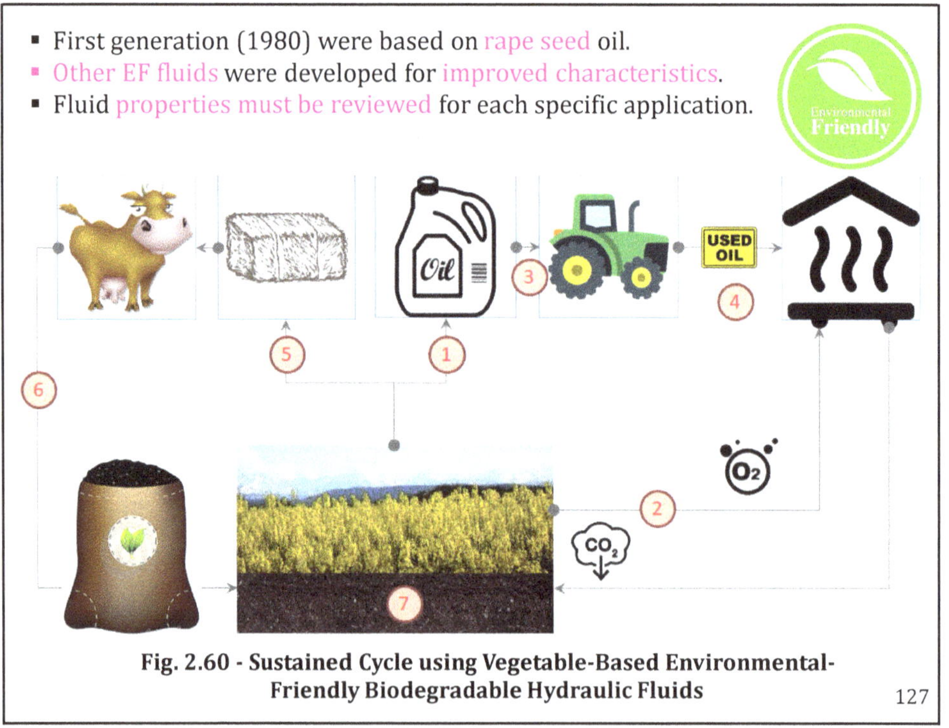

▪ First generation (1980) were based on rape seed oil.
▪ Other EF fluids were developed for improved characteristics.
▪ Fluid properties must be reviewed for each specific application.
Friendly
Oil
USED OIL
1
2
3
4
5
6
7
O_2
CO_2

Fig. 2.60 - Sustained Cycle using Vegetable-Based Environmental-Friendly Biodegradable Hydraulic Fluids

127

127

2.9.2- Composition and Standard Designations for Environmental-Friendly Hydraulic Fluids

Environmental-Friendly

Bio-Based

Synthetic-Based

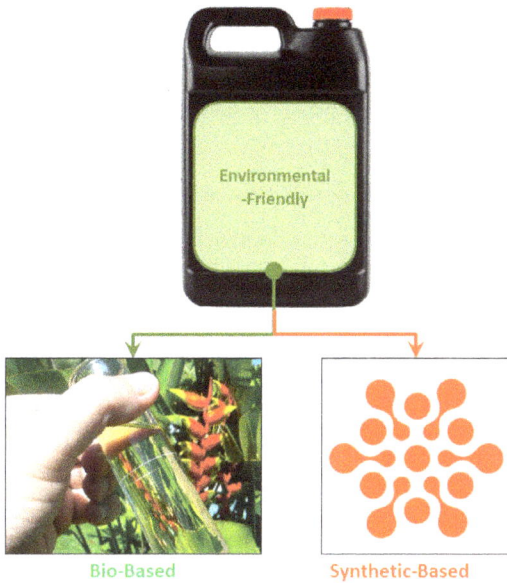

EF hydraulic fluids are:

- Vegetable-Based: e.g. Rape Seed, Canola Oil, and Soya Beans Oil.
- Synthetic-Based: e.g. Glycol-Based and Ester-Based.

Fig. 2.61 - Composition of Environmental-Friendly Hydraulic Fluids

128

128

ISO 6743 Code	Composition
HETG	**Vegetable-Based** (Hydraulic Environmental Triglyceride)
HEPG	**Glycol-based** (Hydraulic Environmental Poly Glycol)
HEES	**Synthetic Easter-Based** (Hydraulic Environmental Ester Synthetic)
HEPR	**Synthetic Hydrocarbons-Based** (HE Poly Alpha Olefin and Related Fluids)

Table 2.24 - Standard Designation for Environmental-Friendly Hydraulic Fluids (According to ISO 6743-4)

129

129

2.9.3- Main Features of Environmental-Friendly Hydraulic Fluids

Advantages:

- ☺ Used for environmental-sensitive applications.
- ☺ Filtration requirements are generally same as for mineral oils.
- ☺ Synthetic-based EF fluids are also fire-resistant and provide good lubrication properties.

Disadvantages:

- ☹ Fluid properties must be reviewed for each specific application (e.g. lubricity, wear resistance characteristics, and compatibility with seals and hydraulic components).
- ☹ Some FE fluids may require reduced pump P, T, and N.
- ☹ Biodegradable specs are highly affected if contaminated by mineral oils.
- ☹ If it replaces mineral oil or engine oil, a thorough flushing is necessary.
- ☹ Some EF fluids have higher specific gravity than mineral fluids.
- ☹ This may require adjusting inlet conditions to avoid cavitation.
- ☹ Expensive.
- ☹ Water contamination starts the biodegrading process.

130

130

Specific Features:

ISO 6743 Code	HETG	HEPG	HEES	HEPR
Biodegradability	Excellent	Good	Good	Acceptable
Viscosity	Limited	Good	Very Good	Excellent
Viscosity Index	Limited Not used Above 65 ^{o}C (149 ^{o}F)	Good Not used Above 65 ^{o}C (149 ^{o}F)	Very Good Used up to 100 ^{o}C (330 ^{o}F)	Excellent Used up to 100 ^{o}C (330 ^{o}F)
Oxidation Stability	Poor	Good	Very Good	Excellent

Table 2.25 - Specific Features of Environmental-Friendly Hydraulic Fluids

131

131

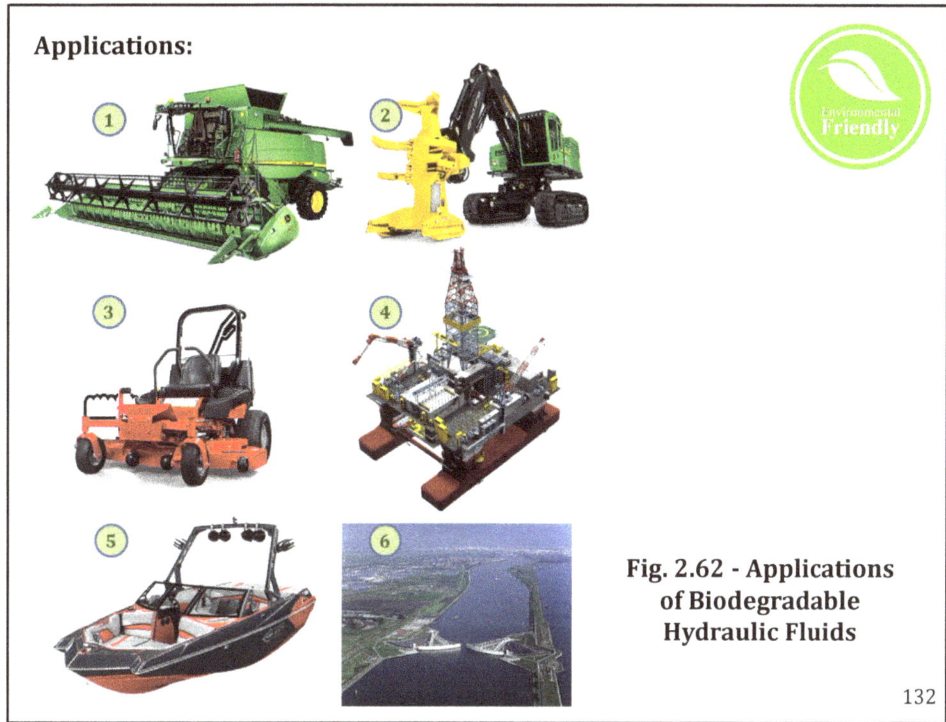

Fig. 2.62 - Applications of Biodegradable Hydraulic Fluids

132

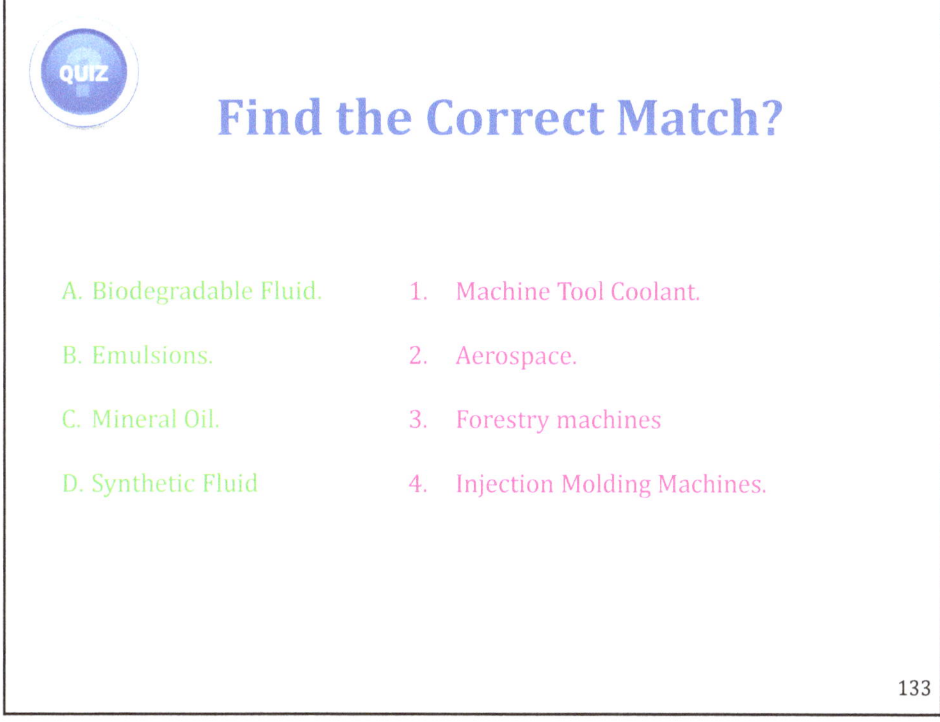

QUIZ
Find the Correct Match?
A. Biodegradable Fluid.
B. Emulsions.
C. Mineral Oil.
D. Synthetic Fluid
1. Machine Tool Coolant.
2. Aerospace.
3. Forestry machines
4. Injection Molding Machines.
133

133

2.10- Best Practices for Hydraulic Fluids Selection

If an ideal hydraulic fluid existed, it would have the following features:

- Be practically Incompressible (large Bulk Modulus).
- Provide proper lubrication (correct Density and Viscosity).
- Have stable viscosity at wide range of temperature (high Viscosity Index).
- Flow easily at low temperature (low Pour Point).
- Does not easily ignite and resists fire (high Flash Point and Ignition Point).
- Improves friction characteristics and reduces wear.
- Resists oxidation, rusting and corrosion.
- Demulsify water.
- Suppresses foam.
- Be compatible with seals and components typical metals.
- Be hydrolytically stable.
- Be Non-toxic, biodegradable, and friend to the environment.
- Be dielectric.
- Be inexpensive.

134

134

Unfortunately, there is no one hydraulic fluid that meets all these requirements together. Therefore, machine builders should compromise the fluid requirements to achieve the best out of their machines.

2.10.1- Manufacturer-Based Fluid Selection

The best advice for selecting the hydraulic fluid is given by the components and systems manufacturers.

2.10.2- Application-Based Fluid Selection

- Most of hydraulic fluids in service are mineral oil based because they generally provide excellent performance at a relatively low cost.
- Fire-Resistant hydraulic fluids are basically used for applications where the operating temperature is extremely high or where fire hazard exists.
- Environmental-friendly hydraulic fluids are mainly used for machines that work nearby the animal life or vegetation life.

135

135

Application	Suitable fluids *)	Max. operating pressure	Ambient temperature	Site
Vehicle construction	1 . 2 . 3	250 bar	- 40 to + 60 °C	inside & outside
Mobile machines	1 . 2 . 3	315 bar	- 40 to + 60 °C	inside & outside
Special vehicles	1 . 2 . 3 . 4	250 bar	- 40 to + 60 °C	inside & outside
Agriculture and forestry machines	1 . 2 . 3	250 bar	- 40 to + 50 °C	inside & outside
Ship building	1 . 2 . 3	315 bar	- 60 to + 60 °C	inside & outside
Aircraft construction	1 . 2 . 5	210 (280) bar	- 65 to + 60 °C	inside & outside
Conveyors	1 . 2 . 3 . 4	315 bar	- 40 to + 60 °C	inside & outside
Machine tools	1 . 2	200 bar	18 to 40 °C	inside
Presses	1 . 2 . 3	630 bar	18 to 40 °C	mainly inside
Ironworks, rolling mills, foundries	1 . 2 . 4	315 bar	10 to 150 °C	inside
Steelworks, water hydraulics	1 . 2 . 3	220 bar	- 40 to + 60 °C	inside & outside
Power stations	1 . 2 . 3 . 4	250 bar	- 10 to + 60 °C	mainly inside
Theatres	1 . 2 . 3 . 4	160 bar	18 to 30 °C	mainly inside
Simulation and testing devices	1 . 2 . 3 . 4	1000 bar	18 to 150 °C	mainly inside
Mining	1 . 2 . 3 . 4	1000 bar	up to 60 °C	outside & underground
Special applications	2 . 3 . 4 . 5	250 (630) bar	- 65 to 150 °C	inside & outside

*) 1= mineral oil; 2= synthetic hydraulic fluids; 3= ecologically acceptable fluids;
4= water, HFA, HFB; 5= special fluids

Table 2.26 - Application-Based Hydraulic Fluids Selection (Courtesy of Bosch Rexroth)

136

136

2.10.3- Properties-Based Fluid Selection

Step 1: Define the fluid category (Mineral based + Fire Resistant + EF)

Step 2: Select the actual fluid within the selected category based on the requirements of the working conditions.

137

137

2.10.4- Compatibility-Based Fluid Selection

	Fluid Types					
Seal materials	Petroleum oil	Water-in-Oil Emulsion	Water Glycol	Phosphate Ester*	Chlorinated hydrocarbon	Synthetic with petroleum fractions
Buna-N (Acrylonitrile)	Excellent	Excellent	Very Good	Poor	Poor	Poor
Neoprene (Chloroprene)	Good	Good	Good	Poor	Poor	Poor
Butyl	Poor	Poor	Good	Fair to good	Poor	Poor
Silicone	Fair	Fair	Fair to poor	Fair to good	Poor to fair	Fair
Ethelene-Propylene	Poor	Poor	Good to excellent	Excellent	Fair	Poor
Viton® (Fluorocarbon)	Excellent	Excellent	Excellent	Good to Excellent	Good to Excellent	Good to Excellent
Metals	Conventional	Conventional	**	Conventional	Conventional	Conventional
Pipe Sealants	Conventional, Loctite® or Teflon® tape	Conventional, Loctite® or Teflon® tape	Loctite® or Teflon® tape	Loctite® or Teflon® tape	Loctite® or Teflon® tape	Loctite® or Teflon® tape

Review Table 2.15

138

138

2.10.5- Viscosity-Based Fluid Selection

The key requirements for selecting the correct viscosity are to:

- Provide better thermal stability. Hence, maximize the working temperature range.
- Maximize the overall efficiency. Hence, provide efficient power transmission.
- Minimize internal leakage. Hence, improves the productivity of the machine.
- Minimize internal friction. Hence, improves components reliability.

Unfortunately, one viscosity can't accomplish all these needs, so the viscosity itself should be compromised for the best application.

139

139

Hydraulic Fluids Selection For:

Video 600 (0.5 min)

- For Thermal Stability: use multi-grade oil.
- For Maximum Energy Saving: work around optimum range.
- For Maximum Productivity: work at high volumetric efficiency.
- For Maximum Reliability: work at high mechanical efficiency.

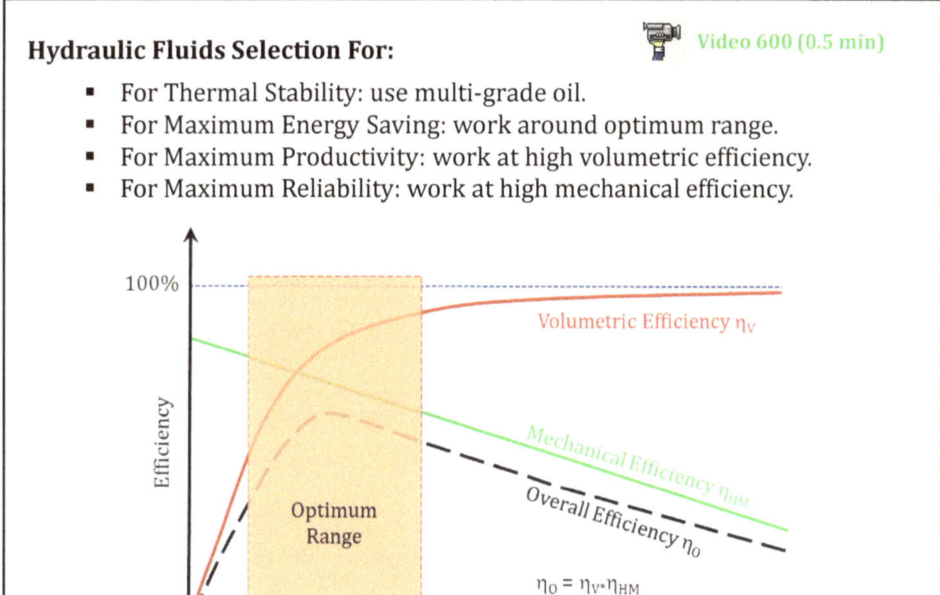

Fig. 2.63 - Pump Efficiencies versus Stribeck Number

140

140

2.10.6- Additives-Based Fluid Selection

An additive package is selected to meet the needs for specific operating conditions. However, some of the additives work against each other. For examples:

- Viscosity index improvers work against foam suppressors.
- Demulsifiers improve the surface tension and the foam suppressant reduces the surface tension.
- Anti-rust blocks anti-wear from the surface and the wear rate will be high.

That's why it is necessary to optimize hydraulic fluid formulation with a balanced additive package.

V703-LB-015-Antifoaming
V704-LB-016-Aeration

2.10.7- Cost-Based Fluid Selection

The following hydraulic fluids are listed with cost ascending order. Each hydraulic fluid is followed by a cost factor with respect to the mineral oil:

- Mineral oil, 1.
- Emulsions, 1.5-2.
- Vegetable-based oil, 2.5.
- Water-Glycol, 4.
- Phosphate Ester, 5.
- Hydrocarbon-based Synthetic fluids, 7.

141

141

2.11- Best Practices for Hydraulic Fluids Replacement

"oils are not similar and can't just replace each other"

- **Fluid Change Intervals:** Nowadays, with the available fluid analysis tools, fluid changing is decided based on the fluid conditions. However, unless otherwise stated, the old school suggests that:
- First change of hydraulic fluid is at 500 hours.
- Subsequent change:
  - Every 2000 hours or once a year (for mineral and synthetic oils).
  - Every 1000 hours or once a year (for water-based and bio-based).
- **Mixing Hydraulic Fluids:**
- Mixing hydraulic fluid could void the machine warranty.
- NO mixing hydraulic fluids of different types.
- NO mixing hydraulic fluids of same type but different additive packages.
- NO mixing of oil and greases.
- NO mixing of an environmental-friendly hydraulic fluid and a mineral oil. Biodegradability characteristics are highly affected by the mineral oil.
- If two fluids are mixed accidentally:
  - Frequent inspections for symptoms of sludge, foaming, oxidations, etc.
  - If any symptoms appeared, flushing is required immediately.
- Making up fluids must be carried out using recommended filtration unit. 142

142

Buying Oil Recommendations: When buying oil in bulk, buyers have a right to set specific certified requirements to ensure the quality.

Sampling of New Oil

- Representative samples must be drawn from purchased fluid.
- An analysis certificate must include at least the following items:
  - Visual inspection.
  - Viscosity @ 40°C.
  - Density.
  - Total Acid Number.
  - Air bubble separation time.
  - Contaminants, gravimetric or ISO cleanliness code.

Claims

- If the oil supplied does not fulfill requirements, returning the consignment might be considered.
- If the problem can be corrected, new samples must be approved.
- The supplier must pay all costs, including machinery failure and downtime.

143

143

Which of the hydraulic fluids listed below are the most commercially used?

A. Petroleum-based (Mineral) oils.

B. Synthetic oil.

C. Fire-resistant fluid.

D. Bio-degradable fluid.

144

144

Which of the hydraulic fluids listed below has custom-made specification?

A. Petroleum-based (Mineral) oils.

B. Synthetic oil.

C. Fire-resistant fluid.

D. Bio-degradable fluid.

145

145

Which of the hydraulic fluids listed below are most recommended for mining applications?

A. Petroleum-based (Mineral) oils.

B. Synthetic oil.

C. Fire-resistant fluid.

D. Bio-degradable fluid.

146

146

Which of the hydraulic fluids listed below are most recommended for agricultural applications?

A. Petroleum-based (Mineral) oils.

B. Synthetic oil.

C. Fire-resistant fluid.

D. Bio-degradable fluid.

147

147

2.12- Best Practices for Hydraulic Fluids Storage

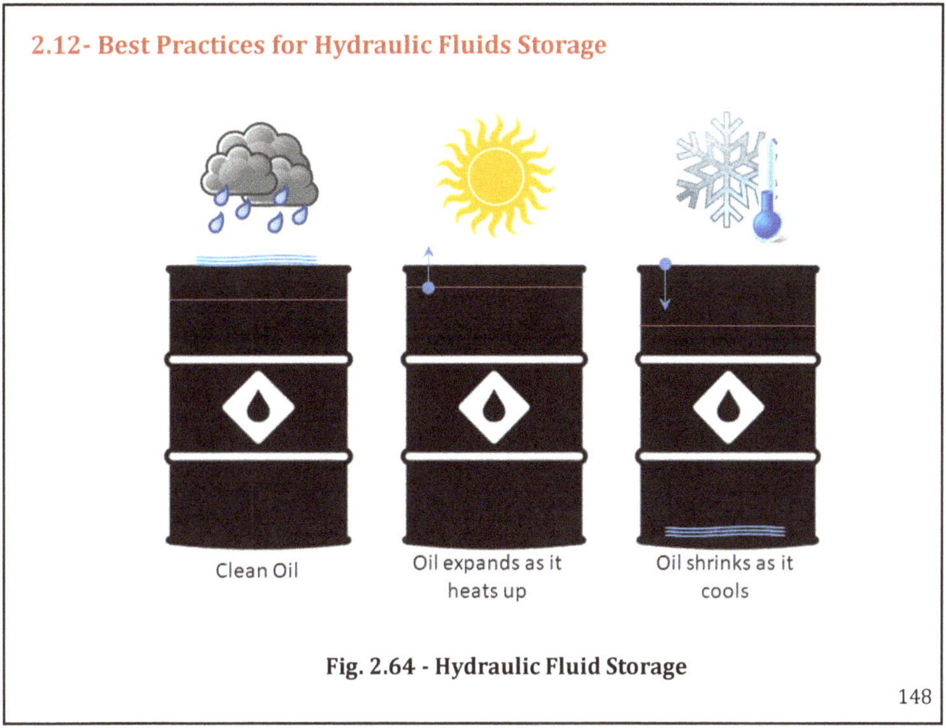
Clean Oil
Oil expands as it heats up
Oil shrinks as it cools

Fig. 2.64 - Hydraulic Fluid Storage

148

148

Best Practices for Hydraulic Fluids Storage

- Avoid storing containers outdoors in harsh weather.
- Store hydraulic fluid drums indoors in a cool, clean, dry, and well-ventilated storage space.
- Storage space must be properly designed and centralized for all lubes.
- Stock the hydraulic fluid drums in rotation, i.e. First-In, First-Out (FIFO).
- Top covers of the hydraulic fluid drums must be cleaned before opening.
- Use fluid filtration unit, with appropriate filtration level for the application, to transfer fluid from a drum to the system. Each filtration unit must be labeled and dedicated for one fluid type.
- For safety of oil transfer, storage space must be designed with spill-control system equipment, eye-wash station, fire-control, and fire-emergency plan logged with local fire department.

149

149

Hydraulic fluids should be stored?

A. In covered containers.
B. In a dry environment.
C. Indoors to avoid direct sunlight and rain.
D. All the above.

V700-LB-012-Hydraulic Fluids Review 1

V708-LB-019-LB-019-Hydraulic Fluids Review 2

150

150

Fig. 2.65 - Proper Hydraulic Fluid Storage

151

151

Chapter 2 Reviews

1. Which of the following statements describes liquid?
   A. A substance that can't overcome shear forces.
   B. A substance that takes the interior shape of a container with a flat surface.
   C. A substance that has internal friction is represented by a property named "viscosity"
   D. All the above.

2. Main function of hydraulic fluid is to?
   A. Resist foaming.
   B. Transmit energy between the pump and the actuator
   C. Resist fire.
   D. Lubricate, cool, clean and seal inside the hydraulic devices.

3. Auxiliary Functions of hydraulic are to?
   A. Resist foaming.
   B. Transmit energy between the pump and the actuator
   C. Resist fire.
   D. Lubricate, cool, clean and seal inside the hydraulic devices.

4. Which of the following viscosity ranges results in decreasing reliability, reducing productivity, and increasing power consumption of a hydraulic system?
   A. Viscosity lower than the recommended range.
   B. Viscosity higher than the recommended range.
   C. Viscosity lower or higher than recommended range.
   D. None of the above.

5. Hydraulic fluids with higher viscosity index can?
   A. Work better at high level of contamination.
   B. Maintain viscosity for wider range of working temperature.
   C. Maintain viscosity for wider range of working pressure.
   D. All the above.

6. The critical temperature, *Pour Point*, of a hydraulic fluid is defined as?
   A. The lowest temperature at which the fluid can be poured + 5.4 F (3 °C).
   B. The lowest temperature at which a liquid gives off vapor sufficient to ignite momentarily or flash when a flame is applied.
   C. The lowest temperature at which the fluid continues to burn even if the flame is removed.
   D. The lowest temperature at which a liquid ignites without an external flame or spark.

7. The critical temperature, *Flash Point*, of a hydraulic fluid is defined as?
   A. The temperature at which the fluid can be poured + 5.4 F (3 °C).
   B. The temperature at which a liquid gives off vapor sufficient to ignite momentarily or flash when a flame is applied.
   C. The temperature at which the fluid continues to burn even if the flame is removed.
   D. The temperature at which a liquid ignites without an external flame or spark.

8. The critical temperature, *Fire Point*, of a hydraulic fluid is defined as?
   A. The temperature at which the fluid can be poured + 5.4 F (3 °C).
   B. The temperature at which a liquid gives off vapor sufficient to ignite momentarily or flash when a flame is applied.
   C. The temperature at which the fluid continues to burn even if the flame is removed.
   D. The temperature at which a liquid ignites without an external flame or spark.

9. The critical temperature, *Ignition Point*, of a hydraulic fluid is defined as?
   A. The temperature at which the fluid can be poured + 5.4 F (3 °C).
   B. The temperature at which a liquid gives off vapor sufficient to ignite momentarily or flash when a flame is applied.
   C. The temperature at which the fluid continues to burn even if the flame is removed.
   D. The temperature at which a liquid ignites without an external flame or spark.

10. Results of a hydraulic fluid high pour point are?
    A. Chemical decomposition because of water contents in the oil.
    B. Orifice blockage, spool sticking, filter clogging and possible pump cavitation.
    C. Difficult machine starting and possible pump cavitation.
    D. Deterioration of rubber seals and gaskets.

11. Results of a hydraulic fluid high oxidation resistance are?
    A. Chemical decomposition because of oil contamination by water.
    B. Orifice blockage, spool sticking, filter clogging and possible pump cavitation.
    C. Difficult machine starting and possible pump cavitation.
    D. Deterioration of rubber seals and gaskets.

12. Results of a hydraulic fluid incompatibility with seals are?
    A. Chemical decomposition because of oil contamination by water.
    B. Orifice blockage, spool sticking, filter clogging and possible pump cavitation.
    C. Difficult machine starting and possible pump cavitation.
    D. Deterioration of rubber seals and gaskets.

13. Results of a hydraulic fluid low hydrolytic stability are?
    A. Chemical decomposition because of oil contamination by water.
    B. Orifice blockage, spool sticking, filter clogging and possible pump cavitation.
    C. Difficult machine starting and possible pump cavitation.
    D. Deterioration of rubber seals and gaskets.

14. Which of the following hydraulic fluids has better foam suppression ability?
    A. Hydraulic fluids with higher oxidation resistance.
    B. Fire resistant hydraulic fluids.
    C. Hydraulic fluid with higher bulk modulus.
    D. Hydraulic fluid with lower viscosity and blended with de-foaming additives.

15. How many broad categories of hydraulic fluids are available?
    A. 1.
    B. 2.
    C. 3.
    D. 4.

16. Which of the hydraulic fluids listed below are the most commercially used?
    A. Petroleum-based (Mineral) oils.
    B. Synthetic oil.
    C. Fire-resistant fluid.
    D. Bio-degradable fluid.

17. Which of the hydraulic fluids listed below has custom-made specification?
    A. Petroleum-based (Mineral) oils.
    B. Synthetic oil.
    C. Fire-resistant fluid.
    D. Bio-degradable fluid.

18. Which of the hydraulic fluids listed below are most recommended for mining applications?
    A. Petroleum-based (Mineral) oils.
    B. Synthetic oil.
    C. Fire-resistant fluid.
    D. Bio-degradable fluid.

19. Which of the hydraulic fluids listed below are most recommended for agricultural applications?
    A. Petroleum-based (Mineral) oils.
    B. Synthetic oil.
    C. Fire-resistant fluid.
    D. Bio-degradable fluid.

20. Hydraulic fluids should be stored?
    A. In covered containers.
    B. In a dry environment.
    C. Indoors to avoid direct sunlight and rain.
    D. All the above.

Chapter 2 Assignment

Student Name: -- Student ID: ------------------

Date: --- Score: -----------------------

Assignment 1: Fill in the shown below table.

Fluid	SG @ [70 ^{0}F = 21 ^{0}C]	Density ρ in $\left[\frac{kg_m}{m^3}\right]$	Density ρ in $\left[\frac{lb_m}{ft^3}\right]$
Mineral Oil	0.87 – 0.9		
Water-Based Oil	0.92		
Vegetable-Based Oil	0.93		
Water/Glycol	1.06		
Phosphoric Esters	1.150		

Assignment 2:

Find the Dynamic Viscosity for an oil that has a Kinematic Viscosity of 150 SUS and a Specific Gravity of 0.85?

Assignment 3:

In the figure shown below, A 1 Liter sample of oil fills a hydraulic cylinder. find the load drifting distance x.

Given:

- Oil has bulk modulus = case A (10,000 bar) and case B (15,000 bar).
- Pressure is increased by 100 bar.
- Cylinder effective area = 10 cc.

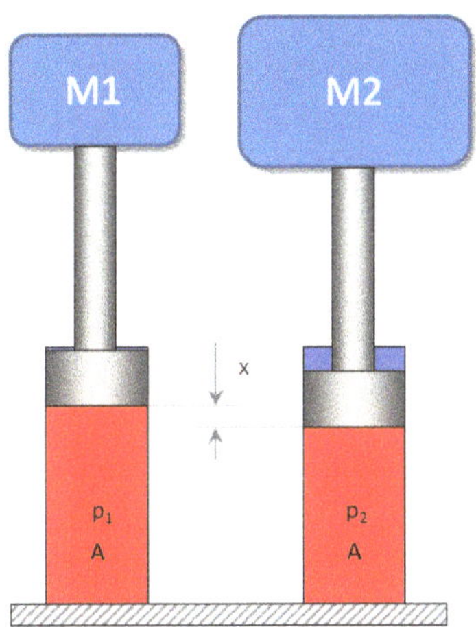

Assignment 4:

In the previously shown figure, if the cylinder is fully extended, find the cylinder pressure increase due to oil thermal expansion.

Given:

- Oil has bulk modulus = case 10,000 bar
- Temperature Change = case A (20 ^{O}F) and case B (40 ^{O}F).
- Coefficient of thermal expansion = 0.0005 (1/ ^{O}F).

Assignment 5:

"SAND" can't resist shear forces. Why it is not defined as "fluid"?

Assignment 6:

What is the liquid that has the highest surface tension?

A. Mineral Oil.
B. Mercury.
C. Fire-Resistant Hydraulic Fluids.
D. Biodegradable Hydraulic Fluids.

Assignment 7:
For a hydraulic system that drives a landing gear system in an aircraft, which fluid do you recommend??

A. Biodegradable hydraulic fluid.
B. Hydraulic fluid with low VI.
C. Hydraulic fluid with high VI.
D. Fire-resistant hydraulic fluid.

Assignment 8:

For each of the following two fluids

1. Hydraulic Fluid with Low Viscosity
2. Hydraulic Fluid with High Viscosity

Which of the following properties are indicated?

A. Better Foam Submission.
B. Lower Pour Point.
C. High Flash Point

Chapter 3
Energetic Contamination

Objectives:

This chapter presents the sources hydraulic fluids energetic contamination. For each source, the chapter explains how the system performance will be affected and possible recommendations to minimize such consequences.

Brief Contents:

0

0

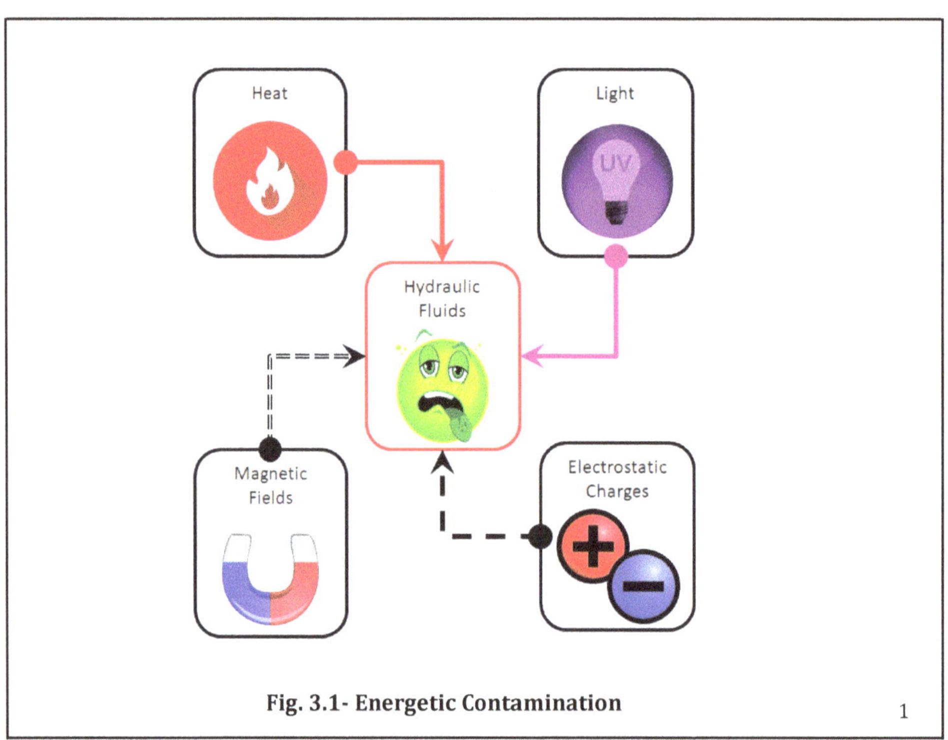

Fig. 3.1- Energetic Contamination

1

1

What is the most common type of energetic contamination for hydraulic fluids?

A. Heat.

B. Static Electricity.

C. Magnetic Field.

D. Light.

2

2

3.1- Contamination by Heat
3.1.1- Sources of Contamination by Heat

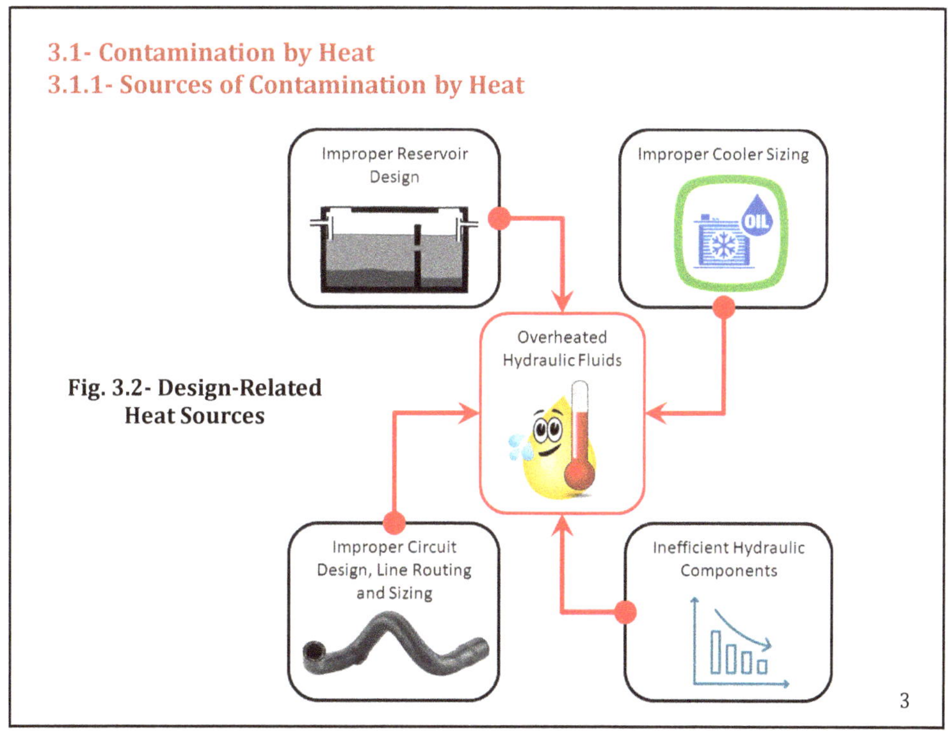

Fig. 3.2- Design-Related Heat Sources

3

3

QUIZ

How can an improper reservoir design contribute to heating up the system?

A- Under sizing the reservoir.

B- Placing suction line beside return line.

C- Not using a baffle plate between the suction and return side.

D. All the above.

4

4

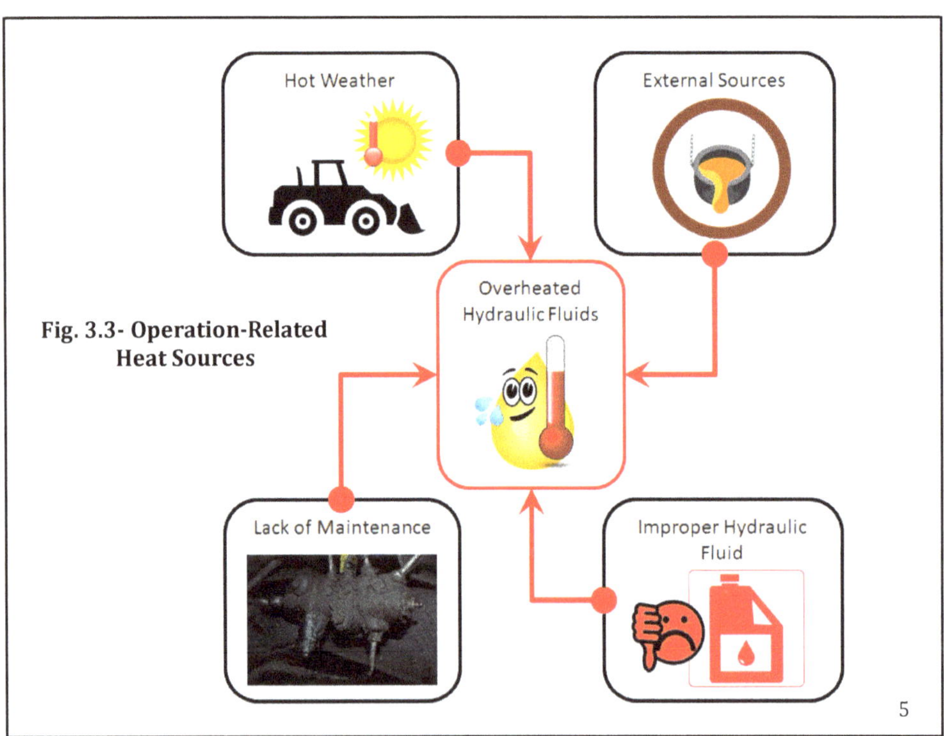

Fig. 3.3- Operation-Related Heat Sources

5

5

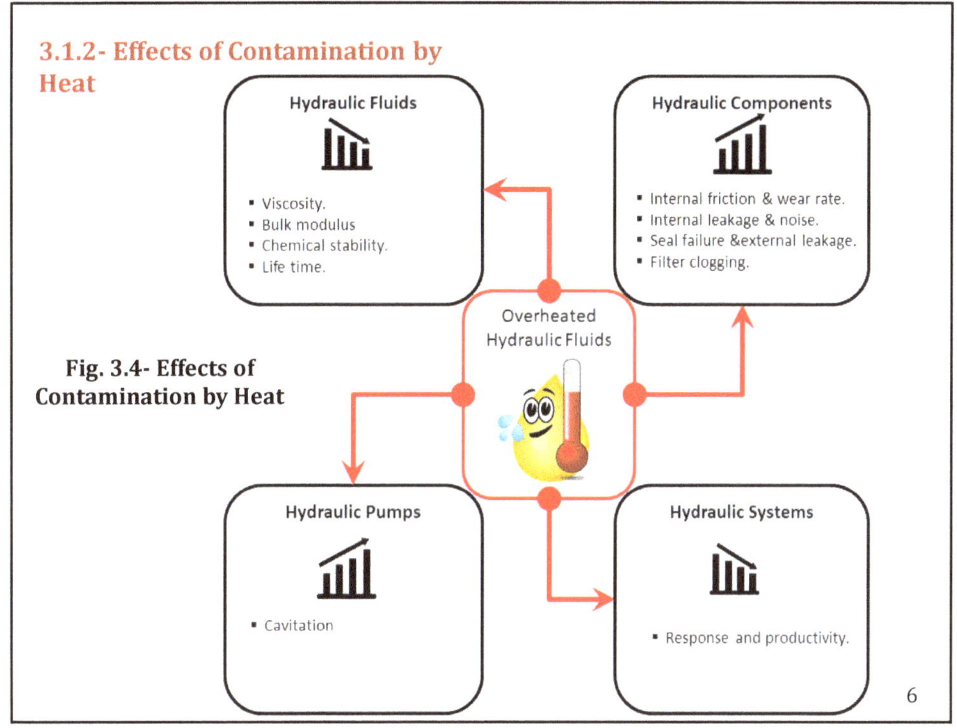

Fig. 3.4- Effects of Contamination by Heat

6

QUIZ

When a hydraulic system that is using petroleum-based hydraulic fluid is overheated, which of the following observations is True ?

A- Viscosity increase and better lubrication.

B- System experiencing less leakage and noise.

C- System experiencing more leakage and becomes sluggish and less productive.

D. Pump avoids cavitation.

7

7

As per most fluid manufacturers specifications:

- Optimum range of working temperature, typically From 38°C to 54°C (100°F to 130°F)
- Even though many fluids are operated above this temperature range.

Petroleum-Based Hydraulic Fluid:

- Critical working temperature is 70°C (158°F).
- Every 10°C (18°F) above the critical temperature
- → Oxidation rate is doubled.
- → Thus, cutting oil life to half.
- For example, running a system at a consistent 80°C (176°F) would reduce the fluid life by 75%.

8

8

Water-Based Hydraulic Fluids:

- Fluid **overheating** causes water evaporation.
- Water loss changes the ratio of water to base fluid.
- Water loss increases the fluid viscosity.
- Water loss lowers fluid fire resistivity.
- Water loss increases concentration of additives in the fluid.

- On the other hand, **when temperatures are too low**:
- The fluid thickens.
- increasing the energy required to move it though the system.
- This can cause pump cavitation and/or sluggish behavior of actuators.
- Water based fluids will freeze and shut down the whole system if the temperature is low enough.

9

9

When a hydraulic system that is using water-based hydraulic fluid is overheated, which of the following observations is False ?

A- Additives concentration in the fluid increases.

B- Fire resistance of the fluid increases.

C- Pump cavitation is likely possible.

D. Fluid viscosity is increased.

10

10

3.1.3- Best Practices to Minimize Contamination by Heat

❑ **During system Design:**

- Proper design and sizing of hydraulic reservoirs and heat exchangers.
- Proper sizing and routing of hydraulic transmission lines.
- Proper hydraulic circuit design.
- Selection of good quality branded products.
- Selection of hydraulic fluids based on the manufacturer recommendations.

❑ **During System Commissioning:**

- Placing hydraulic reservoirs in a well-ventilated area rather than in a dead zone.
- Shielding transmission lines that are nearby heat sources or exposed to direct sunlight.
- If needed, use industrial type fan to create air flow around the reservoir.

11

11

During System Operation:

- Frequently changing hydraulic fluids based on the manufacturer recommendations and/or fluid analysis.
- Continually cleaning the outer surfaces of the components and transmission lines to improve the heat dissipation.
- Performing preventive maintenance, particularly heat exchanger.
- Continuously monitoring the working temperature of the cooling water source, make sure it meets the design requirements.
- Apply routine heat exchanger maintenance.

12

12

3.2- Contamination by Magnetic Fields
3.2.1- Sources of Contamination by Magnetic Fields

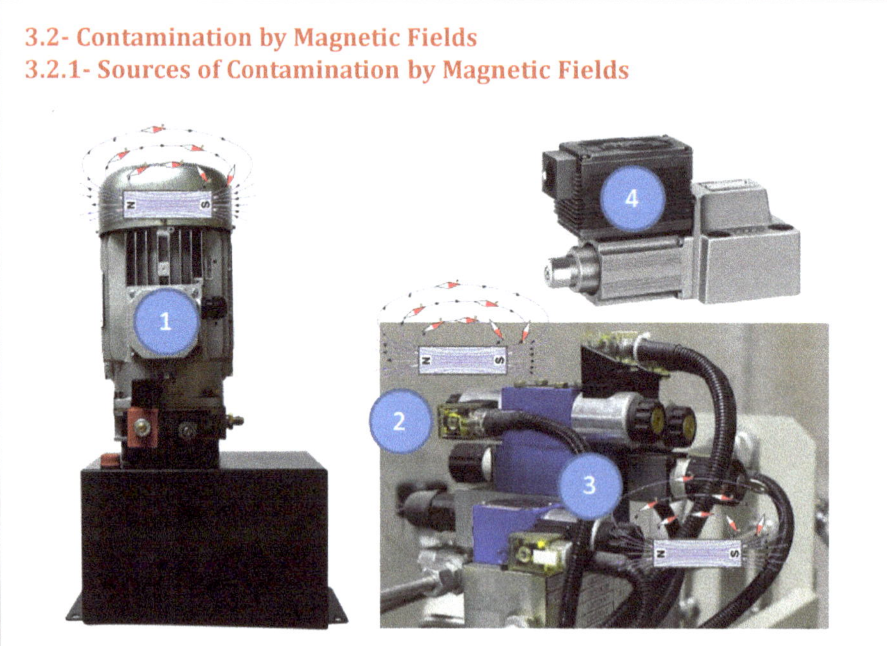

Fig. 3.5- Sources of Contamination by Magnetic Fields

13

13

3.2.2- Effect of Contamination by Magnetic Fields

Magnetic fields attract metallic impurities and fine particles that find their way into the hydraulic components.

3.2.3- Best Practices to Minimize Contamination by Magnetic Fields

Magnetic fields around hydraulic components can be minimized by:

- Use of good quality branded solenoids and
- Shield electrical cables.
- Separate the electrical lines apart from the hydraulic lines.

14

14

QUIZ

As best practices to minimize the effect of magnetic fields around hydraulic components

A- Maintain high working pressure.

B- Maintain working temperature within recommended limits.

C- Keep the outer surfaces cleaned.

D. Separate hydraulic transmission lines apart from the electrical cables.

15

15

3.3- Contamination by Electrostatic Charges

3.3.1- Sources of Contamination by Electrostatic Charges

- Static Electricity is increased by:
  - fine-filtration due to the triboelectric effect (transfer of electrons between the fluid and the filter media).
  - Hydraulic fluid rubbing with the inside surface of a hydraulic reservoir & transmission lines
- The magnitude of charge depends on:
  - Flow speed.
  - Fluid conductivity, and
  - Material surrounding the fluid flow.

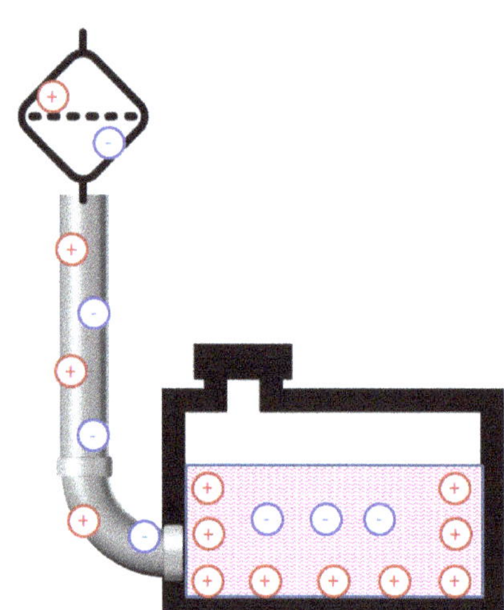

Fig. 3.6- Sources of Contamination by Electrostatic Charges

16

16

3.3.2- Effects of Contamination by Electrostatic Charges

- Electrostatic charges are repeatedly discharged to another surface of lower voltage (usually earth or other metal).
- Discharging process is accompanied by a clicking sound and a spark.
- The discharging frequency depends on the charging rate.
- This can result in:
  - Microscopical etching of the surfaces
  - Carbon deposits on the surface
  - Burn holes in filter media,
  - Damage electronic components
  - Accelerate oil aging
  - Expedite oil oxidation forming varnish.
  - Clearly, if the discharge occurs in a flammable atmosphere the effect can be serious, but these instances are rare.

Fig. 3.7- Static Discharge by Tank-Mounted Filter

17

17

3.3.3- Best Practices to Minimize Contamination by Electrostatic Charges

The hazard of the electrostatic discharge in hydraulic systems is minimized by:

- Applying special synthetic Anti-Static filter media.
- Proper bonding of reservoirs and transmission lines.
- Designing the hydraulic reservoir to limit fluid movement inside it.
- Use of hydraulic fluids with high electrolyte conductivity.
- Hydraulic fluids that contain metallic-based additives, like zinc, have high conductivity. So, the charge carried by the oil is generally dissipated as it passes around the system.
- Using dielectric hydraulic hoses.

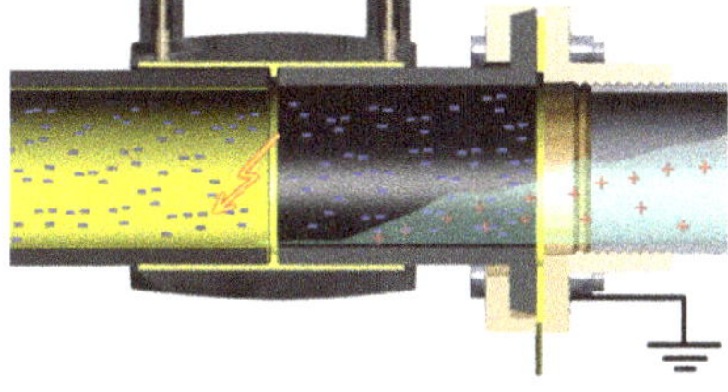

Fig. 3.8- Hydraulic Transmission Line Bonding

18

18

3.4- Contamination by Light

- When hydraulic fluids are exposed to direct *sunlight* or *ultraviolet light*, chemical decomposition will occur.
- Therefore, hydraulic fluids must be stored indoors in covered and sealed containers.

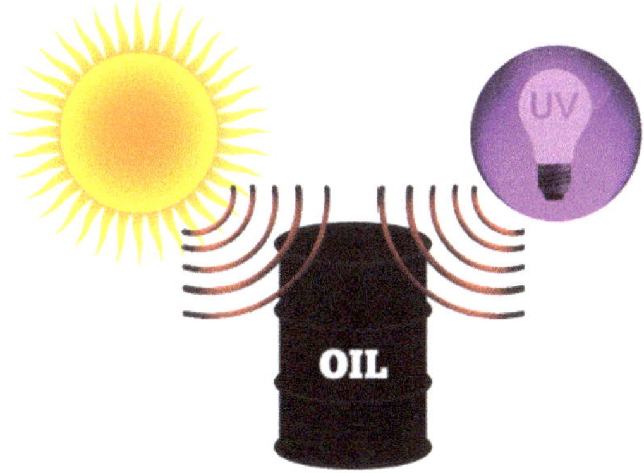

Fig. 3.9- Sunlight and Ultraviolet Light Cause Hydraulic Fluid Decomposition

19

19

Chapter 3 Reviews

1. What is the most common type of energetic contamination for hydraulic fluids?
   A. Heat.
   B. Static Electricity.
   C. Magnetic Fields.
   D. Light.

2. How can an improper reservoir design contribute to heating up the system?
   A. Under-sizing the reservoir.
   B. Placing suction line beside return line.
   C. Not using a baffle plate between suction side and return side.
   D. All the above.

3. When a hydraulic system that is using petroleum-based hydraulic fluid is overheated, which of the following observations is TRUE?
   A. Viscosity increase and better lubrication.
   B. System experiencing less leakage and noise.
   C. System experiencing more leakage and becomes sluggish and less productive.
   D. Pump avoids cavitation.

4. When a hydraulic system that is using water-based hydraulic fluid is overheated, which of the following observations is False?
   A. Additives concentration increases.
   B. Fire resistance of the fluid increases.
   C. Pump cavitaion is likely possible.
   D. Fluid viscosity increases.

5. As best practices to minimize the effect of magnetic fields around hydraulic components.
   A. Maintain high working pressure.
   B. Maintain working temperature within recommended limits.
   C. Keep the outer surfaces cleaned.
   D. Separate hydraulic transmission lines apart from the electrical cables.

Chapter 3 Assignment

Student Name: -- Student ID: ------------------

Date: --- Score: ------------------------

Assignment 1: State three design-related reasons for system overheating

Assignment 2: State three operation-related reasons for system overheating

Assignment 3: State three actions required to minimize the effect of static electricity in a hydraulic system.

Chapter 4
Gaseous Contamination

Objectives:

This chapter presents the sources of hydraulic fluids gaseous contamination. For each source, the chapter explains how the system performance will be affected and recommendations to minimize such consequences.

Brief Contents:

0

0

Which of the following conditions could result in contaminating hydraulic fluids by air?

A. Bad reservoir design and transmission lines sizing.

B. Driving a pump at high speed.

C. Working at high pressure.

D. Using hydraulic fluid with low viscosity.

1

1

4.1- Sources of Gaseous Contamination

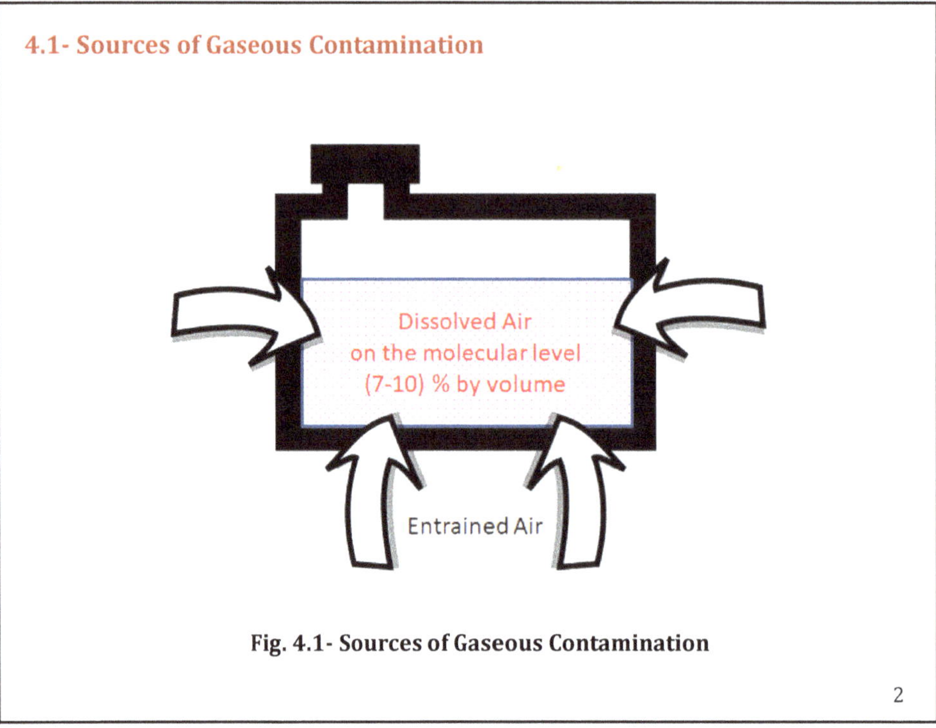

Fig. 4.1- Sources of Gaseous Contamination

2

2

- **Suction Line:** Leaking air into the system or low oil level.
- **Return Line:** Not properly submerged in the fluid.
- **Dissolved Air Separation:** Due to negative pressure.
- **Hydraulic Fluid Evaporation:** due to overheating the fluid.
- **Pump:** improper priming (pre-filling).
- **Reservoir:** improper design, sizing, baffles, and line placement.
- **Transmission Lines:** Improper sizing results in turbulent flow.
- **Accumulator:** Nitrogen leaking to fluid side due to bladder or seal failure.
- **Commissioning:** Systems startup with improper pre-filling or air bleeding.
- **Maintenance:** Oil abused by extended recommended working hours.
- **Flow Surges:** such as during cylinder retraction or fast cyclic motion.
- **Hydraulic Fluid:** Poor hydraulic fluid quality.

3

3

Air can be present in hydraulic fluids in which of the following forms?

A. Dissolved air.
B. Entrained air.
C. Foam.
D. All the above.

4

4

4.2- Forms of Air in Hydraulic Fluids

Air present in hydraulic fluids in three forms as follows:

Dissolved Air: In normal operating conditions, oil contains 7 to 10% by volume homogeneously *dissolved air*. Dissolved air does not significantly affect hydraulic system performance and is not visible to the naked eye.

Entrained Air: Entrained air (*Aeration*) appears as tiny emulsified bubbles below the surface of the fluid. Highly aerated fluids have a milky appearance and can cause a variety of performance problems.

Foaming: *Foaming* is a surface phenomenon and readily identified by accumulation of bubbles on top of the fluid surface.

5

5

4.3- Standard Test Methods for Measuring Air Content in Hydraulic Fluids

- The technical paper (FPMC2014-7823) reported in the Proceedings of the ASME/BATH 2014 Symposium
- In order to obtain accurate results, tests must be conducted immediately after obtaining the sample.

Measurement Method		
Mechanical	**Optical**	**Electrical**
▪ Density ▪ Change in Volume ▪ Compressibility ▪ Speed of Sound	▪ Translucency ▪ Photography ▪ Light Scattering ▪ Radio-metricity	▪ Electrical Conductivity ▪ Electrical Impedance ▪ Permittivity

Table. 4.1- Overview of Different Measurement Techniques

Table 4.2- Evaluation of the Commonly used Measurement Methods

Method	Effort	Accuracy
Compressibility	☹	☹
Photography	☺	☺
Density via Orifice Flow	☹	😐
Sensor	☺	☺

6

6

QUIZ

When a hydraulic fluid is contaminated by air, which of the following consequences can occur?

A. Fluid appearance change and fluid loses the ability for lubrication.

B. System becomes noisy and actuators move erratically.

C. Fluid becomes more compressible and system becomes sluggish.

D. All the above.

7

7

4.4- Effects of Gaseous Contamination

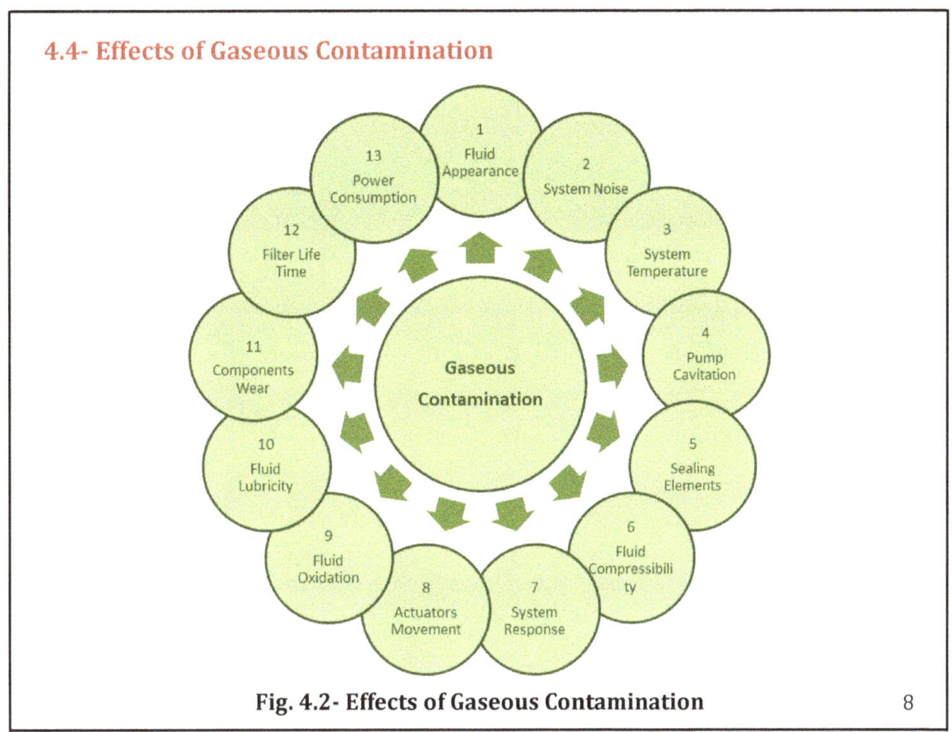

Fig. 4.2- Effects of Gaseous Contamination 8

8

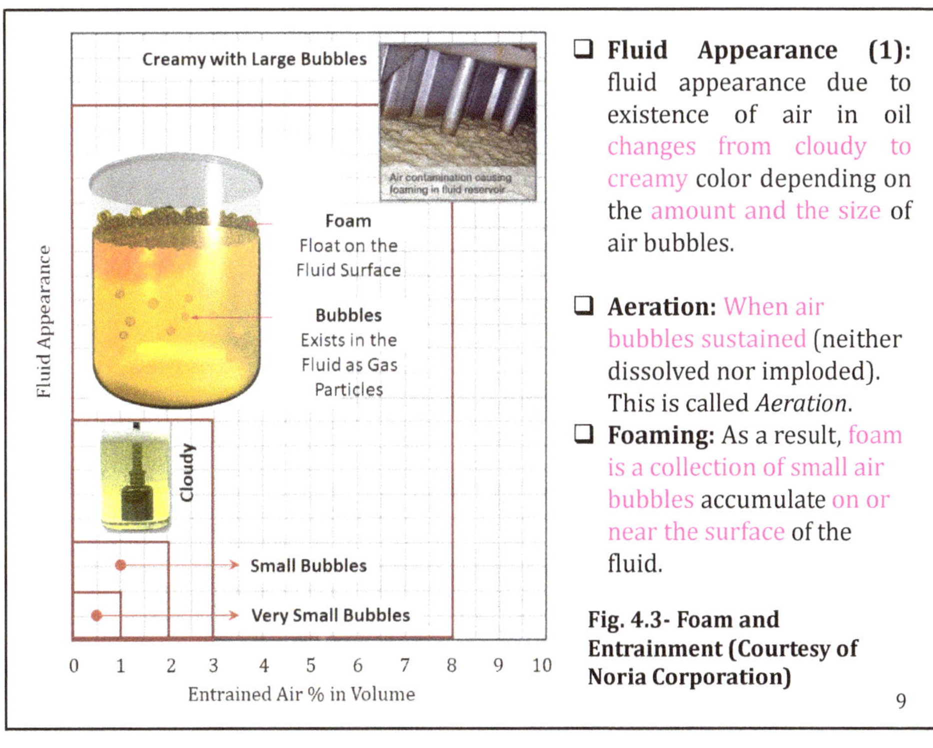

- **Fluid Appearance (1):** fluid appearance due to existence of air in oil changes from cloudy to creamy color depending on the amount and the size of air bubbles.
- **Aeration:** When air bubbles sustained (neither dissolved nor imploded). This is called *Aeration.*
- **Foaming:** As a result, foam is a collection of small air bubbles accumulate on or near the surface of the fluid.

Fig. 4.3- Foam and Entrainment (Courtesy of Noria Corporation) 9

9

- **System Noise (2):** Increased system noise and vibration due to air bubble explosions.
- **System Temperature (3):** Difficult fluid temperature control because foam is an efficient thermal insulator. Therefore, system temperature typically is increased by 18-36 °F (10-20 °F) above normal operating temperature.
- **Pump Cavitation (4):** Presence of foam resists pump suction increasing pump cavitation.
- **Sealing Elements (5):** Increased chance of seal damages due to either sudden expansion or self-Ignition (diesel effect).

10

10

Which one of the following hydraulic fluid properties is most affected by entrained air?

A. Antiwar.

B. Antirust.

C. Bulk Modulus.

D. Fire Resistivity.

11

11

- **Fluid Compressibility (6):** As shown in Table 4.3, *Bulk Modulus* for hydraulic fluids is severely reduced as a result of the increase in undissolved air.

- **System Response (7):** Hydraulic fluid becomes spongey and the system becomes less responsive (more sluggish).

- **Actuators Movement (8):** Erratic movement of cylinders and motors.

Air content - %	Temperature - °F	Adiabatic Bulk Modulus - psi
0.0	80	268,000
0.1	80	250,000
1.0	80	149,000
0.0	180	163,000
1.0	180	106,000

Table. 4.3- Effect of Air Content on Bulk Modulus

12

12

Fluid Oxidation (9): Increased rate of oil oxidation due to increased oxygen content.

Fluid Lubricity (10): Oil performs as two-phases losing its ability to lubricate.

Components Wear (11): Increased wear and rate of failure due to poor lubrication.

Filter Service Life (12): Reduce filter service life time because of increased component wear.

Power Consumption (13): Bubbles increases the pressure drop through transmission lines and valves.

13

13

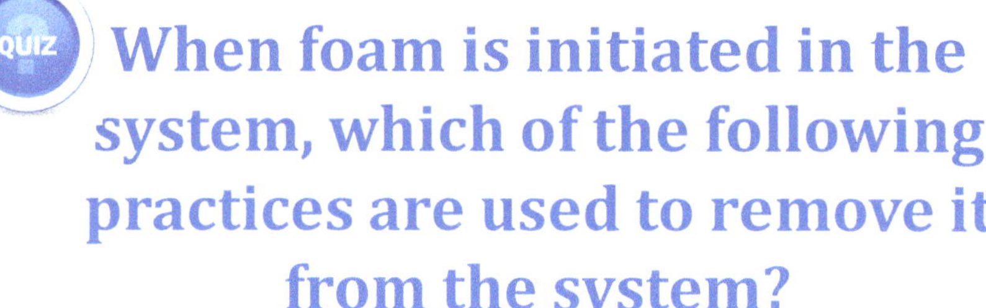

When foam is initiated in the system, which of the following practices are used to remove it from the system?

A. Foam will be self-removed by just circulating the fluid in the system.

B. Shut down the system for some time and bleed the foam from the bottom of the reservoir.

C. Shutdown the system for time long enough for the foam to be released and dissipated at the surface. Otherwise use bubble eliminators devise.

D. Just use antifoaming additives.

14

14

4.5- Best Practices to Minimize Gaseous Contamination
4.5.1- Preventive Practices to Minimize Gaseous Contamination

During System Design:

- Properly size, layout, and assemble pump suction line.
- Properly size hydraulic transmission lines for laminar flow.
- Properly design reservoirs to help dissipate air.

During System Commissioning:

- Properly prime pumps and motors before starting the system.
- Properly bleed air out of the system.
- Use of proper additive package such as foam suppressors.

During System Operation:

- Monitor pump inlet pressure, fluid temperature, accumulator charge pressure on a continuous basis.
- Help reduce foaming by using oil with anti-foaming additives.
- Respect time intervals for replacing hydraulic fluids.

Next volume of this textbook's series will provide detailed discussions about the above-mentioned bullets.

15

15

4.5.2- Curative Practices to Remove Gaseous Contamination

Simple Method to Remove Foam:

- Foam can be simply removed by shutting down the system long enough to allow the air to collect and dissipate at the fluid surface in the reservoir.
- Air may collect in the highest elevation points of the system piping. In such a case, the system should be bled out on startup.

Advanced Method to Remove Foam:

- If this simple method does not help, bubble removal device can be used to mechanically remove bubbles from fluids.

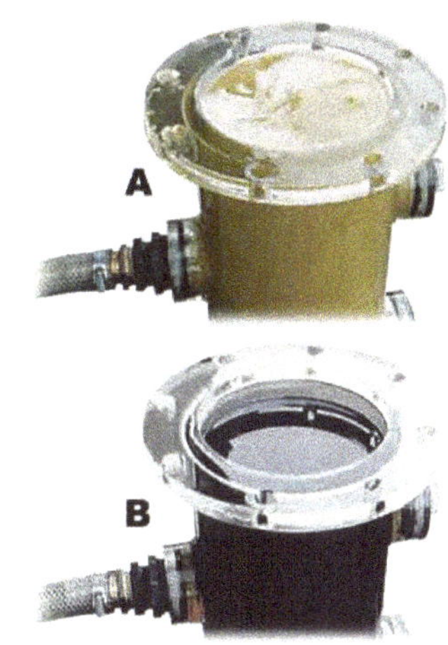

Fig. 4.4- Air Bubble Removal,
A) Aerated Fluid,
B) De-Aerated Fluid
(Courtesy of Noria Cooperation)

16

16

Bubble Eliminator.

- Consists of a tapered tube connected with a cylindrical tube chamber.
- Fluid containing bubbles flows tangentially into the tapered tube from an inlet port and generates a swirling flow that circulates the fluid through the flow passage.
- The swirling flow accelerates as the radius decreases, reducing the fluid pressure along the central axis as the fluid moves downstream by Bernoulli's equation.
- At the end of the tapered tube, the swirl flow decelerates downstream and the pressure recovers as the fluid moves through the cylindrical tube.
- In a sense, this is a fluid-flow driven centrifugal action for bubble removal.

 Video 483 (2.5 min)

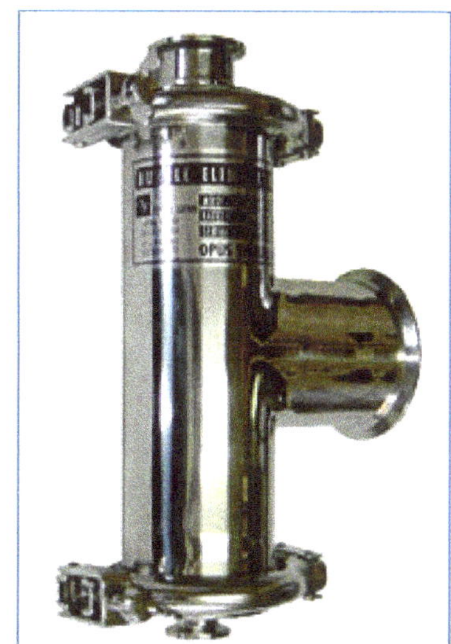

Fig. 4.5- Bubble Eliminator
(www.opussystem.com)

17

17

Chapter 4 Reviews

1. Which of the following conditions could result in contaminating hydraulic fluids by air?
   A. Bad reservoir design and transmission lines sizing.
   B. Driving a pump at high speed.
   C. Working at high pressure.
   D. Using hydraulic fluid with low viscosity.

2. Air can be present in hydraulic fluids in which of the following forms?
   A. Dissolved air.
   B. Entrained air.
   C. Foam.
   D. All the above.

3. When a hydraulic fluid is contaminated by air, which of the following consequences can occur?
   A. Fluid appearance change and fluid losses the ability for lubrication.
   B. System becomes noisy and actuators move erratically.
   C. Fluid becomes more compressible and system becomes sluggish.
   D. All the above.

4. Which one of the following hydraulic fluid properties is affected by entrained air?
   A. Antiwar.
   B. Antirust.
   C. Bulk Modulus.
   D. Fire Resistivity.

5. When foam is initiated in the system, which of the following practices are used to remove it from the system?
   A. Foam will be self-removed by just circulating the fluid in the system.
   B. Shut down the system for some time and bleed the foam from the bottom of the reservoir.
   C. Shutdown the system for time long enough for the foam to be released and dissipated at the surface. Otherwise use bubble eliminators devise.
   D. Just use antifoaming additives.

Chapter 4 Assignment

Student Name: -- Student ID: ------------------

Date: --- Score: -----------------------

Assignment 1: List 5 reasons for air to entrain into a hydraulic system.

Assignment 2: List 5 symptoms for a hydraulic system that is contaminated by air.

Chapter 5
Fluidic Contamination

Objectives:

This chapter covers the sources of hydraulic fluids fluidic contamination. For each source, the chapter explains how the system performance will be affected and possible recommendations to minimize such consequences.

0

0

Brief Contents:

1

1

Generally speaking, fluidic contamination means oil contaminated by?

A. Solid particles.

B. Air.

C. Water.

D. Sludge.

2

2

5.1- Sources of Fluidic Contamination

Fig. 5.1- Sources of Contamination by Free Water

Other *Fluidic Contaminations* may result from:

- Mixing of incompatible hydraulic fluids.
- Residual fluids from flushing or pickling process.
- Other fluids used in the vicinity of the hydraulic system such as paints, cleaning solvents, metal working fluids and coolants.

3

3

Absorptive capacity of a hydraulic fluid depends on?

A. Fluid temperature.

B. Fluid's molecular structure.

C. Additive packages of the fluid.

D. All of the above.

4

4

5.2- Forms of Water Contamination in Hydraulic Fluids

- Different hydraulic fluids have different water absorptive capacity.
- Absorptive capacity depends on:
  - Fluid Temperature.
  - Fluid's molecular structure.
  - Additive packages of the fluid.
- Water content measurement unit is *part per million* (ppm).
- For example, 10,000 ppm =1% water.
- Saturation level of a hydraulic fluid is the *Maximum Water Content* that can dissolve within the molecular structure of the hydraulic fluid at an identified *Critical Temperature*.

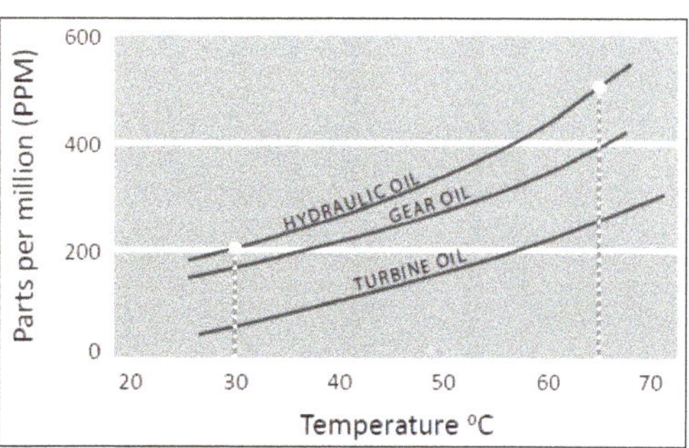

Fig. 5.2- Saturation Level of Different Hydraulic Fluids (Courtesy of C.C. Jensen Inc.)

5

5

QUIZ

Above saturation point, residual water separates from the fluid forming?

A. Rust.

B. Wax.

C. Varnish.

D. Free water.

6

6

- All numbers in this table are only rough guides, which strongly differ in dependency with the used base oil, additive packages and the application of the hydraulic system.

Fluid Type	Critical Water Content (ppm)
Mineral oil (HLP)	200 - 500
Biodegradable oil (HEES)	700
Fire resistant fluid (HFC=Water in Glycol Emulsion)	> 4000

Table 5.1- Saturation Level of Different Hydraulic Fluids at 20 ^{o}C (68 ^{o}F)

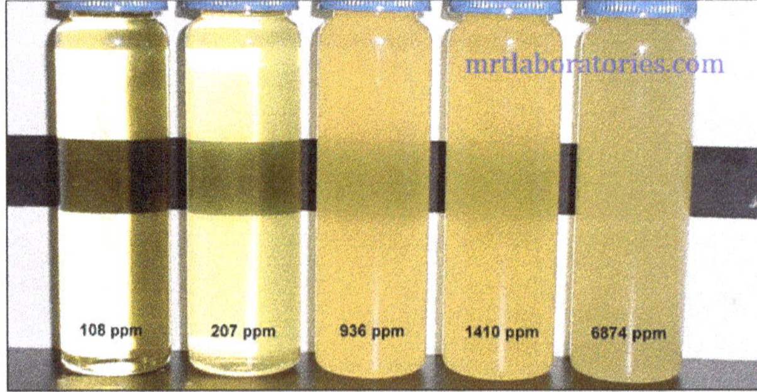

Fig. 5.3- Various Levels of Contamination by Water in Oil

7

7

Dissolved (Emulsified) Water:

- Water droplets dispersed at a molecular level below the saturation level.
- Is not visible when in solution.
- Appears as a cloud in the oil as temperature is lowered to the critical temperature that begins to force the water out of solution.

Free Water:

- Above saturation point, residual water separates forming free water.
- Settles to the bottom of the tank and should be drained periodically.
- Free water is more harmful than dissolved water.

Fig. 5.4- Forms of Water in a Hydraulic Fluid

8

8

5.3- Standard Test Methods for Measuring Water Content in Hyd. Fluids
5.3.1- Karl-Fischer Method (ISO760 - ASTM D6304 – DIN 51777)

- The *Karl-Fischer* method is based on titration (chemical analysis) using electrochemical device consisting of two components, the Karl Fischer titrator and an integrated oven.
- Used to determine the total water content of oils.
- Can't distinguish between the dissolved and free water.
- Difficult to interpret the results when the concentration is < 500 ppm.

Fig. 5.5- KF Titrator (www.metrohm.com)

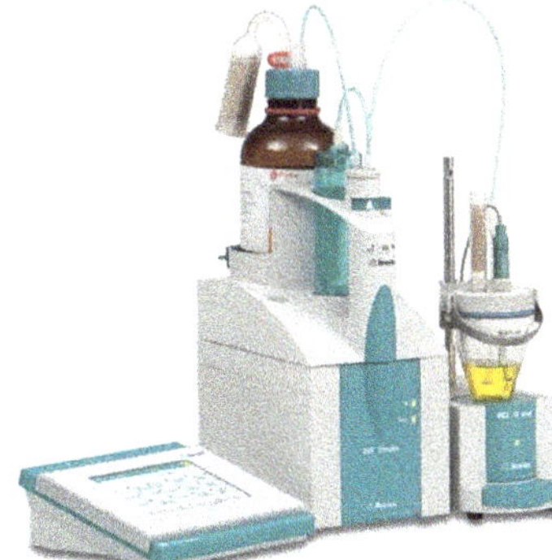

9

9

5.3.2- Fourier Transform Infrared (FTIR) (ASTM E2412)

- IR analysis is a chemical-free method.
- IR analysis is based upon the same principle as a microwave oven.
- Microwave ovens transmit radiation through food.
- Water molecules absorb the "micro" wavelengths and so heat up.
- Carbohydrates, fat, protein, plastic, paper and glass do not absorb microwave radiation.
- FTIR *spectrometer* device consists of a radiative source of infrared (IR) and a detector.
- The FTIR device compares the spectrum of the contaminated oil sample versus fresh oil sample.
- Water content is determined by calculating the area between the two spectrums along the wave number range.

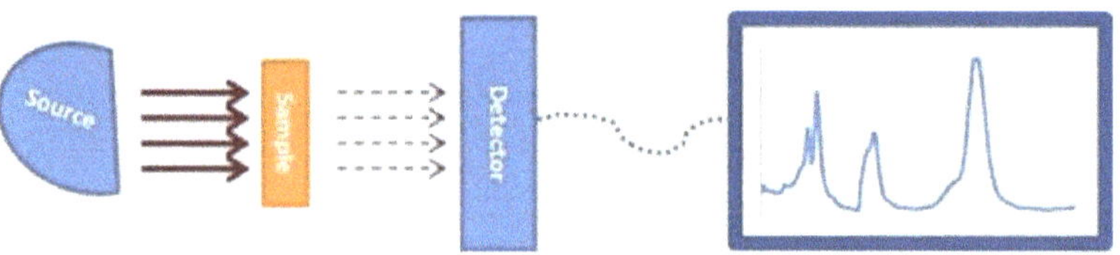

Fig. 5.6- Schematic of Typical Spectrometer (Courtesy of Spectro Scientific) 10

10

- The IR light absorbed by pure water can be identified by a peak in the IR spectrum at about wavelength 3400cm^{-1}.

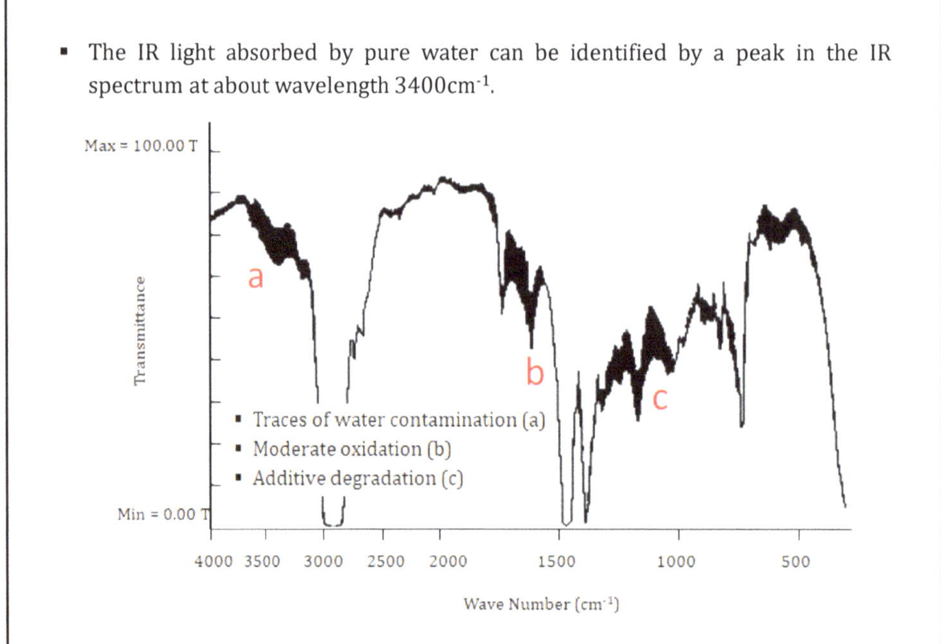

Fig. 5.7- Measurement of Water using FTIR method (Courtesy of MSOE) 11

11

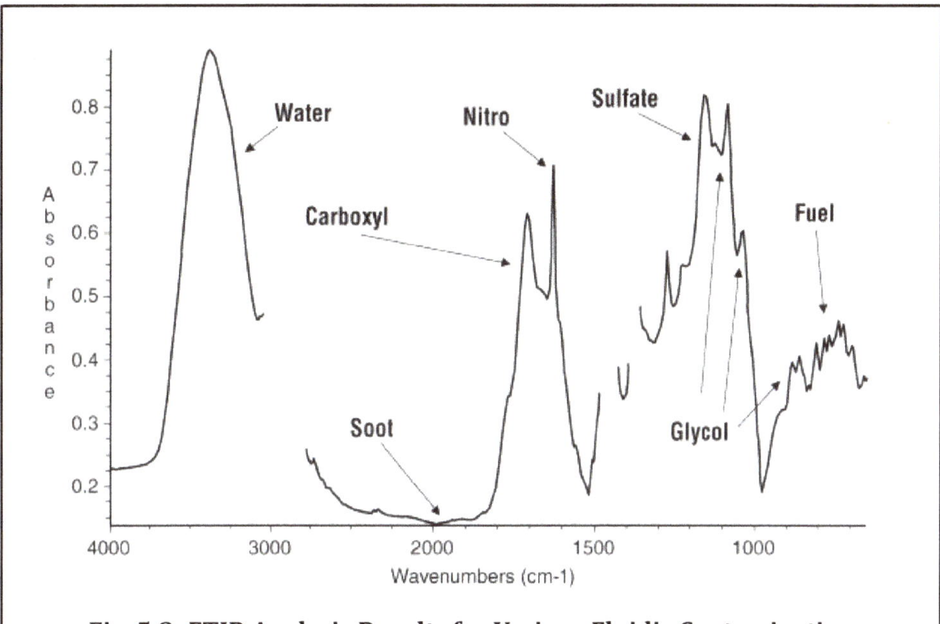

Fig. 5.8- FTIR Analysis Results for Various Fluidic Contamination (Courtesy of Spectro Scientific)

12

12

5.3.3- Centrifuge

- Applicable for water contents > 0.1% (1000 ppm).
- The spinning of the sample in the centrifuge causes the higher density water to collect at the bottom of the centrifuge tube.
- The volume of the water is compared to the total volume of sample placed in the centrifuge tube.

5.3.4- Crackle Test

- The "Crackle" test can be done routinely onsite to determine if an oil sample is contaminated with water.
- This test is just a qualitative test to provide a yes or no answer.
- Is based on the fact that water boils at a lower temperature than oil.
- When a contaminated oil sample is heated, water violently changes phase from liquid to vapor creating a popping noise.

13

13

- Maintain surface temperature of a hot plate at 135°C (300°F).
- Using a clean dropper, place a drop of oil on the hot plate.

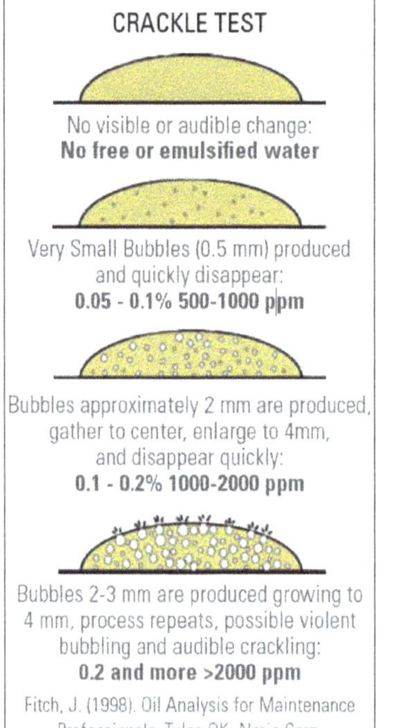

CRACKLE TEST
No visible or audible change:
No free or emulsified water
Very Small Bubbles (0.5 mm) produced and quickly disappear:
0.05 - 0.1% 500-1000 ppm
Bubbles approximately 2 mm are produced, gather to center, enlarge to 4mm, and disappear quickly:
0.1 - 0.2% 1000-2000 ppm
Bubbles 2-3 mm are produced growing to 4 mm, process repeats, possible violent bubbling and audible crackling:
0.2 and more >2000 ppm
Fitch, J. (1998). Oil Analysis for Maintenance Professionals, Tulsa OK, Noria Corp.

Fig. 5.9- Crackle Test (Courtesy of Spectro Scientific)

14

14

Karl-Fischer method is based on?

A. Visual Fluid Appearance.

B. Titration (chemical analysis) using electrochemical device.

C. Transmitted microwave through the fluid and a wave detector.

D. Water boils at a temperature higher than oil.

15

15

5.4- Effects of Fluidic Contaminants

- Oil is considered contaminated when the water content exceeds the saturation level.
- However, water more than 0.5% by volume in a hydrocarbon-based fluid accelerates degradation.
- The degree of damage depends on:

- Form of the water.
- % of water content.
- For how long.

Fig. 5.10- Factors Affect the Degree of Damage due to Fluidic Contamination

16

16

- Water has the same set of effects like gaseous contamination.
- Unlike gaseous contamination, system damage and loss of performance due to fluidic contaminants occurs over an extended period.

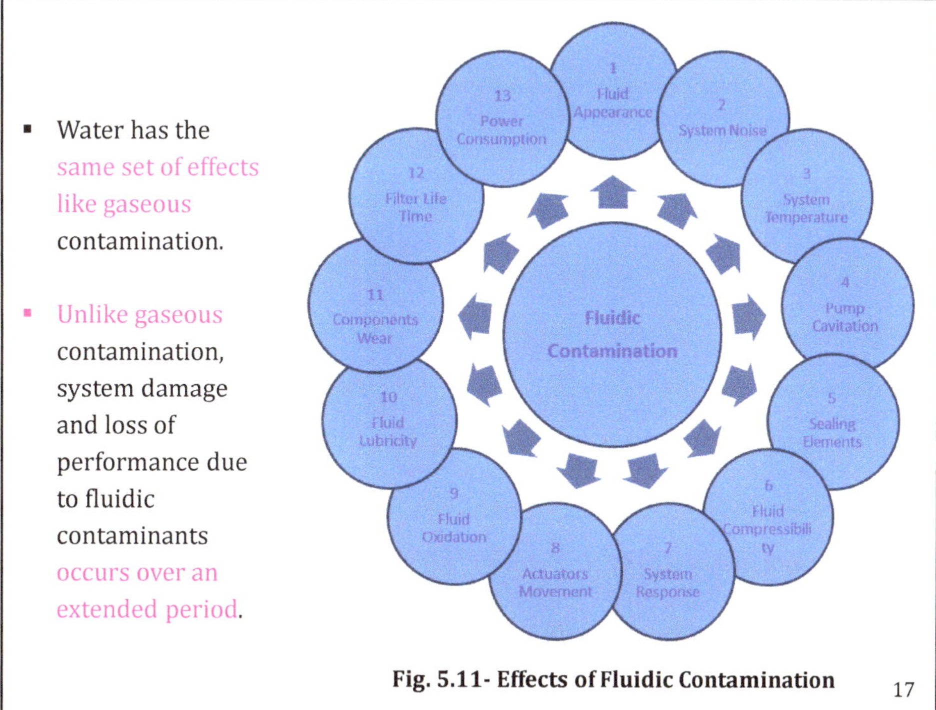

Fig. 5.11- Effects of Fluidic Contamination

17

17

- **Fluid Appearance (1):**
  - (on the left) oil with small amount of free water in the bottom
  - (on the right) after shacking by hand for 30 seconds.
- **System Noise (2):** Increased system noise and vibration due to lack of lubrication.
- **System Temperature (3):** water boils at a lower temperature than oil.
- **Pump Cavitation (4):** Water evaporation increases possibility of pump cavitation.

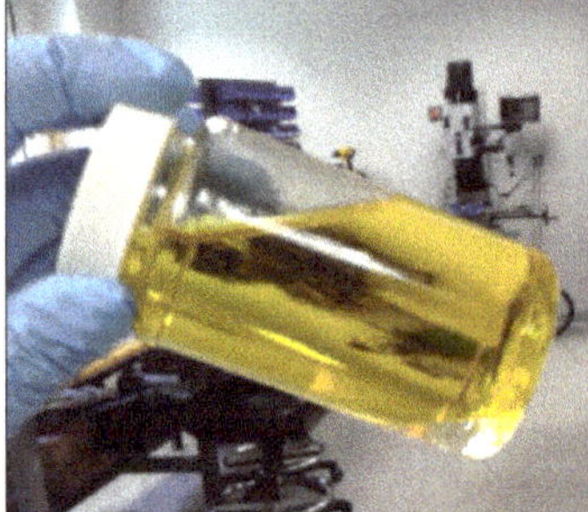

Fig. 5.12- Milky/Cloudy Appearance of Hydraulic Fluid Contaminated by Water

18

18

- **Sealing Elements (5):** Water content affects sealing performance.
- **Fluid Compressibility (6):** Contaminated oil has higher equivalent bulk modulus because water has higher bulk modulus than oil.
- **System Response (7):** affected due to oil degradation.
- **Actuators Movement (8):** Stick-Slip movements.
- **Fluid Oxidation (9):** Increased rate of oil oxidation.
- **Fluid Lubricity (10):** Oil loses its ability to lubricate.
- **Components Wear (11):** Increased wear rate and failure due to lack of lubrication, rust, and corrosion.
- **Filter Service Life (12):** Reduced because of sludge formation.
- **Power Consumption (13):** Higher power consumption due to loss of system performance.

19

19

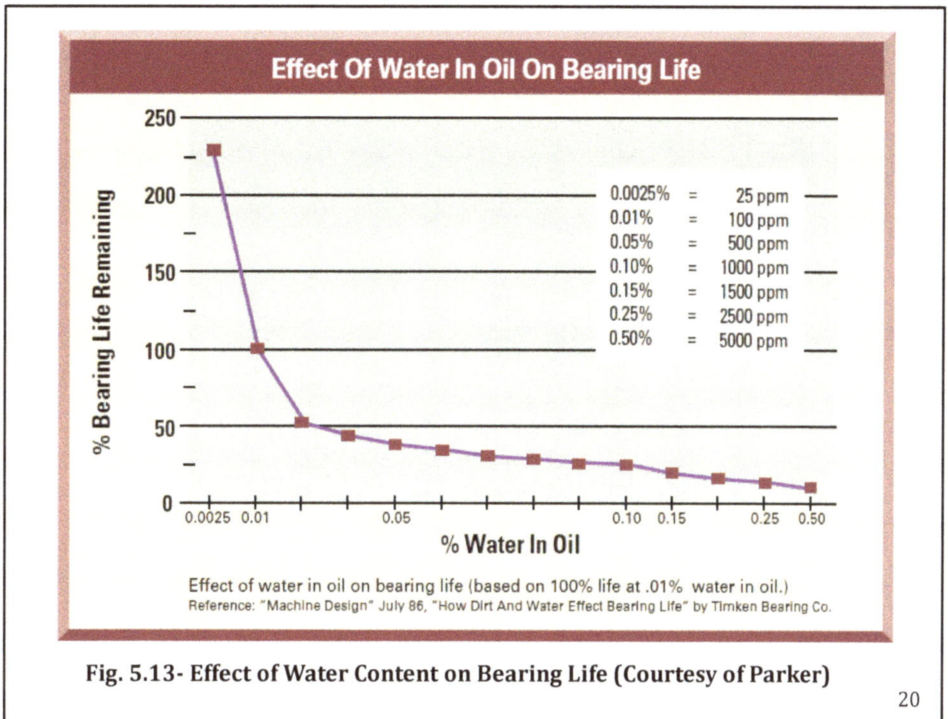

Fig. 5.13- Effect of Water Content on Bearing Life (Courtesy of Parker)

20

20

Contamination by water has the following additional effects:

Water Icing: Water icing in cold weather results in forming hard crystals.

Bacterial Contamination:

- In biodegradable fluids, water supports biological growth
- Generates organic contamination and microbes.
- It will be seen as accumulation of green microbes sticking on the inside surfaces of the reservoir.

Additives: Water content results is:

- Decrease of additive performance and
- Increase of additive depletion.

21

21

Oil Degradation: Mixing of incompatible hydraulic fluids can result in:

❑ Phosphate Ester or brake fluid + Mineral oil:

- → Acids and sludge.
- → Seals will swelling,
- → Filters clogging,
- → Critical orifices plugged, and
- → Spool valves become sluggish.

❑ Volatiles (such as diesel fuel, gasoline or solvents) + hydraulic oil:

- → Reduce oil's viscosity.
- → Damage may occur to the system due to lack of lubrication.

❑ Water + automatic transmission fluids:

- → Sludge and small hard crystalline particles to form.

22

22

5.5- Best Practices to Minimize Fluidic Contamination

5.5.1- Preventive Practices to Minimize Fluidic Contamination

preventing practices are much more cost effective than removing the contamination.

❑ **New Hydraulic Fluid:**

- DO NOT mix oils without previously investigating compatibility.
- DO NOT use oils additives that are not necessary for the application.
- Use fluid with high hydrolytic stability to minimize fluid chemical degradation when contaminated by water.
- Compare the oil in operation to fresh oil regularly in order to discover any sudden appearance of water, air or other contaminants.

23

23

- **In Service Hydraulic Fluid:** Continuous removal of water out of a hydraulic fluid can improve hydraulic system reliability and is considered as a *Life Extension Method (LEM)*.

LEM - Moisture Level

Current moisture level, ppm	Life Extension Factor								
	2	3	4	5	6	7	8	9	10
50,000	12,500	6,500	4,500	3,125	2,500	2,000	1,500	1,000	782
25,000	6,250	3,250	2,250	1,563	1,250	1,000	750	500	391
10,000	2,500	1,300	900	625	500	400	300	200	156
5,000	1,250	650	450	313	250	200	150	100	78
2,500	625	325	225	156	125	100	75	50	39
1,000	250	130	90	63	50	40	30	20	16
500	125	65	45	31	25	20	15	10	8
260	63	33	23	16	13	10	8	5	4
100	25	13	9	6	5	4	3	2	2

1% water = 10,000 ppm. | Estimated life extension for mechanical systems utilizing mineral-based fluids

Example: By reducing average fluid moisture levels from 2,500 ppm to 156 ppm, machine life (MTBF) is extended by a factor of 5

Table 5.2- Life Extension of a Machine (Courtesy of C.C. Jensen Inc.) 24

24

- **After Flushing:** After flushing and pickling process, system must be dried by blowing clean hot dry air into the transmission line.

- **Water Content Sensors:** *Relative Humidity* (RH) sensors are used to provide early warning about the water content in the hydraulic fluids.

Water content sensors are available in different styles.

- Left: typical low-cost, in-line sensor for measuring dissolved water content in hydraulic, lubricating and insulating fluids.
- Right: Offline sensor for checking during routine maintenance.

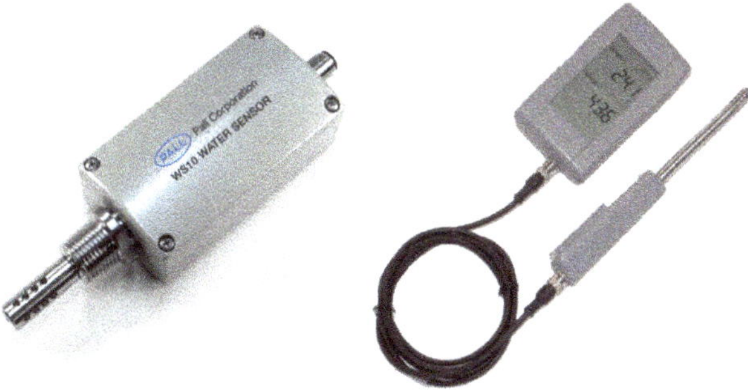

Fig. 5.14- Water Content Sensors (Courtesy of Pall Corporation) 25

25

- **Operational Actions:**
  - Avoid high-pressure sprays around seals, shafts, fill ports and breathers when washing machines.
  - Maintain seals in steam and heating/cooling water systems.
  - Chanel water to divert water flow away from reservoir breathers and top covers.
  - Use and maintain high-quality rod wiper seals in hydraulic cylinders.
  - Prevent water from entering new oil by storing drums indoors.
  - Periodically drain water from low points in system.

- **Closed and Pre-Pressurized Reservoir:** Closed reservoirs may be a solution in highly humid environments such as offshore and marine applications.

26

26

- **Desiccant Filter Breather:** Absorbs the moisture from the air entering the reservoir.

Fig. 5.15- Desiccant Filter Breather (www.descase.com)

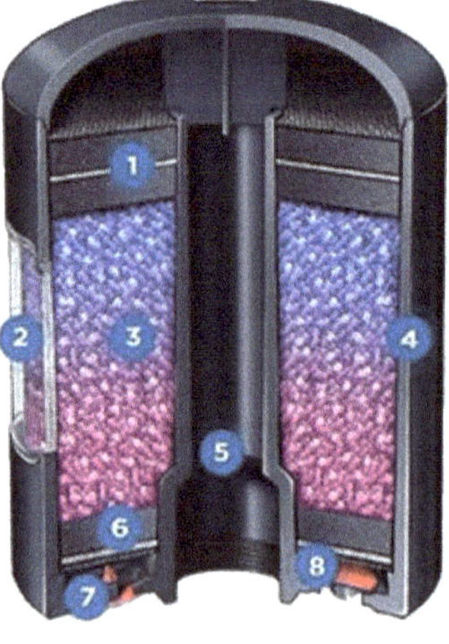

1. Secondary Filter Element.
2. Visual Indicator
3. Water Vapor Adsorbent
4. Rugged Housing
5. Integrated Stand pipe
6. Foam Pad
7. Quad Check-Valves.
8. Filter Element removes airborne contamination to 0.3-micron absolute and stops free water.

27

27

5.5.2- Curative Practices to Remove Fluidic Contamination

- Normal filtration will not remove water.

- Depends on:
  - Volume of the contaminated oil.
  - Form of water content whether dissolved or free water.
  - Level of contamination by water.

- Water removal methods can be
  - Simple less expensive.
  - High cost advanced techniques.

- For example,
  - Small amount of water content can be removed by using absorptive breathers or active venting systems.
  - Large quantities of water, vacuum dehydration, coalescence, and centrifuges are appropriate techniques for its removal.

28

28

Comparative information on these techniques and their relative effectiveness. Care should be taken to apply the best technique to a given situation and its demands for water removal.

	Usage	Prevents Humidity Ingression	Removes Dissolved Water	Removes Free Water	Removes Large Quantities of Free Water	Limit of Water Removal
Adsorptive Passive Breather	prevention	Y				n/a
Active Venting System	prevention and removal	Y	Y	Y		down to <10% saturation
Water Absorbing Cartridge Filter	removal			Y		only to 100% saturation
Centrifuge	removal			Y	Y	only to 100% saturation
Coalescer	removal			Y	Y	only to 100% saturation
Vacuum Dehydrator	removal		Y	Y	Y	down to ~20% saturation

Table 5.3 - Water Prevention and Removal Techniques (Courtesy of Donaldson)

29

29

5.5.2.1- Water Removal Techniques for Small Water Contents

❑ **Fluid Replacement:**

- Water can't be removed 100%
- For quantity (< 500 gallon =2000 liter), it is recommended to replace it and flush the system.

❑ **Periodic Disposal of Free Water by Gravity:**

- For large oil quantity, keep the machine at rest (min 2 hours).
- Drain the settled water at the bottom of the tank.
- Oil is heated in an open tank to help evaporate the residual water.

❑ **Active Venting System:** The method of *Active Venting System* is

- Also known as *Head Space Dehumidification*.
- Circulating dehumidifying desiccant hot air (with low dew point) from the reservoir head space.
- Small air flow [~ 4 standard cubic feet per minute (SCFM)] is required. The side effect is air in oil increases.
- Water in the oil migrates to the dry air.
- The side effect is air in oil increases.

30

30

❑ **Adsorption:**

- *Adsorption* means accumulating the free water on the surface of some adsorbent material such as silica.
- Contaminated oil is circulated through desiccant filters that adsorb water.
- Filter can be replaced quickly and easily.
- This method is expensive and not effective for large quantity of oil.

31

31

- Alternatively, a special water adsorbent is placed in the reservoir.
- This method is applicable for small-sized reservoirs.
- Side effect of that is some of the absorbing material may migrate to the fluid and fine filtration is required to remove it.

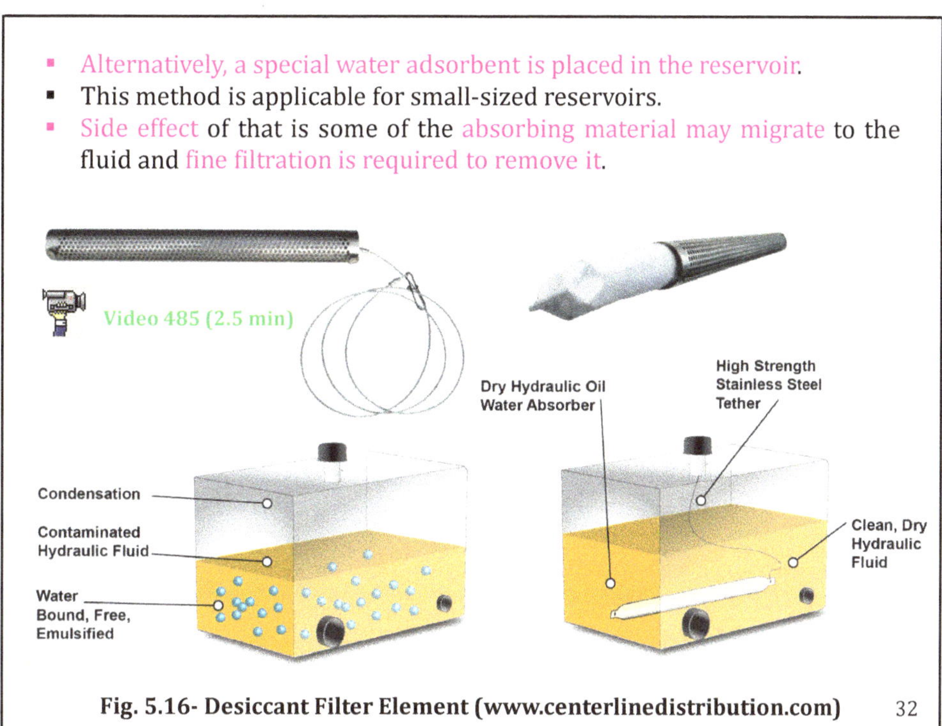

Fig. 5.16- Desiccant Filter Element (www.centerlinedistribution.com) 32

32

❑ Absorption (Coalescence):

- It means trapping and accumulating water into a special filter separator.
- Oil is pumped through a special filter separator from the lowest point.
- Water aggregates in droplets sinking down in the bottom of the filter.
- Water is automatically removed through a water discharge system.
- Filter separator removes particles, oxidation and water in one operation.
- The clean and dry oil is returned to the system.

Video 481 (2.5 min)

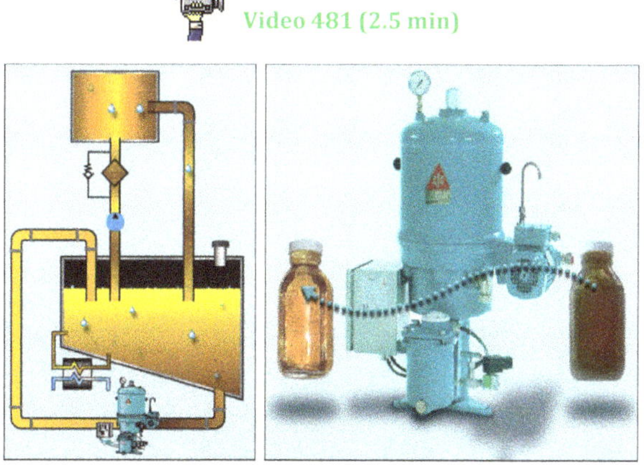

Fig. 5.17- CJC Coalescence Filter Separator (Courtesy of C.C. Jensen Inc.) 33

33

- Such a filter element must be replaced based on:
  - Manufacturer recommendations or
  - At least once a year or
  - When pressure drop exceeds 2 bar

Video 484 (2.5 min)

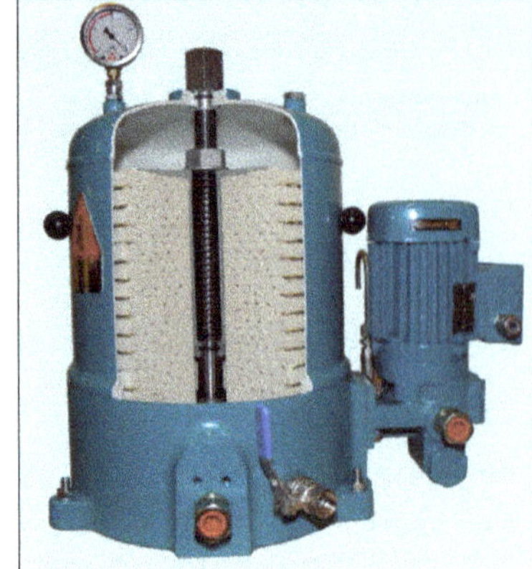

Fig. 5.18- CJC Filter Separator and Filter Elements (Courtesy of C.C. Jensen Inc.)

34

34

5.5.2.2- Water Removal Techniques for Large Water Contents

- **Centrifugal Water Separators:**
  - *Centrifugal Water Separators* is based on the fact water has higher mass density than hydraulic fluids.
  - Water is separated and collected through the wall while the cleaned oil is directed back to the reservoir.
  - Side effect is that some oil additive may be removed. Oil should be tested to verify additive package is acceptable for continued use.

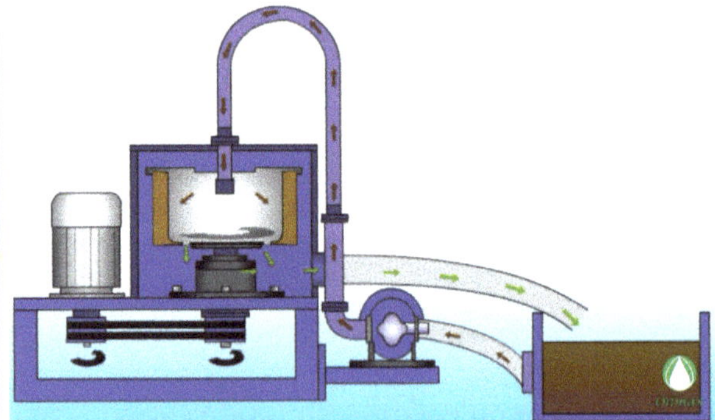

Video 486 (1 min)

Video 487 (1 min)

Fig. 5.19- Concept of Operation of Centrifugal Water Separator (www.oilmax.com)

35

35

- **Mass Transfer Vacuum Dehydrator:**
  - Mass Transfer Vacuum Dehydrators is used for medium to large oil systems, particularly where high viscosity fluids are employed.
  - Vacuum dehydration removes 100 % free water and as much as 90 % of dissolved water at minimum cost and ease of use.

Fig. 5.20- HNP075 Series Oil Purifier (Courtesy of Pall Corporation)

36

36

Removal of water to levels below the saturation curve ensures that free water will not be reformed after cooling the hydraulic fluid as follows:

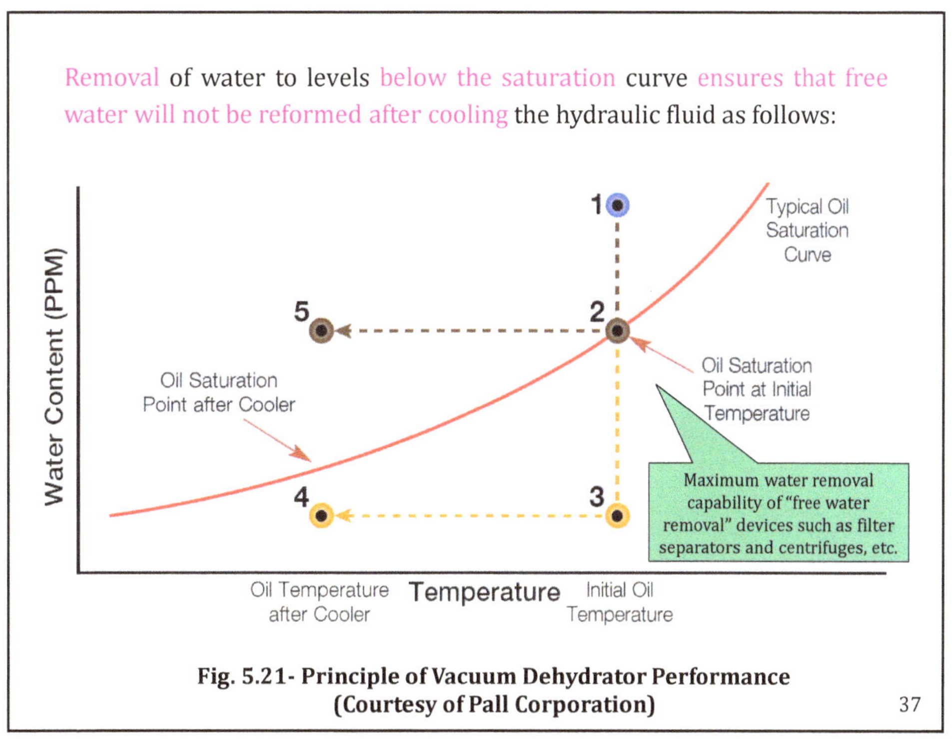

Fig. 5.21- Principle of Vacuum Dehydrator Performance (Courtesy of Pall Corporation)

37

37

Procedure of water removal and purifying hydraulic fluid using Mass Transfer Vacuum Dehydrators .

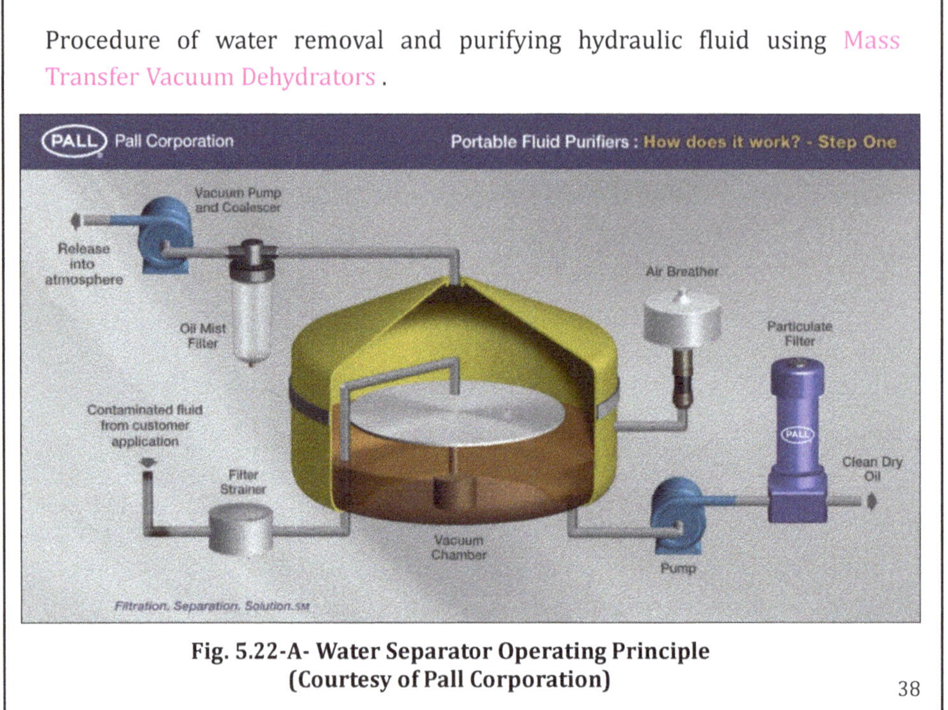

Fig. 5.22-A- Water Separator Operating Principle (Courtesy of Pall Corporation)

38

38

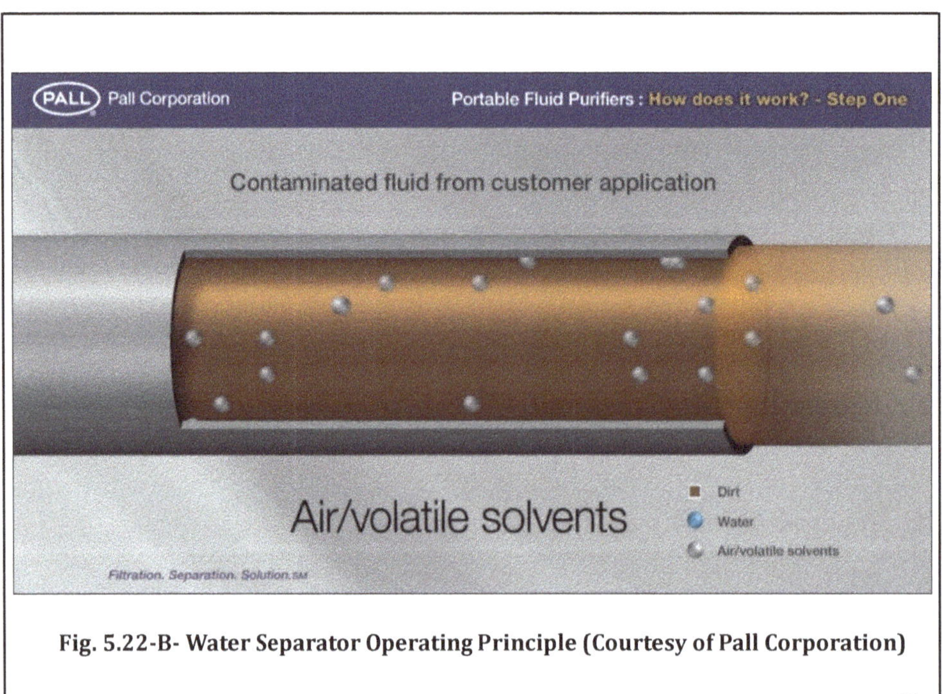

Fig. 5.22-B- Water Separator Operating Principle (Courtesy of Pall Corporation)

39

39

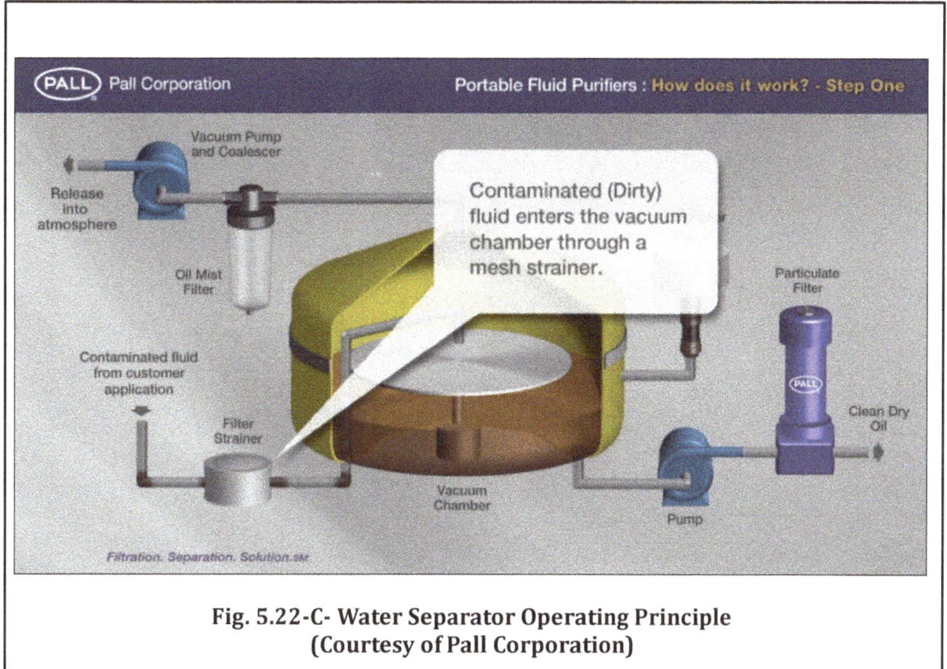

Fig. 5.22-C- Water Separator Operating Principle (Courtesy of Pall Corporation)

40

40

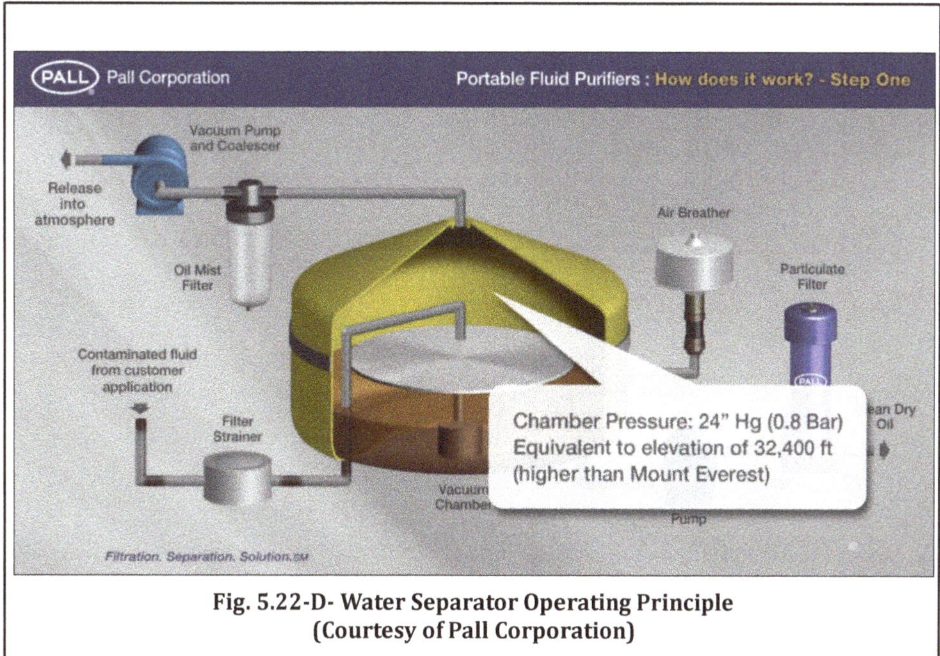

Fig. 5.22-D- Water Separator Operating Principle (Courtesy of Pall Corporation)

41

41

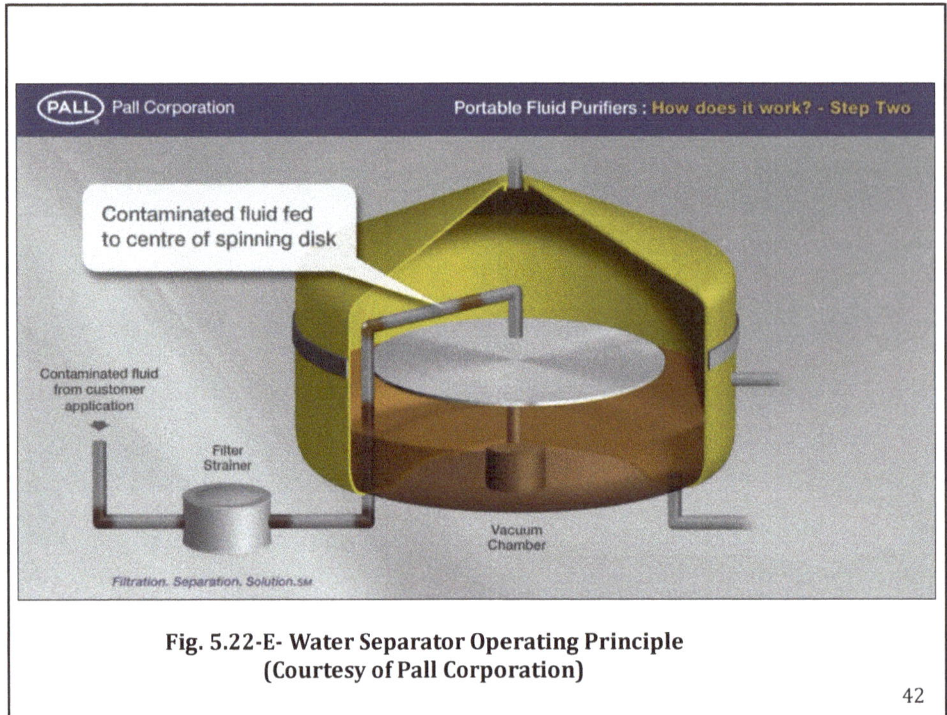

Fig. 5.22-E- Water Separator Operating Principle
(Courtesy of Pall Corporation)

42

42

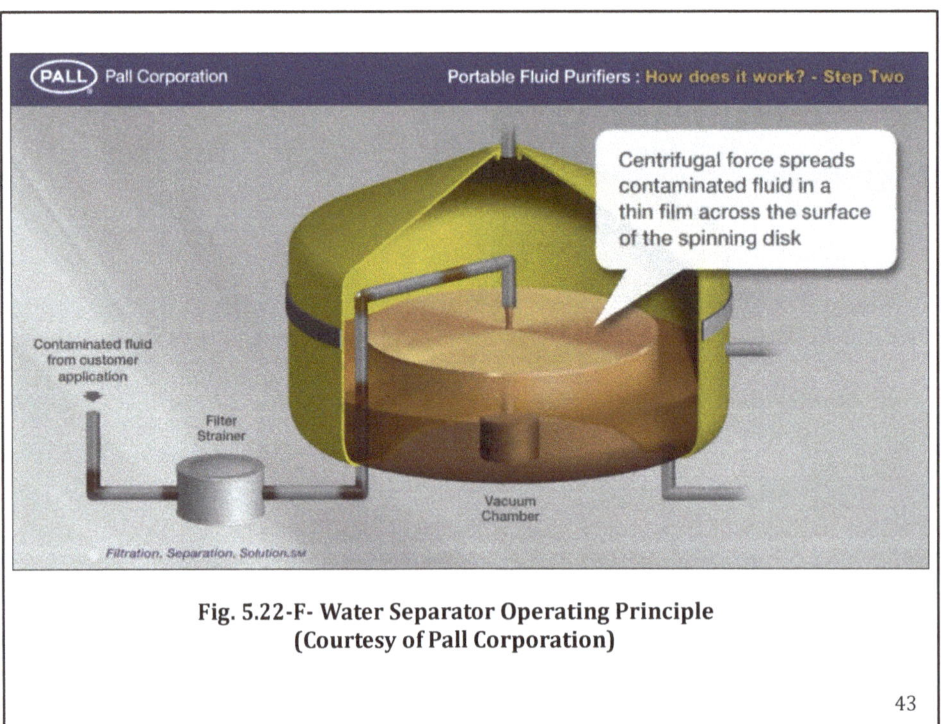

Fig. 5.22-F- Water Separator Operating Principle
(Courtesy of Pall Corporation)

43

43

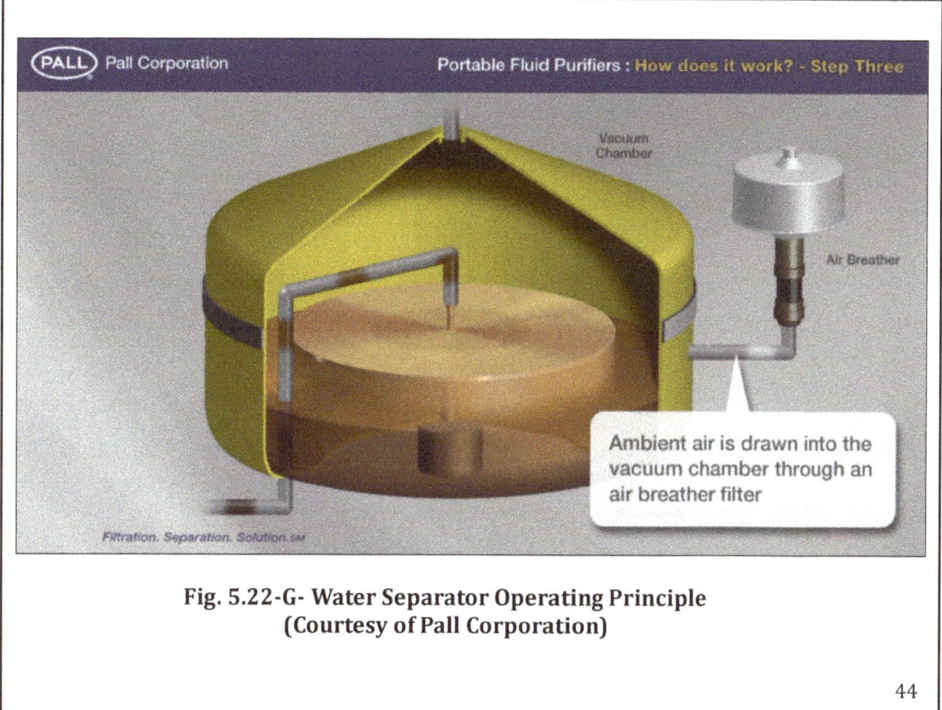

Fig. 5.22-G- Water Separator Operating Principle (Courtesy of Pall Corporation)

44

44

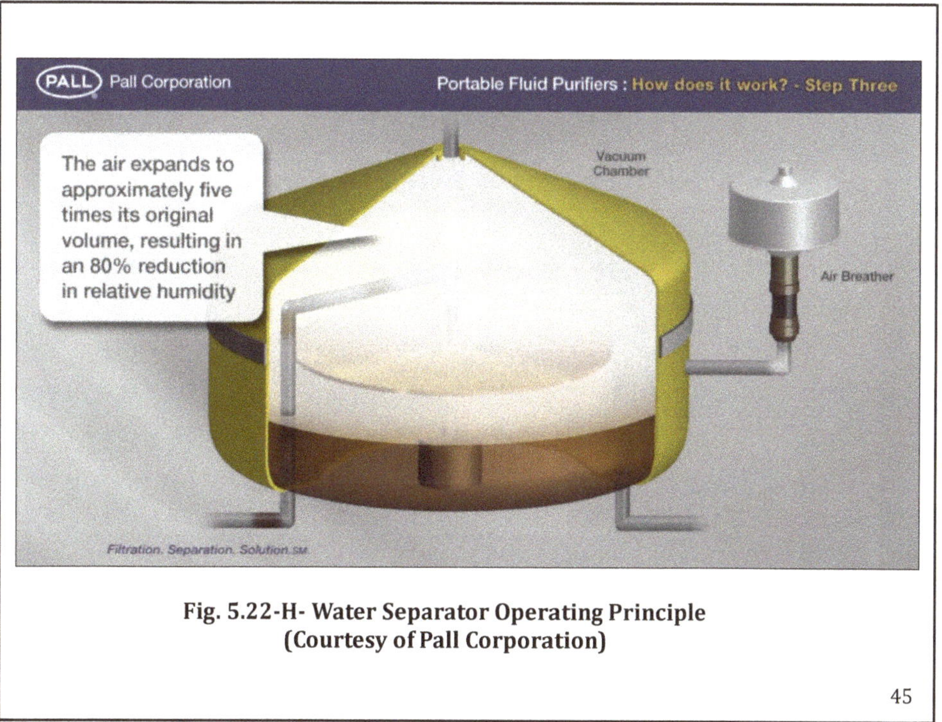

Fig. 5.22-H- Water Separator Operating Principle (Courtesy of Pall Corporation)

45

45

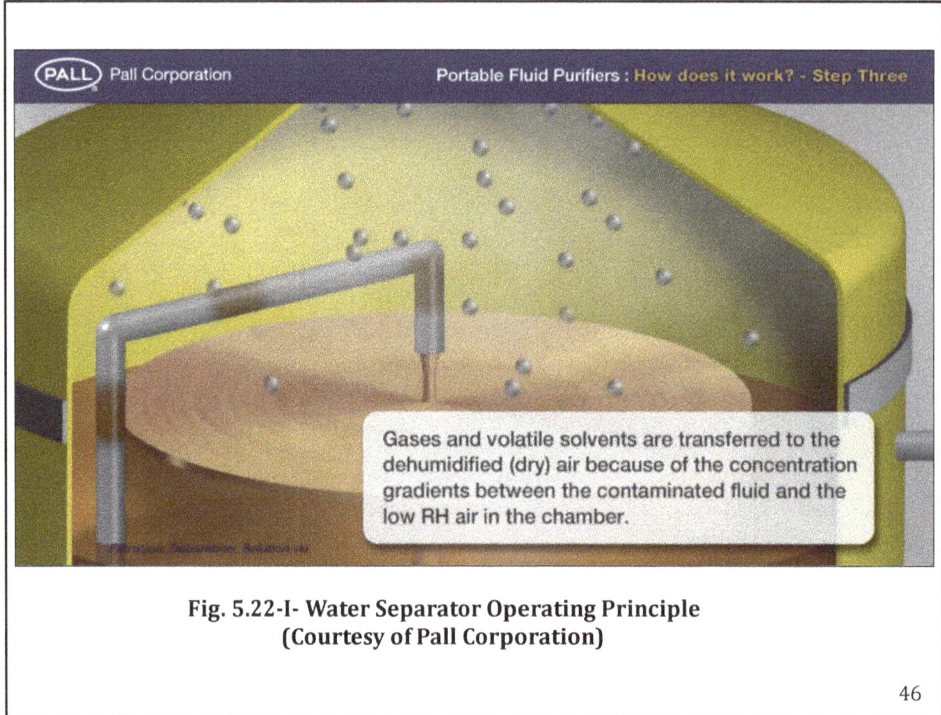

Fig. 5.22-I- Water Separator Operating Principle (Courtesy of Pall Corporation)

46

46

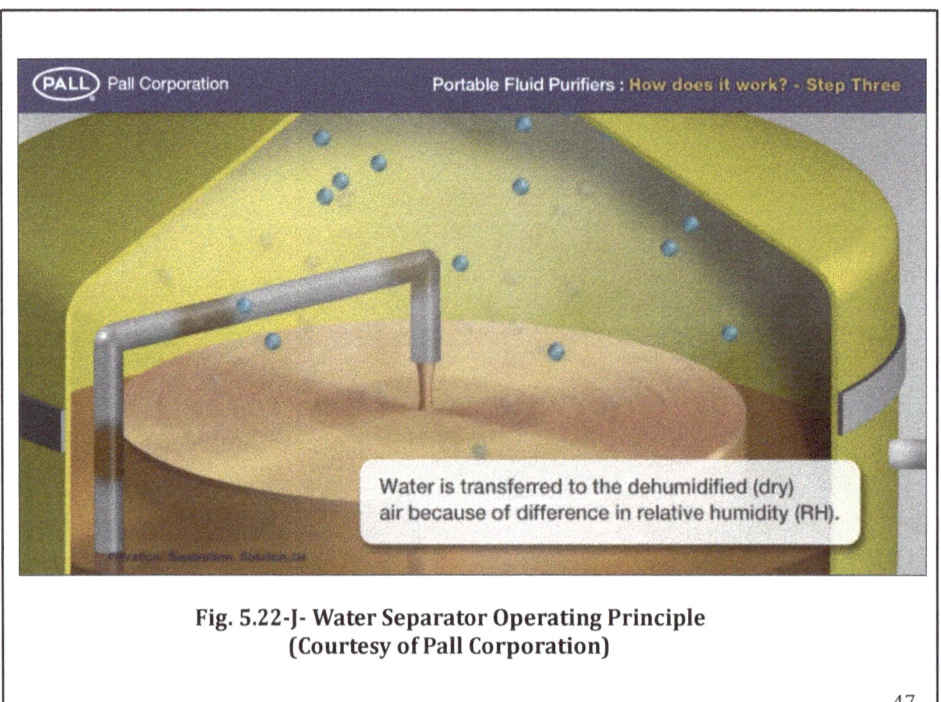

Fig. 5.22-J- Water Separator Operating Principle (Courtesy of Pall Corporation)

47

47

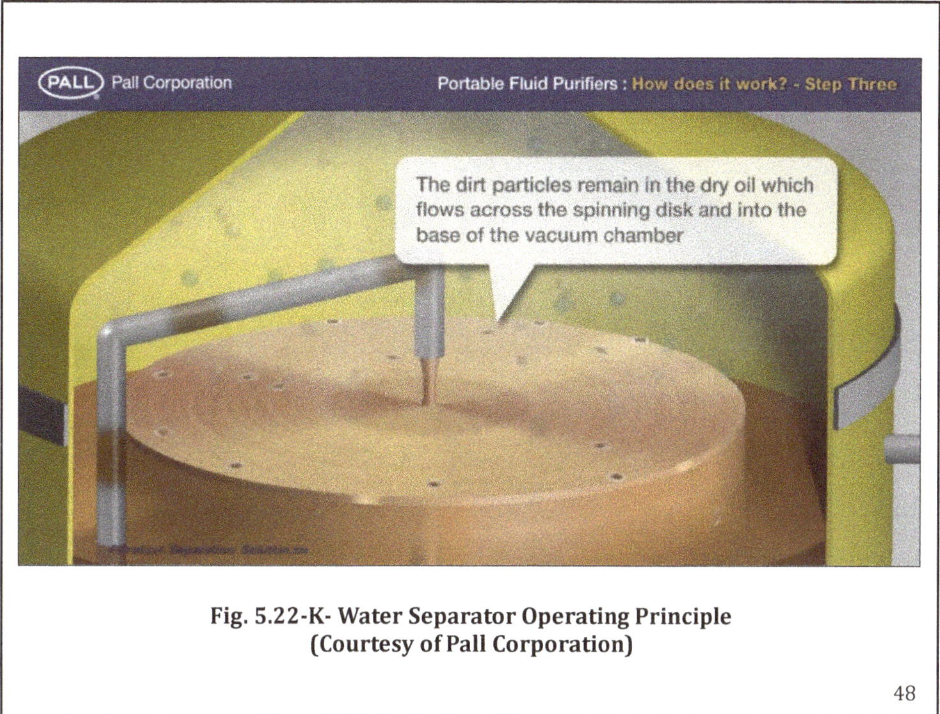

Fig. 5.22-K- Water Separator Operating Principle (Courtesy of Pall Corporation)

48

48

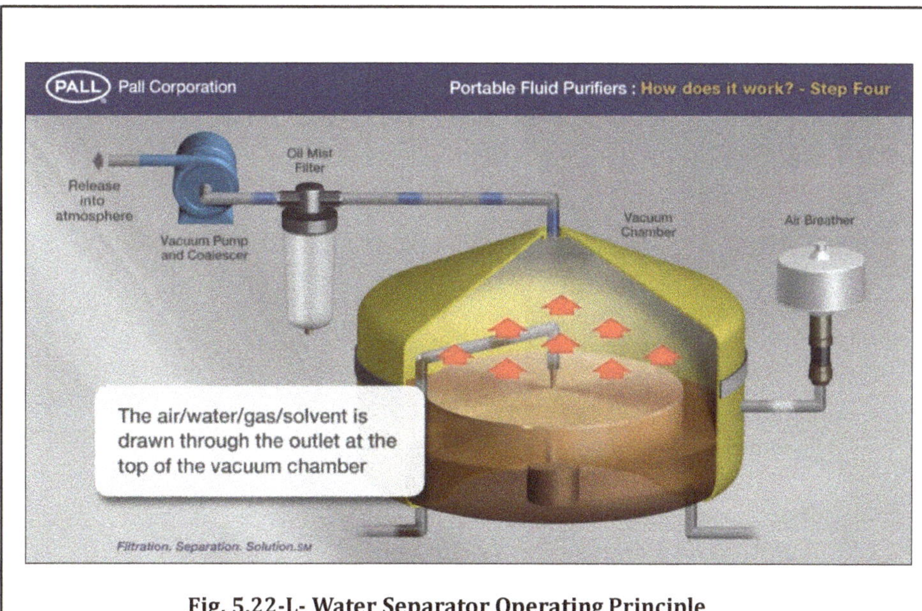

Fig. 5.22-L- Water Separator Operating Principle (Courtesy of Pall Corporation)

49

49

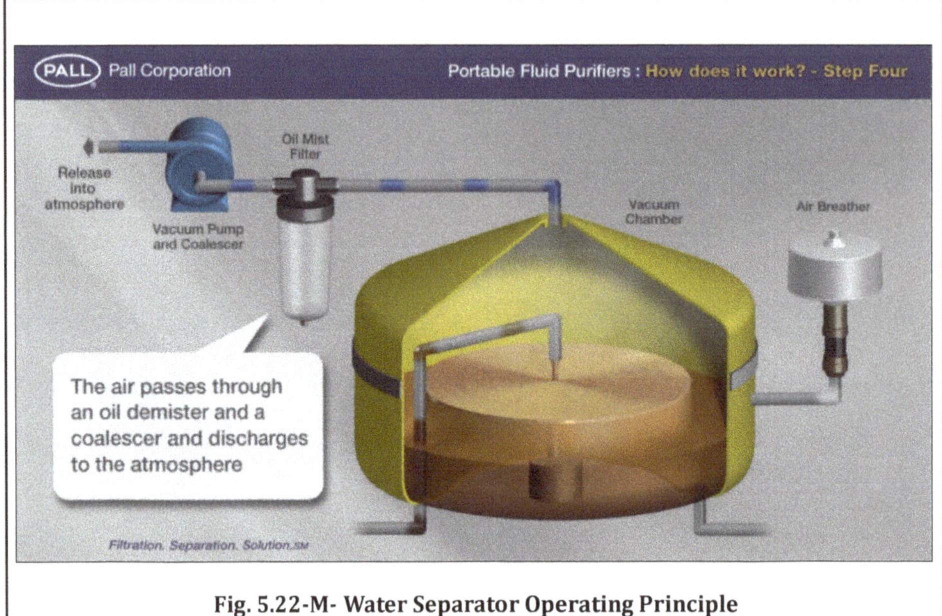

Fig. 5.22-M- Water Separator Operating Principle (Courtesy of Pall Corporation)

50

50

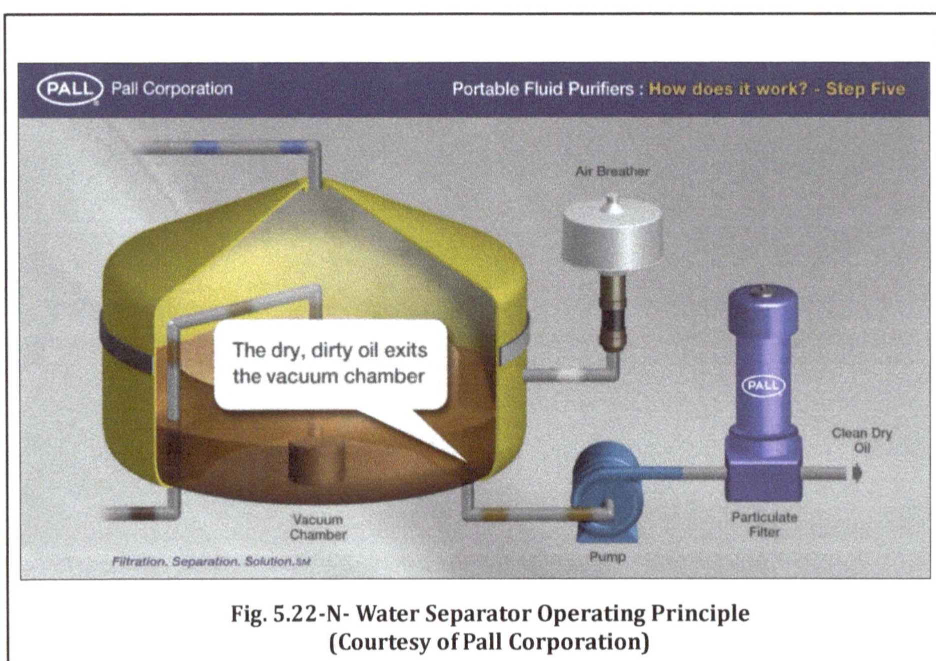

Fig. 5.22-N- Water Separator Operating Principle (Courtesy of Pall Corporation)

51

51

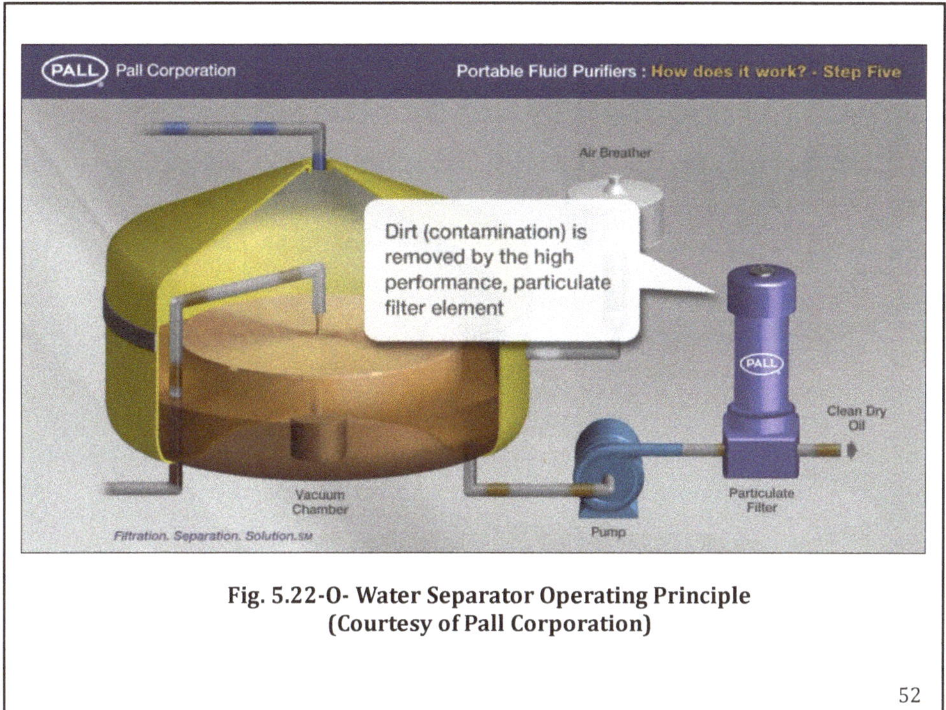

Fig. 5.22-O- Water Separator Operating Principle (Courtesy of Pall Corporation)

52

52

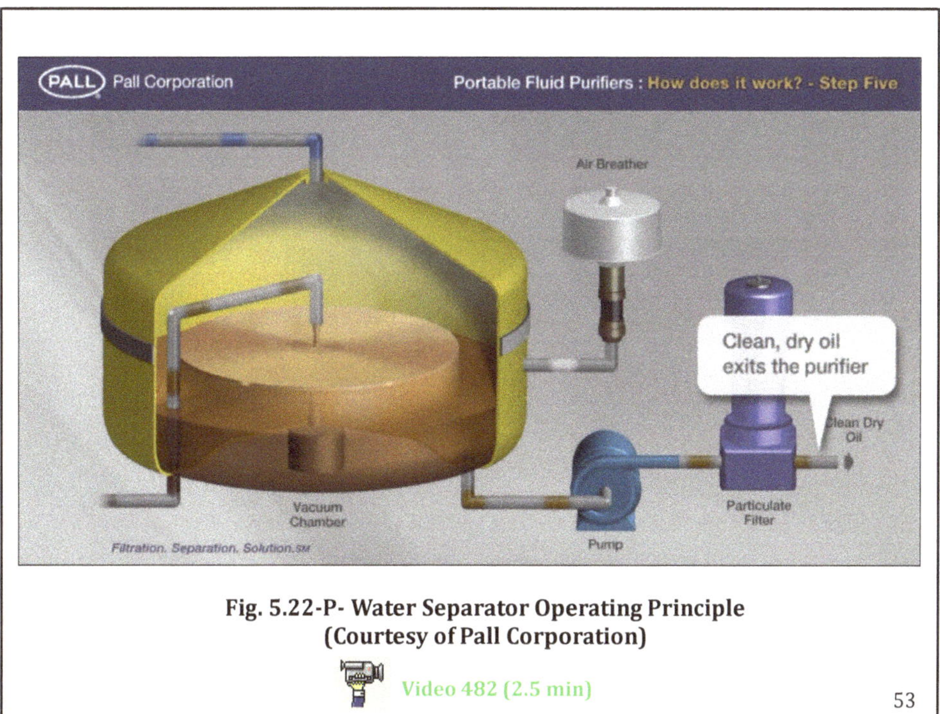

Fig. 5.22-P- Water Separator Operating Principle (Courtesy of Pall Corporation)

Video 482 (2.5 min)

53

53

If 10 gallons (40 liters) are highly contaminated by water, most suitable solution is to ?

A. Pass the oil through Desiccant Filter Element.

B. Pass the oil through centrifuge device.

C. Pass the oil through Mass Transfer Vacuum Dehydrator.

D. Drain the whole amount of fluid, flush the system, and fill in the system with new fresh oil through 10-micron filter.

54

54

Chapter 5 Reviews

1. Generally speaking, fluidic contamination means oil contaminated by?
   A. Solid particles.
   B. Air.
   C. Water.
   D. Sludge.

2. Absorptive capacity of a hydraulic fluid depends on?
   A. Fluid temperature.
   B. Fluid's molecular structure.
   C. Additive packages of the fluid.
   D. All of the above

3. Above saturation point, residual water separates from the fluid forming?
   A. Rust.
   B. Wax.
   C. Varnish.
   D. Free water.

4. Karl-Fischer method is based on?
   A. Visual Fluid Appearance.
   B. Titration (chemical analysis) using electrochemical device.
   C. Transmitted microwave through the fluid and a wave detector.
   D. Water boils at a temperature higher than oil.

5. If 10 gallons (40 liters) are highly contaminated by water, most suitable solution is to
   A. Pass the oil through Desiccant Filter Element.
   B. Pass the oil through centrifuge device.
   C. Pass the oil through Mass Transfer Vacuum Dehydrator.
   D. Drain the whole amount of fluid, flush the system, and fill in the system with new fresh oil through 10-micron filter.

Chapter 5 Assignment

Student Name: -- Student ID: ------------------

Date: --- Score: -----------------------

A: List 5 symptoms for a hydraulic system that is contaminated by water.

B: State the recommended method to remove water and varnish from a large quantity of oil and explain way this method is selected?

Chapter 6
Chemical Contamination

Objectives:

This chapter presents the sources of chemical contamination. For each source, the chapter explains how the system performance will be affected and possible recommendations to minimize such consequences.

0

0

Brief Contents:

1

1

Which of the following conditions expedites oil degradation?

A. High operating temperature.
B. High water and air contents.
C. Presence of catalyzing metals.
D. All of the above.

2

2

6.1- Sources of Chemical Contamination

Gaseous+ Fluidic + Thermal → Oil degrading → forms of oil degradation (*Oxidation*, *Hydrolysis*, and *Thermal Degradation*).

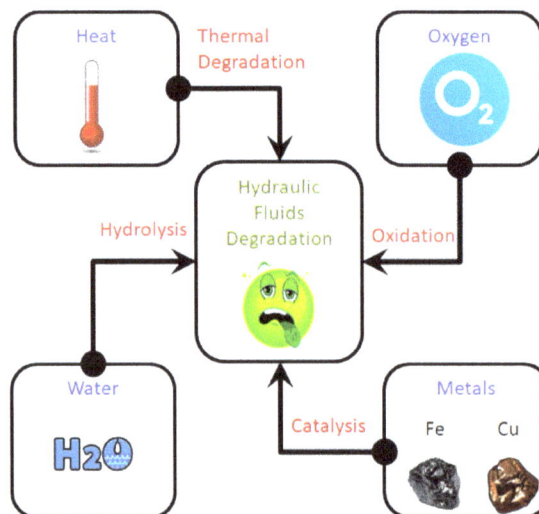

Fig. 6.1- Sources of Chemical Contamination

3

3

Which combination of products result from hydraulic fluid chemical degradation?

A. Oxygen + Acids + Sand + Rust.

B. Sludge + Acids + Microbes + Water.

C. Sludge + Acids + Varnish + Rust.

D. All of the above.

4

4

→ Degradation Products (Rust, Varnish, Acids, and Sludge.).

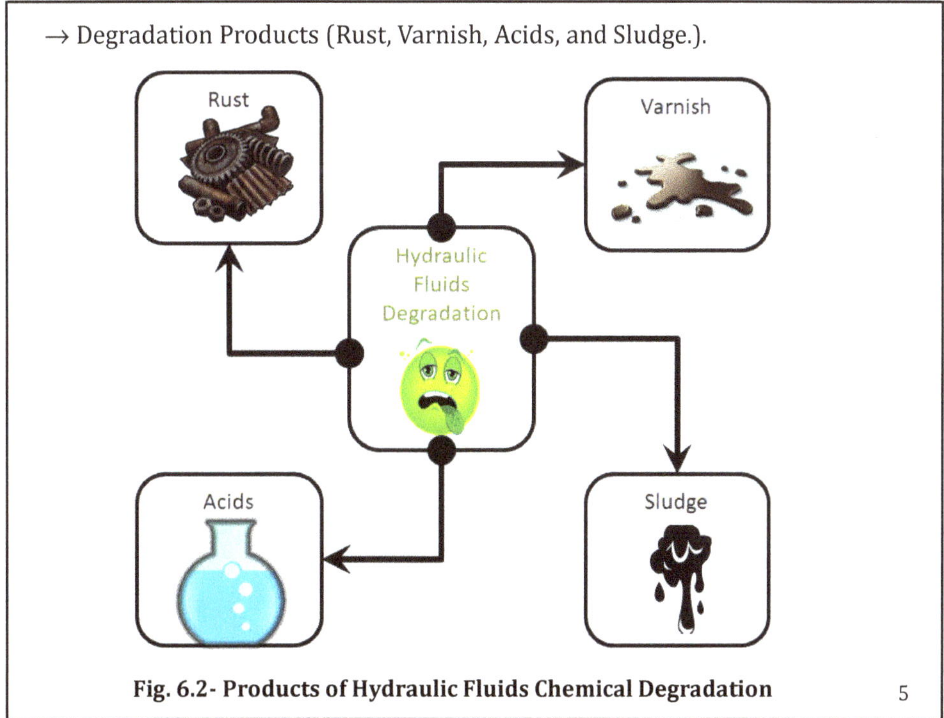

Fig. 6.2- Products of Hydraulic Fluids Chemical Degradation

5

5

6.2- Products of Hydraulic Fluid Degradation

6.2.1- Rust

- *Rust* is a surface degradation of iron metallic components due to oxidation.
- Presence of (Water moisture + Heat + Copper & Iron particles)
- Oxidation can occur twice as fast for every 10 ^{0}C (18°F) temperature rise.
- → Oxidation is accelerated.
- → Rust formation within the fluid.
- → Corrosion and metal fatigue are seen within the components.

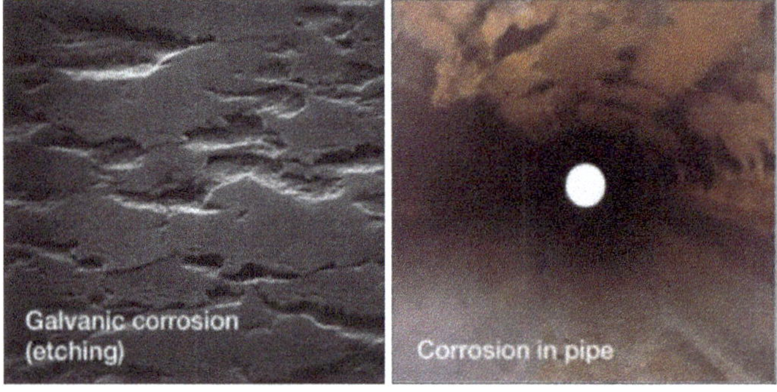

Fig. 6.3- Effect of Rust on Hydraulic Pipes (Courtesy of Pall Corporation) 6

6

Rust is also formed on the outer surfaces of the components and affect hydraulic lines (1), valves (2), pumps (3), cylinder rods (4), etc.

Fig. 6.4- Effects of Rust on Hydraulic Components 7

7

6.2.2- Acids

- ❑ Hydrolysis of ester-based fluids + Reaction of additives like zinc and calcium sulfonate:
- → Acid formation.
- ❑ Increased acidity:
- → promotes corrosion,
- → shortens fluid and components service life, and
- → leads to increased wear in the internal surfaces of machine.

Fig. 6.5- Corrosion in a Machine Component due to Acid Formation 8

8

6.2.3- Sludge

❑ Sludge is a thick polymerized compound.

❑ Hydraulic fluid exposure to high temperatures

→ Fluid break-down forming resinous material.

→ Sludge is formed.

→ Plugs small openings and orifices causing sudden system malfunction and/or failure.

→ Reduces heat transfer.

→ Cloggs filters and strainers.

Fig. 6.6- Sludge in Hydraulic Fluids

9

9

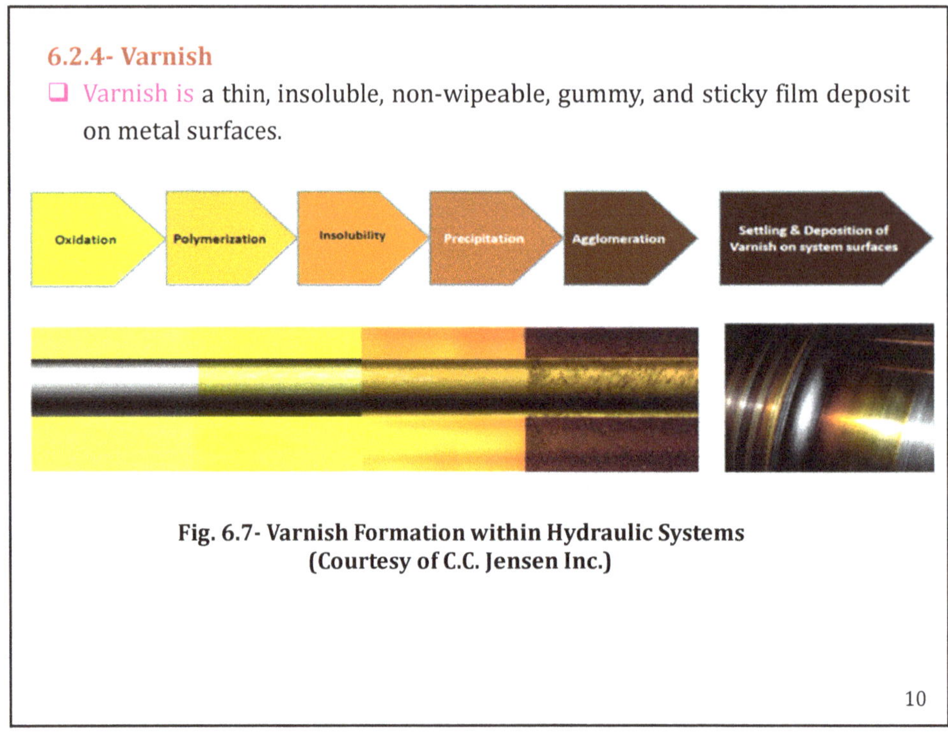

6.2.4- Varnish
❑ Varnish is a thin, insoluble, non-wipeable, gummy, and sticky film deposit on metal surfaces.
Oxidation
Polymerization
Insolubility
Precipitation
Agglomeration
Settling & Deposition of Varnish on system surfaces
10

Fig. 6.7- Varnish Formation within Hydraulic Systems (Courtesy of C.C. Jensen Inc.)

10

11

As a result of varnish formation in a hydraulic fluid?

A. Actuators move erratically in stick-slip mode.

B. Water content in the oil increased.

C. Bearing life increase.

D. Oil viscosity reduced.

12

12

- ❑ Varnish creates a sticky layer.
- → This layer attracts the abrasive particles of all sizes.
- → Creating a sand-paper grinding surface.
- → Radically speeds up machine wear.

Fig. 6.8- Varnish Sticky Layer Attracts Abrasive Particles (Courtesy of C.C. Jensen Inc.)

13

13

- ❑ Varnish creates a sticky layer.
- → Block fine tolerances,
- → Make spool valves seize.
- → Clogs filters.
- → Acts as an insulator reducing the effect of the heat exchangers.

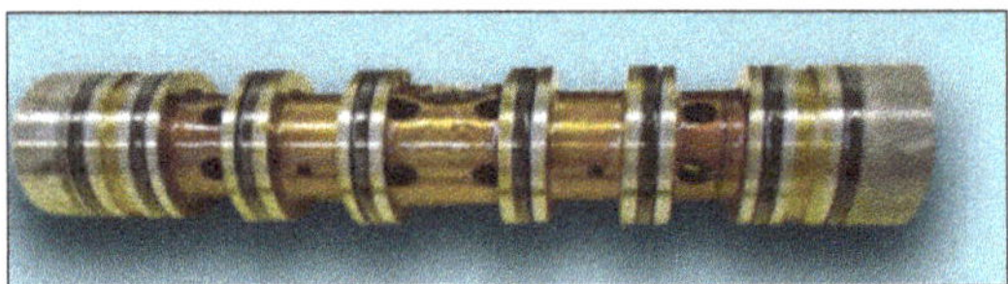

Fig. 6.9- Varnish Sticky Layer Seizes Valve Spools (Courtesy of C.C. Jensen Inc.)

Fig. 6.10- Varnish Sticky Layer Clogs Filters (Courtesy of C.C. Jensen Inc.) 14

14

Which of the following symptoms can be used as an indication of hydraulic oil degradation?

A. Oil has increased ability for lubrication.

B. Oil has sour, putrid, and acidic smell.

C. Oil has increased viscosity index.

D. All of the above.

15

15

6.3- Effects of Chemical Contamination

- **Fluid Appearance:** As shown in Fig. 6.11, as compared to a sample of new oil, oil that is degraded has dark color.
- **Fluid Odor** As shown in Fig. 6.12, as compared to a sample of new oil, oil that is degraded has sour, putrid, and acidic smell.
- **Increased Oil Viscosity:** As shown in Fig. 6.13, increasing oil viscosity results in a higher pressure drop in valves and transmission lines.

Fig. 6.11- Fluid Appearance (C. of Noria Corp.)

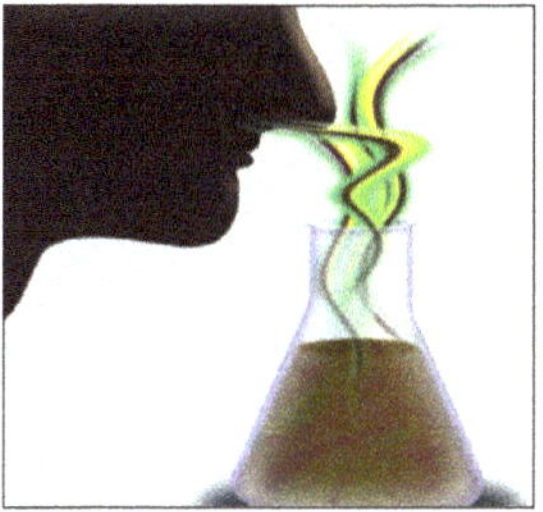

Fig. 6.12- Fluid Odor (C. of C.C. Jensen Inc.)

Fig. 6.13- Fluid Viscosity (C. f C.C. Jensen Inc.)

16

16

- **Reduced System Performance:** stick-slip actuators motion.
- **Decreased Additive Performance:** Some additives react with the degrading products and consequently lose their effect.
- **Shorter Oil Life:** Oil life is significantly reduced.
- **Filter Life:** Reduce filter life because of sludge formation.
- **Component Life:** Reduced service life because of corrosive wear.
- **Reduced Productivity:** Productivity reduced due to increased downtime and filter change frequency.
- **Increased Maintenance Costs:** Maintenance cost increased due to shorter oil and component service life.
- **Environmental Pollution Consequences:** Environmental pollution increased due to frequent disposals and possible leakage.

17

17

6.4- Standard Test Methods for Measuring Oil Degradation

1. **Ultracentrifuge Test:**

- Uses the centrifugal forces to extract and settle the contaminants of the oil.
- The sediments are compared with a sedimentation rating system to determine the degradation of the oil.

2. **Fourier Transformation Infrared Spectroscopy Analysis (FTIR):**

- *FTIR* analysis is same as used in water content measuring.

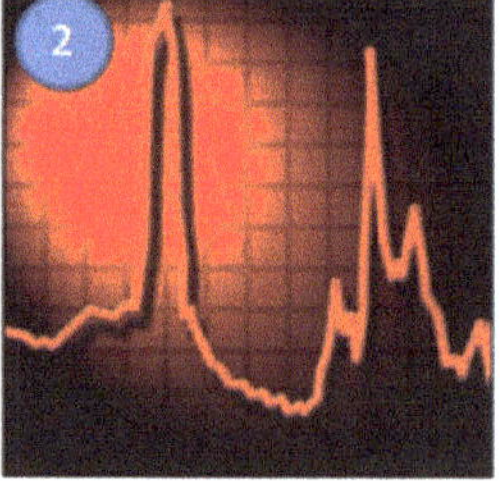

Fig. 6.14- Standard Test Methods for Measuring Oil Degradation (Courtesy of C.C. Jensen Inc.)

18

18

3. **Membrane Patch Colorimetry (MPC)**

- Is an analysis that indicates that the oil contains degradation products.
- Varnish is captured in a white 0.45-micron cellulose membrane.
- Varnish shows as a yellow, brownish or dark color depending on the amount of varnish present in the oil.

4. **QSA Test:**

- QSA method identifies the varnish potential rating on a scale 1-100.
- Is based on colorimetric analysis by
- Comparing the result to a large database.

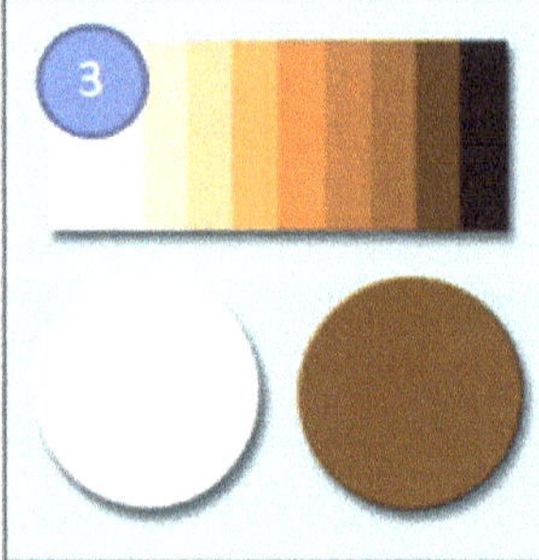

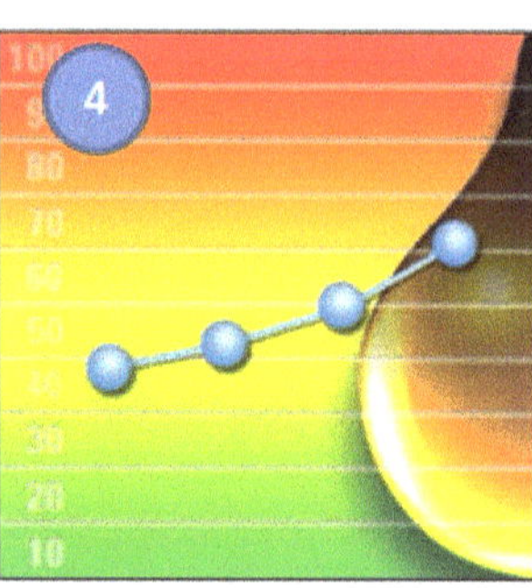

19

19

1. **Gravimetric Analysis:** Measures the weight of residual components to determine the level of oil degradation by.

2. **Viscosity Test:** Used as an indicator of oil degradation.

3. **Remaining Useful Life Evaluation Routine (RULER) Test:** Measures the remaining amount of anti-oxidants (oil additives). When the additives get depleted due to oil degradation, RULER number decreases.

4. **Total Acid Number (TAN):** Measures the level of acidic compounds. used as an indicator of oil degradation, since acidity is a product of degradation.

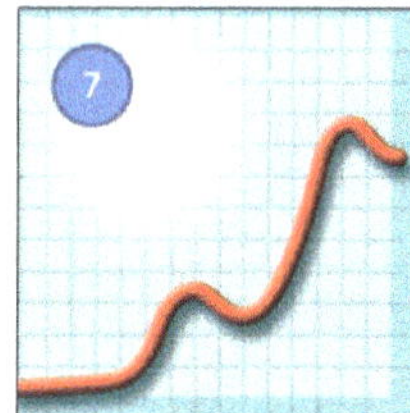

20

20

Which of the following statements is TRUE?

A. Water and gaseous contamination do not affect hydraulic oil degradation.

B. Acidity of a hydraulic fluid should be limited to maximum of +0.5 TAN higher than that of new oil.

C. Varnish is not a product of hydraulic oil degradation.

D. All of the above is TRUE.

21

21

6.5- Best Practices to Minimize Chemical Contamination

6.5.1- Preventive Practices to Minimize Chemical Contamination

- ❑ **Control contents of water and air in the hydraulic fluid.**

- ❑ **Control working fluid temperature.**

- ❑ **Control the acidity level of the hydraulic fluid.**
  - Acidity of a hydraulic fluid should be limited to maximum of +0.5 TAN higher than that of new oil.
  - If +1 TAN is measured, an immediate action is required (i.e. if new oil has 0.5 TAN, then 1.0 TAN is alert and 1.5 TAN is alarming value).

- ❑ **Hydraulic Fluid Analysis:** Periodic testing for measuring fluid conditions such as TAN, varnish and sludge formation, oil viscosity, etc. are key information for predictive maintenance.

- ❑ **Hydraulic Fluid Additives:** Use of proper additive package such as anti-oxidation, rust inhibitors, emulsifiers, and foam suppressors.

22

22

- ❑ **As an example of new technology:**

DuraClean™ is an ultra-premium multi-grade hydraulic oil provided exclusively by Parker.

- Has a unique additive chemistry designed:
- ISO 46, all season, multi-grade hydraulic fluid.
- Replaces ISO 32, 46, and 68 mono-grades.
- High viscosity index for wide operating temperature ranges.
- Outstanding oxidation life to maximize component life.
- Formulated to help extend the life of hoses and seals.
- Prevents varnish formation.
- Clean, as packaged, to ISO 17/15/12 cleanliness level.
- Special formulation that allows for rapid air release and water separation.
- Excellent filterability to minimize filter blockage.
- Outstanding acrylate anti-foam agent contains no silicones, which can lead to inaccurate particle counts.
- Excellent shear stability for stable viscosity over time.
- Superior thermal stability for uncompromised performance at high temperatures.

23

23

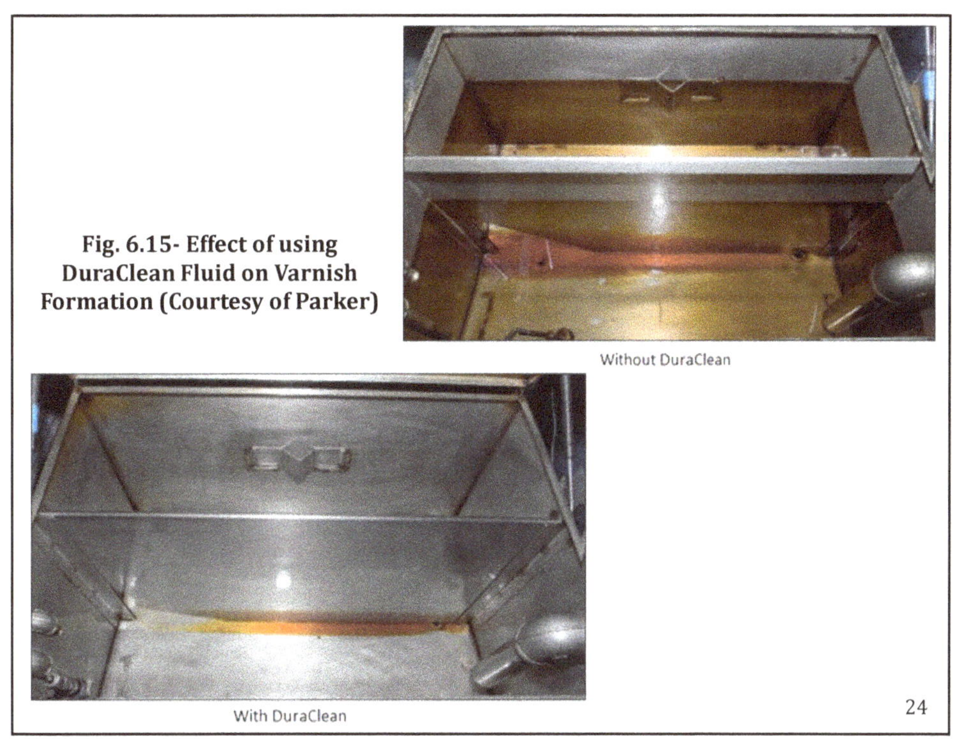

Fig. 6.15- Effect of using DuraClean Fluid on Varnish Formation (Courtesy of Parker)

24

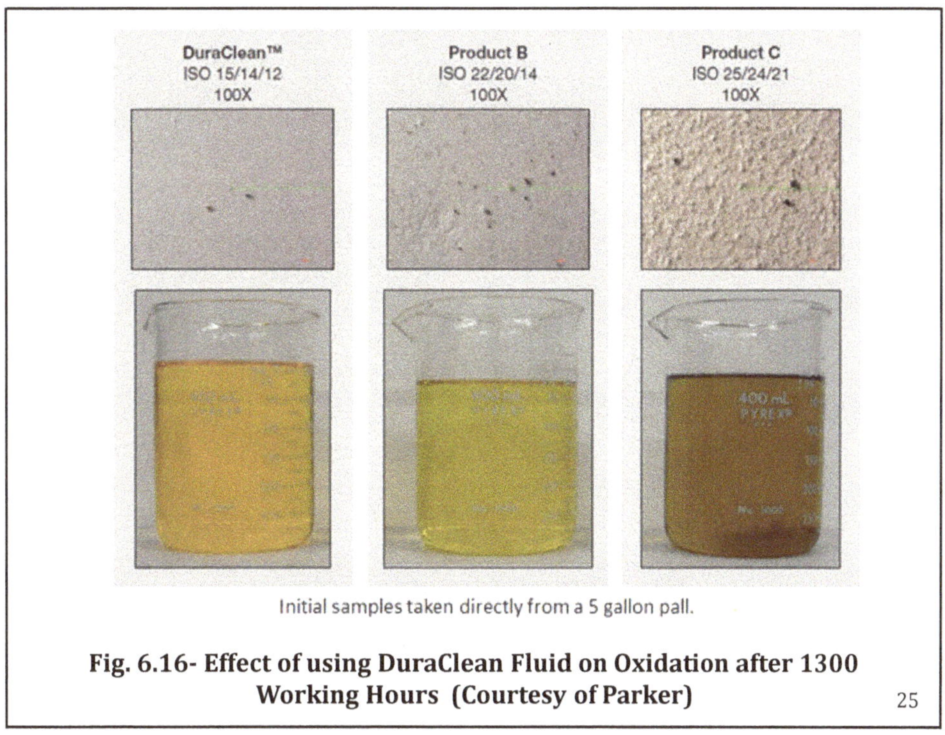

Fig. 6.16- Effect of using DuraClean Fluid on Oxidation after 1300 Working Hours (Courtesy of Parker)

25

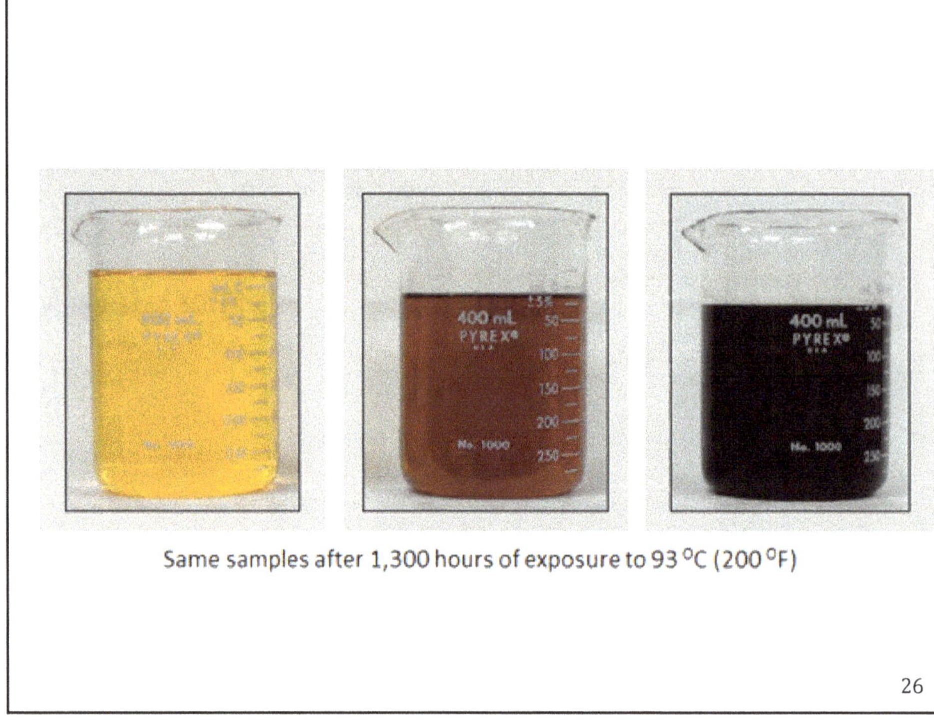

Same samples after 1,300 hours of exposure to 93 °C (200 °F)

26

26

6.5.2- Curative Practices to Remove Chemical Contamination

If hydraulic fluid degrades, serious and immediate actions are required. The following sections provide most common methods of water removal.

❑ **Fluid Replacement:**

- As it has been previously mentioned, if the quantity of the contaminated oil is small (< 500 gallon =2000 liter), it is recommended to replace it and flush the system.

❑ **Acidity Neutralization:**

- The alkalinity of the oil is supposed to neutralize incoming acidity.
- Acid number 3-5 times higher than that of new oil results in severe acidic corrosion of system components.
- In such fluids the acid number can be lowered and maintained by changing the fluid.

27

27

❑ **Varnish and Sludge Removal:**

- Oil degradation products cannot be removed with conventional mechanical filters because they are submicron particles and a fluid in a fluid, like when sugar is dissolved in water.
- These degradation products can be removed by fine filters through a combination of adsorption and absorption processes.

Fig. 6.17- Oil Degradation Products (Courtesy of C.C. Jensen Inc.)

28

28

Adsorption is the physical or chemical binding of molecules to a surface (like getting a cake thrown into your face).

Absorption molecules are absorbed into the media.

Fig. 6.18- Difference between Absorption and Adsorption (Courtesy of C.C. Jensen Inc.)

29

29

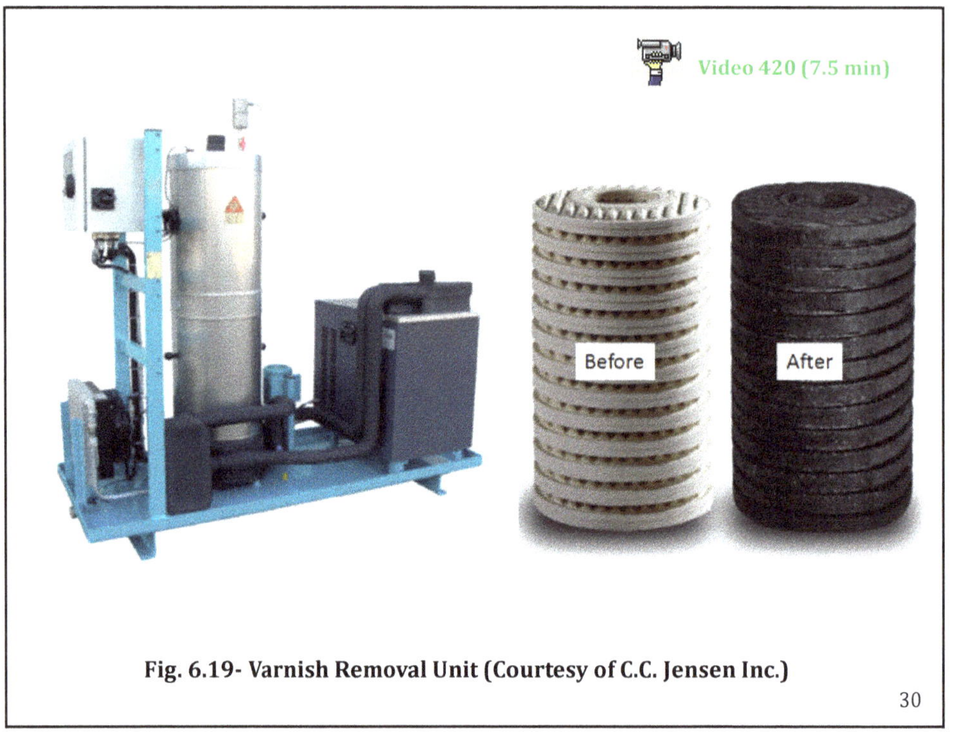

Fig. 6.19- Varnish Removal Unit (Courtesy of C.C. Jensen Inc.)

30

30

TECHNICAL DATA

Varnish Removal Unit		VRU 27/108 380 - 420V @ 50 Hz & 440 - 480V @ 60 Hz
Pump inlet pressure max.	bar/psi	0.5/7
Power consumption aver.	kW	2
Full load current max.	A	4
Filter Insert VRi 27/27	pcs.	4
Oil reservoir volume max. *)	ltr/gal	45,000/11,900
Oil viscosity **)		<ISO VG68
Oil temperature max *)	°C/°F	105/221
Varnish holding capacity up to	kg/lb	8/18
Total weight	kg/lb	290
Design pressure, filter	bar/psi	4/58
Dimensions lxwxh incl. + free height	mm inches	1600x650x1598+575 63x25.6x62.9+22.6

*) For more than 45,000 L or higher temperatures, please contact us
**) For viscosities higher than ISO VG68, please contact us

Table 6.1- Technical Data of the Varnish Removal Unit (Courtesy of C.C. Jensen Inc.)

31

31

- **As an example of new technology**:,

CJC™ Filter Inserts, made of *Cellulose Fibers*,

- have a high surface area and are effective as adsorbents and absorbents.
- Each cellulose fiber consists of millions of cellulose molecules.
- Each strand of cellulose molecule has a diameter of 10-30 microns.
- Degradation products are adsorbed and absorbed into the cellulose material.

32

32

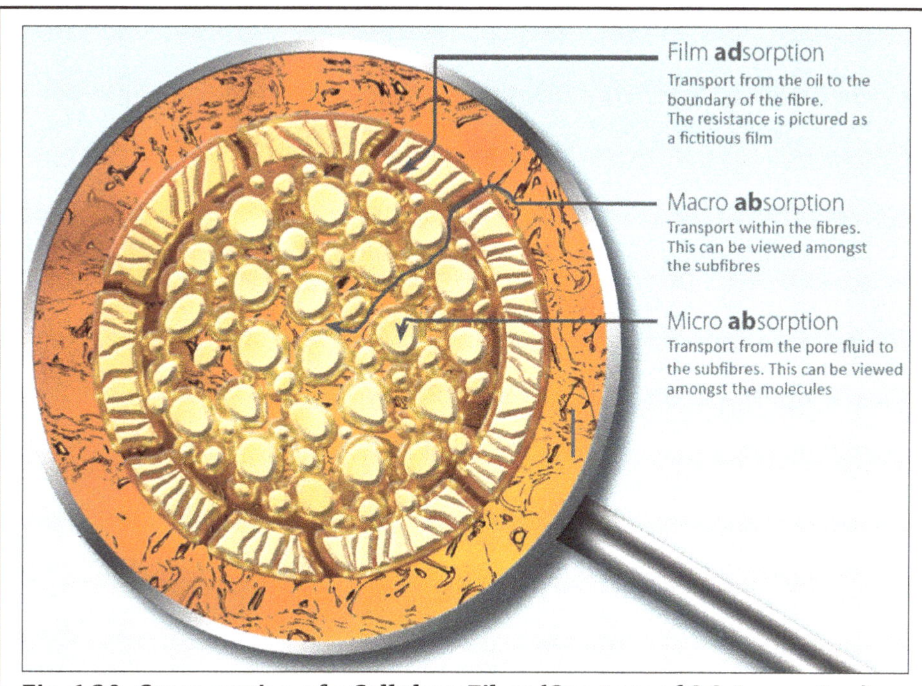
Film **ad**sorption
Transport from the oil to the boundary of the fibre. The resistance is pictured as a fictitious film
Macro **ab**sorption
Transport within the fibres. This can be viewed amongst the subfibres
Micro **ab**sorption
Transport from the pore fluid to the subfibres. This can be viewed amongst the molecules

Fig. 6.20- Cross-section of a Cellulose Fiber (Courtesy of C.C. Jensen Inc.) 33

33

- Contaminated oil approaches the cellulose fibers in an almost new Filter Insert.

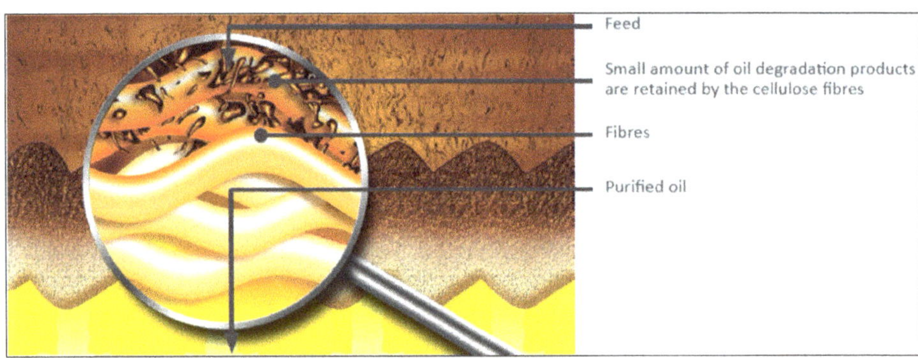
Feed
Small amount of oil degradation products are retained by the cellulose fibres
Fibres
Purified oil

Fig. 6.21- Contaminated Oil Approaching Cellulose Fibers (Courtesy of C.C. Jensen Inc.)

34

34

- This illustration shows that the Filter Insert is still delivering clean oil even though the cellulose fibers are nearly saturated.

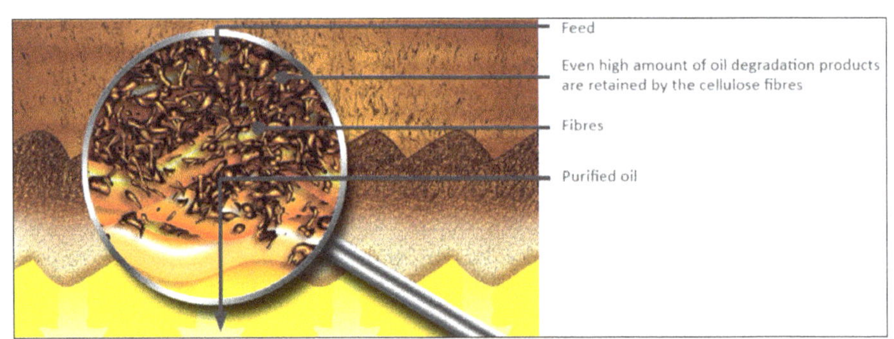
Feed
Even high amount of oil degradation products are retained by the cellulose fibres
Fibres
Purified oil

Fig. 6.22- Filter Inserts Near Saturation (Courtesy of C.C. Jensen Inc.)

Video 513 (2.5 min)

35

35

Chapter 6 Reviews

1. Which of the following conditions expedites oil degradation?
   A. High Operating temperature.
   B. High water and air contents.
   C. Presence of catalyzing metals.
   D. All of the above.

2. Which combination of products result from hydraulic fluid chemical degradation?
   A. Oxygen + Acids + Sand + Rust.
   B. Sludge + Acids + Microbes + Water.
   C. Sludge + Acids + Varnish + Rust.
   D. All of the above.

3. As a result of varnish formation in a hydraulic fluid?
   A. Actuators move erratically in stick-slip mode.
   B. Water content in the oil increased.
   C. Bearing life increase.
   D. Oil viscosity reduced.

4. Which of the following symptoms can be used as an indication of hydraulic oil degradation?
   A. Oil has increased ability for lubrication.
   B. Oil has sour, putrid, and acidic smell.
   C. Oil has increased viscosity index.
   D. All of the above.

5. Which of the following statements is TRUE?
   A. Water and gaseous contamination do not affect hydraulic oil degradation.
   B. Acidity of a hydraulic fluid should be limited to maximum of +0.5 TAN higher than that of new oil.
   C. Varnish is not a product of hydraulic oil degradation.
   D. All of the above is TRUE.

Chapter 6 Assignment

Student Name: -- Student ID: ------------------

Date: -- Score: ------------------------

A: List 4 products that result due to hydraulic fluid degradation.

B: State and explain two quantitative and two qualitative standard methods to evaluate oil degradation.

Chapter 7
Particulate Contamination

Objectives:

This chapters presents the sources of particulate contamination. For each source, the chapter explains how the system performance will be affected and possible recommendations to minimize such consequences.

0

0

Brief Contents:

1

1

Particulate contaminants can be in which one of the following forms?

A. Silt.

B. Nonabrasive particles such as fibers.

C. Gelatinous particles or microorganisms.

D. All of the above.

2

2

7.1- Forms of Particulate Contamination

- **Abrasive Particles:**
  - Most are metallic due to component wear such as (aluminum, chromium, copper, iron, lead, tin, silicon, sodium, zinc, barium and phosphorous).
  - Some other abrasive particles are nonmetallic such as sand.

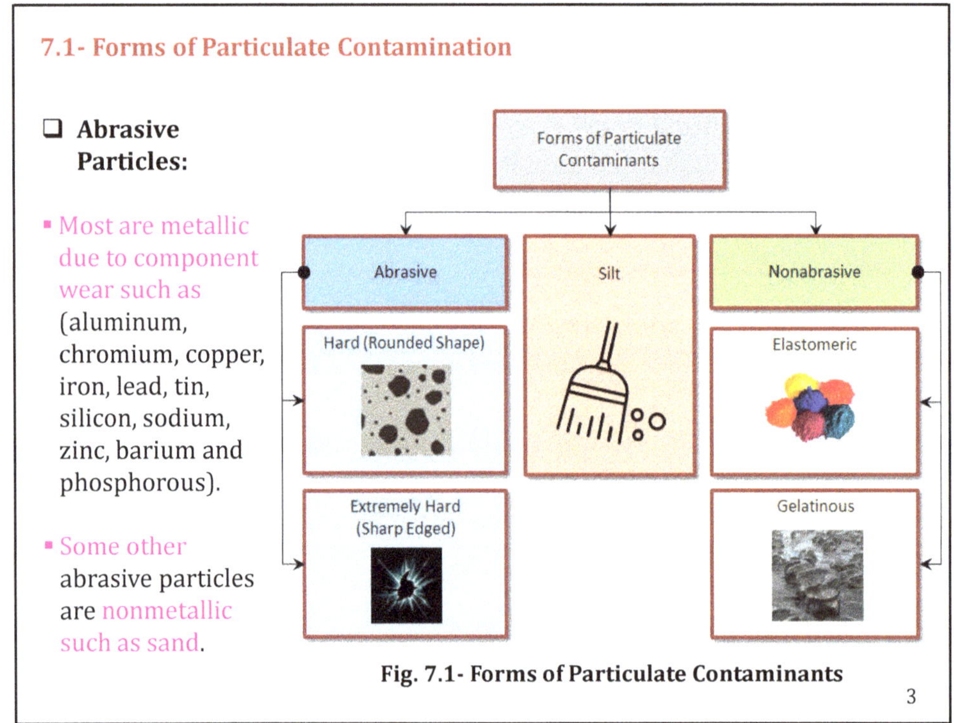

Fig. 7.1- Forms of Particulate Contaminants

3

3

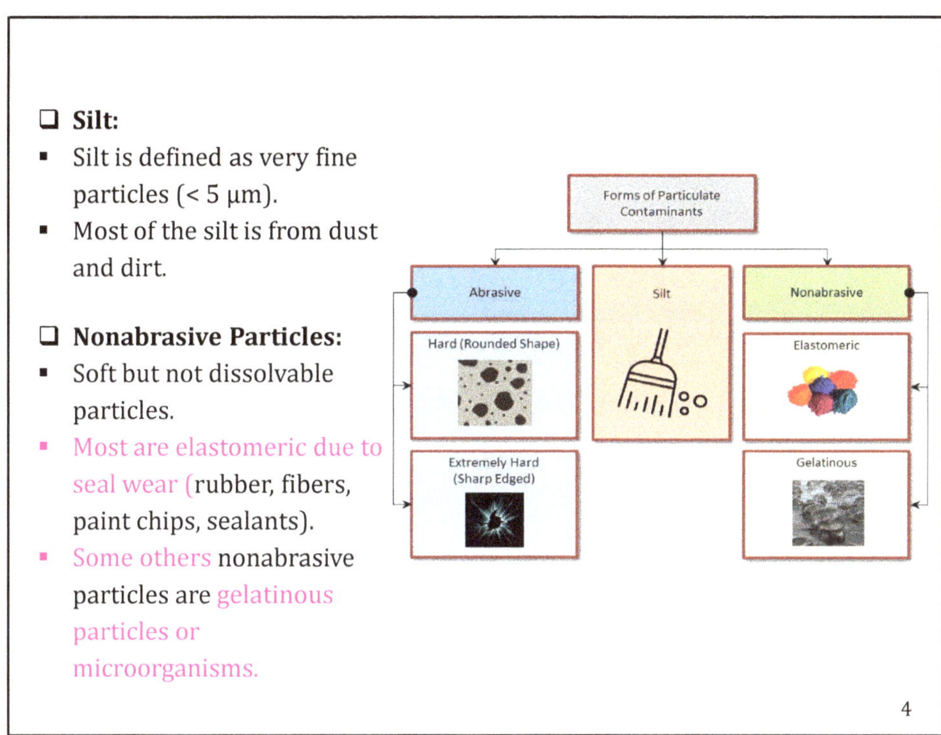

❑ Silt:
▪ Silt is defined as very fine particles (< 5 µm).
▪ Most of the silt is from dust and dirt.
❑ Nonabrasive Particles:
▪ Soft but not dissolvable particles.
▪ Most are elastomeric due to seal wear (rubber, fibers, paint chips, sealants).
▪ Some others nonabrasive particles are gelatinous particles or microorganisms.
Forms of Particulate Contaminants
Abrasive
Silt
Nonabrasive
Hard (Rounded Shape)
Extremely Hard (Sharp Edged)
Elastomeric
Gelatinous
4

4

Residual sand from a foundry during casting a pump housing is considered?

A. Built in contaminates.

B. Introduced (ingested) contaminants.

C. Introduced (induced) contaminates.

D. Generated contaminates.

5

5

Humidity condensed into a reservoir is considered?

A. Built in contaminates.
B. Introduced (ingested) contaminants.
C. Introduced (induced) contaminates.
D. Generated contaminates.

6

6

Debris result from tube flaring is considered?

A. Built in contaminates.
B. Introduced (ingested) contaminants.
C. Introduced (induced) contaminates.
D. Generated contaminates.

7

7

Bearing wear is considered?

A. Built in contaminates.
B. Introduced (ingested) contaminants.
C. Introduced (induced) contaminates.
D. Generated contaminates.

8

8

7.2- Sources of Particulate Contamination

- **Built-in:** during manufacturing, assembly, and storage.
- **Introduced (Ingested):** from the environment.
- **Introduced (Induced):** during system servicing, make up fluid, and cleaning.
- **Generated:** due component wear and normal system operation.

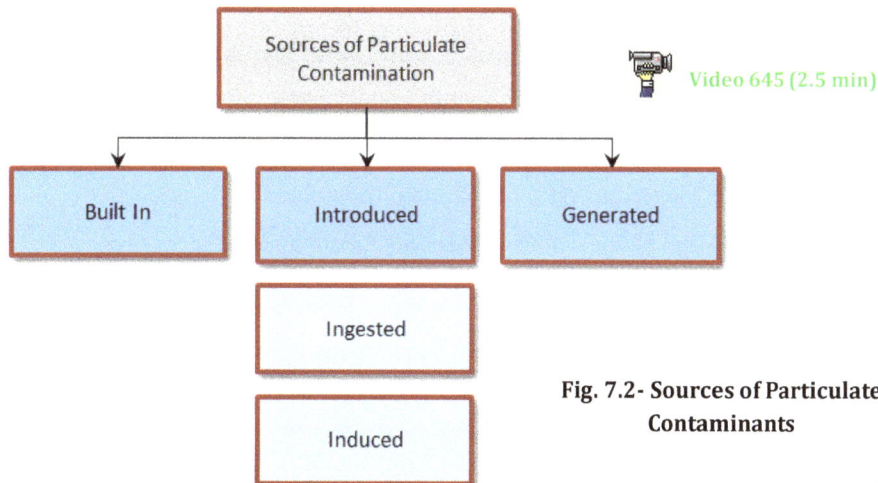

Fig. 7.2- Sources of Particulate Contaminants

9

9

7.2.1-Built-in Particulate Contamination

- Is defined as the particles remaining in the system following initial construction of the hydraulic components and system.

- Built-in contamination is also called *Primary Contamination*.

- End users should not assume that new components and systems are 100% clean.

- It is wise to pre-clean all components prior to assembly and utilize "good housekeeping" techniques in the assembly area.

10

10

- Built-in particulate contamination is a result of, but not limited to:
- **Foundry Operations:** Core sand and dust
- **Machining Operations:** metal chips and weld splatter.
- **Painting:** Paint flakes and overspray particulates.
- **Assembly:** Lubricants, Teflon tape, and other sealing materials.
- **Plumbing:** Hose cutting, tube bending and flaring, pipe threading, and fittings tightening.
- **Testing:** particles from testing fluid and environment.
- **Initial Cleaning:** Sands from sandblasting, fibers and lint from rags.
- **Storage and Handling:** Dust, insects, rust, scale from pipes, and airborne contaminants.
- **Shipping:** Packaging materials.

Video 452 (1 min)

11

11

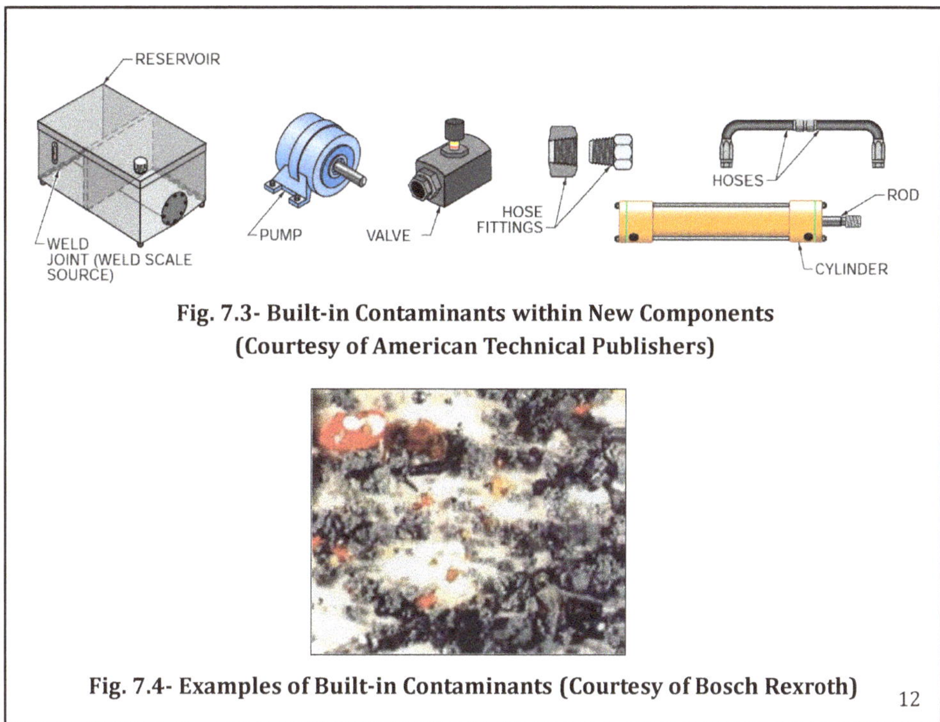

Fig. 7.3- Built-in Contaminants within New Components (Courtesy of American Technical Publishers)

Fig. 7.4- Examples of Built-in Contaminants (Courtesy of Bosch Rexroth) 12

12

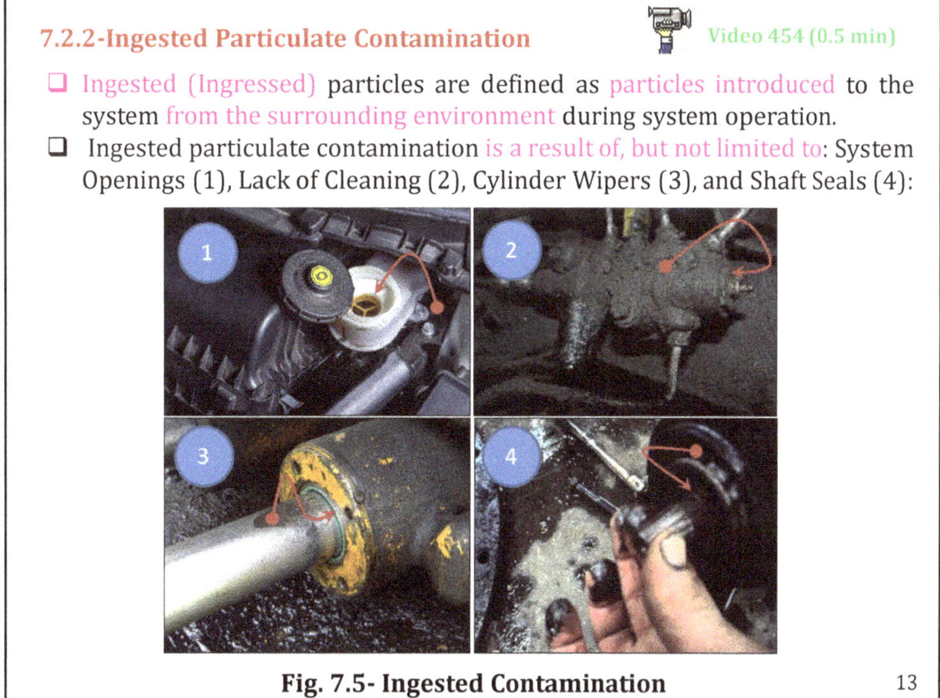

7.2.2-Ingested Particulate Contamination

- ❑ Ingested (Ingressed) particles are defined as particles introduced to the system from the surrounding environment during system operation.
- ❑ Ingested particulate contamination is a result of, but not limited to: System Openings (1), Lack of Cleaning (2), Cylinder Wipers (3), and Shaft Seals (4):

Fig. 7.5- Ingested Contamination 13

13

7.2.3-Induced Particulate Contamination

- ❑ Is defined as the particles introduced into the system during maintenance, repair, and troubleshooting.
- ❑ Whenever a hydraulic system is "opened up" contaminates may be induced into the system.
- ❑ Induced particulate contamination is a result of, but not limited to:
- **Make Up Hydraulic Fluids:**

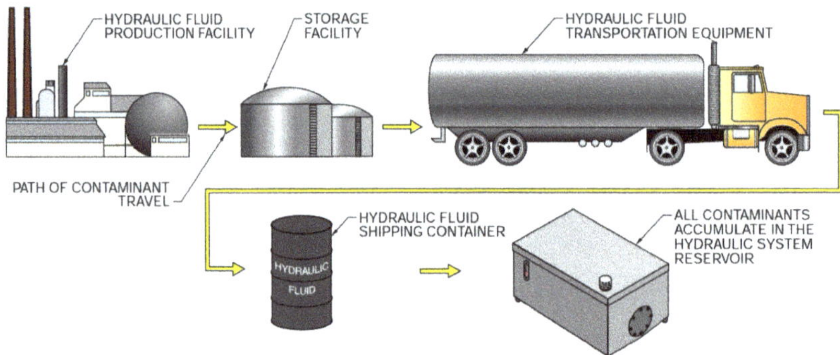

Fig. 7.6- Introduced Contaminants During Hydraulic Fluid Handling (Courtesy of American Technical Publishers) 14

14

- **Filter Change:** Changing a filter element requires opening the filter housing and replacing the current element with a new one that was taken out of package. This process can induce particles into the system
- **Component Rebuilding:** Component overhauling process requires system dissembling, cleaning, possibly machining, reassembling, cleaning, lubrication, testing, packaging, and storage. All these steps can be accompanied by introducing some amount of dirt into the component.
- **Reservoir Clean-out:** Cleaning a reservoir may result in inducing lint from rags or sand if sandblasting is used.
- **Hydraulic Line Replacement:** Cutting hoses by saw blade, bending and flaring tubes, threading and welding pipes, and fitting tightening are accompanied by induced particulate contamination.
- **Open Ports of Components:** Leaving ports of hydraulic components open (such as pump intake and discharge ports) during servicing a hydraulic system provides continuous ingression of particles into the system.

15

15

7.2.4-Generated Particulate Contamination

- Is defined as the particles internally generated during normal system operation. Generated particulate contamination is a result of, but not limited to:

- Metallic particles due to components wear or loss of metal due to other reasons.

- Sludge products due to oxidation.

- Rubber compounds and elastomers degradation due to aging, temperature and high velocity fluid streams. During degradation, the elastomers will start releasing particles into the hydraulic system. Sources include hoses, accumulator bladders and seals, particularly dynamic seals, such as cylinder pistons and shaft seals.

16

16

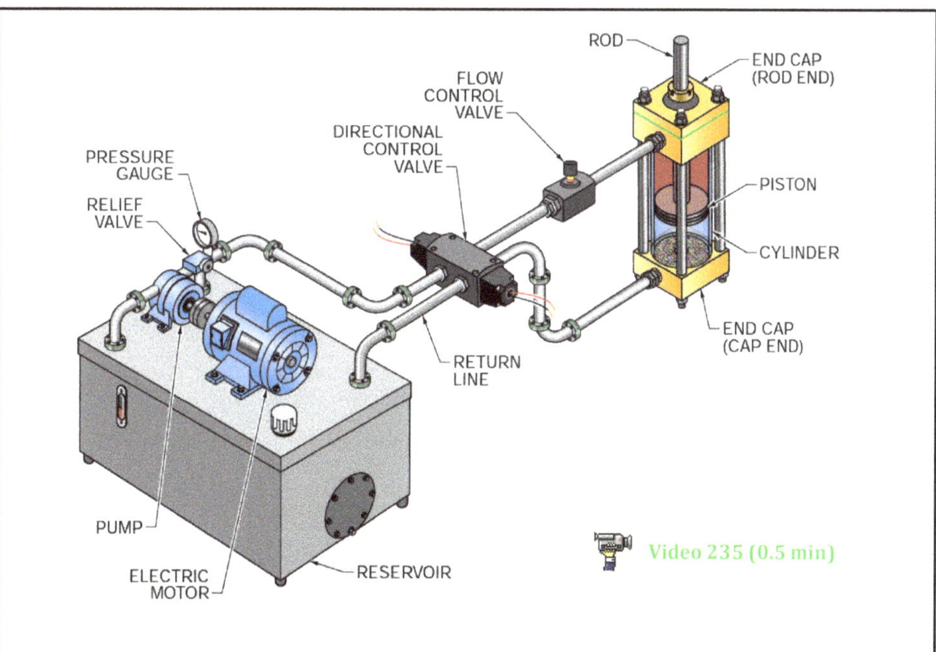

Video 235 (0.5 min)

Fig. 7.7- Particulate Contamination Generated During Normal System Operation (Courtesy of American Technical Publishers)

17

17

7.2.5- Wear Mechanisms in Hydraulic Components

Wear and loss of material is caused by different mechanisms depending on the existing combination of factors causing the wear.

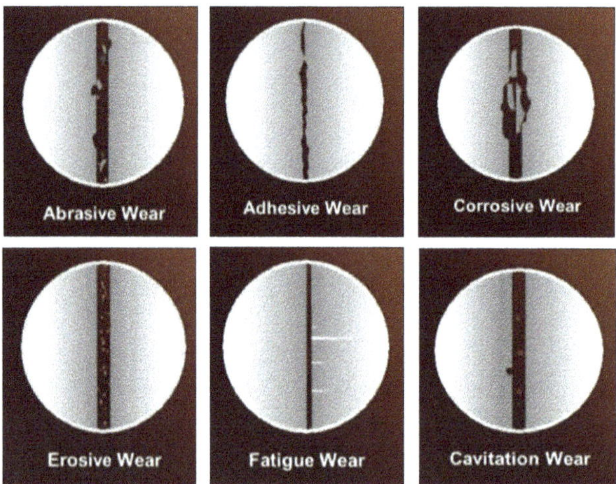

Fig. 7.8- Wear Mechanisms in Hydraulic Components (Courtesy of Parker)

18

18

7.2.5.1- Abrasive Wear Mechanism

- ❑ Is caused when hard particles bridge two moving surfaces, scraping one or both.

- Hydraulic fluid is expected to create a lubricating film to:
  - Separate moving surfaces.
  - prevent metal-to-metal contacts.
  - Allow the silt (small) particles to pass through causing no damage.

- Ideally, the lubricating film is thick enough to:
  - Completely fill the clearance between moving surfaces.
  - Reduce wear rate.
  - Extend a component life to millions of pressurization cycles.

19

19

- **Operating (Dynamic) Clearance** and consequently the actual thickness of a lubricating film depends on:

  - F, Applied load.
  - v, Relative speed of the two surfaces.
  - ν, Fluid viscosity.

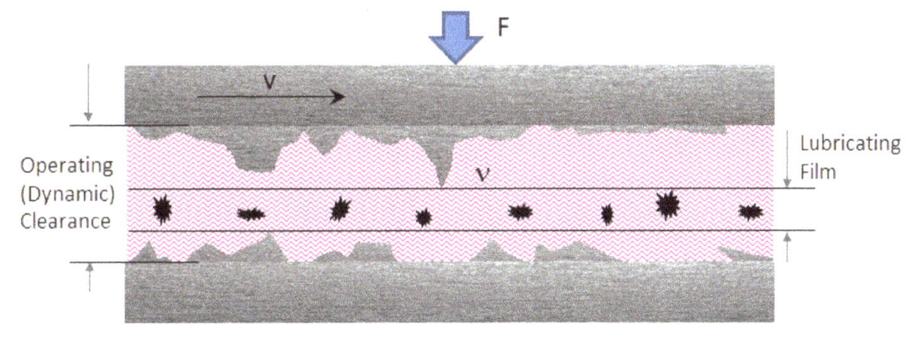

Fig. 7.9- In Normal Conditions, Silt Particles Pass Through Causing No Damage

20

20

- Operating (dynamic) clearance ≠ machine clearance
- It depends upon the load, speed, and lubricant viscosity.

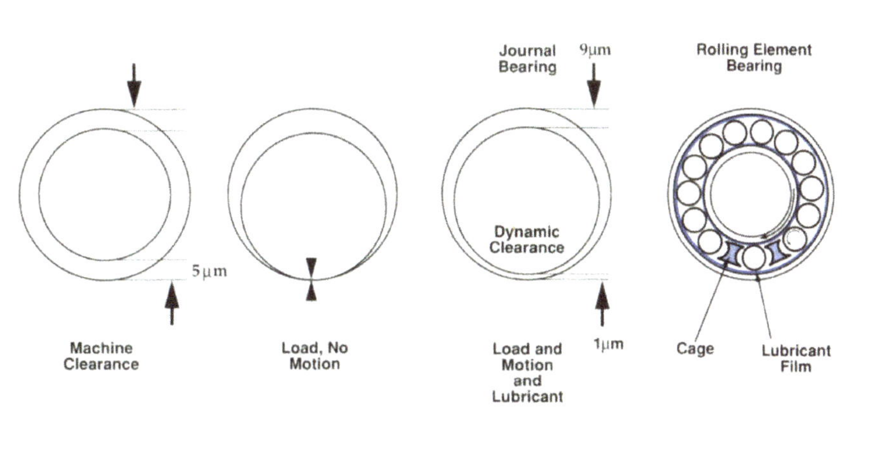

Fig. 7.10- Dynamic Clearances in Bearings (Courtesy of Pall)

21

21

Dynamic oil film

Component	Oil film thickness in micron (μm)
Journal, slide and sleeve bearings	0.5-100
Hydraulic cylinders	5-50
Engines, ring/cylinder	0.3-7
Servo and proportional valves	1-3
Gear pumps	0.5-5
Piston pumps	0.5-5
Rolling element bearings / ball bearings	0.1-3
Gears	0.1-1
Dynamic seals	0.05-0.5

Table 7.1- Typical Dynamic Oil Film Thickness in Various Hydraulic Components (Courtesy of Noria Corporation)

22

22

- Particle size ≈ dynamic clearance → most damaging.
- It acts like grinding → remove material → cause grooving.

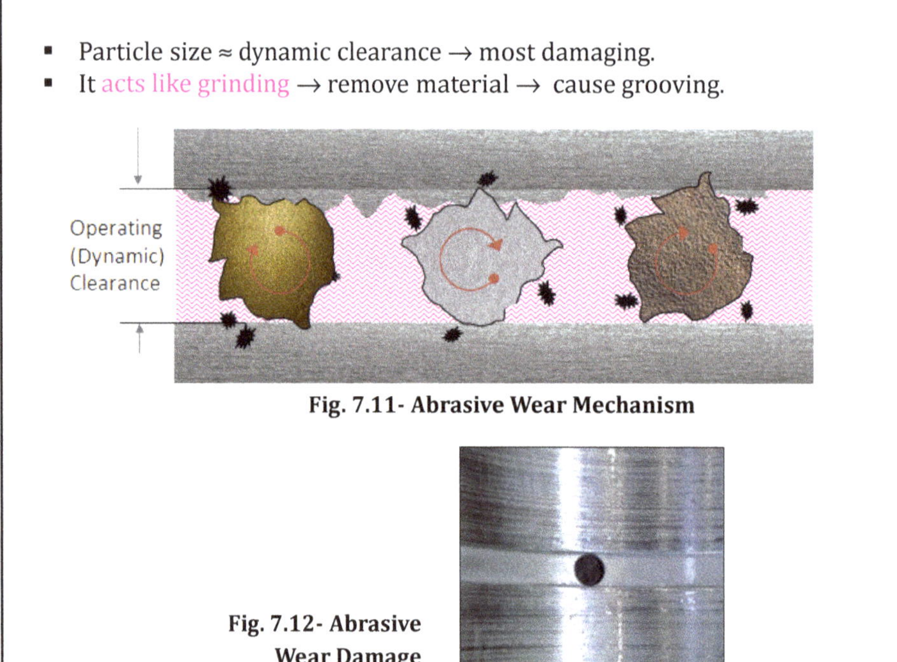

Fig. 7.11- Abrasive Wear Mechanism

Fig. 7.12- Abrasive Wear Damage

23

23

7.2.5.2- Adhesive Wear Mechanism

- *Adhesive* wear results when moving surfaces tend to stick together.

- Increased load + increased relative speed + loss of oil viscosity
- → Oil lubricating film collapsed to a very thin film (< 1 µm thick).
- → Moving metals are squeezed and "cold welded" together.
- → Particles are generated as the surface asperities are sheared off.

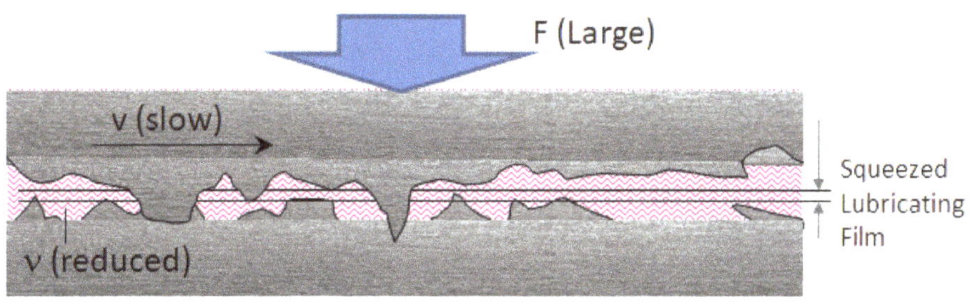

Fig. 7.13- Adhesive Wear Mechanism

24

24

7.2.5.3- Corrosive Wear Mechanism

- Is the loss of material over a large area typically caused by water, chemical, or microbial contamination in the fluid.
- Rust due to oxidation is one form of corrosive wear.
- Rust entered the cylinder through failed rod seals causing more wear.

Fig. 7.14- Corrosive Wear due to Rust (www.gallagherseals.com)

25

25

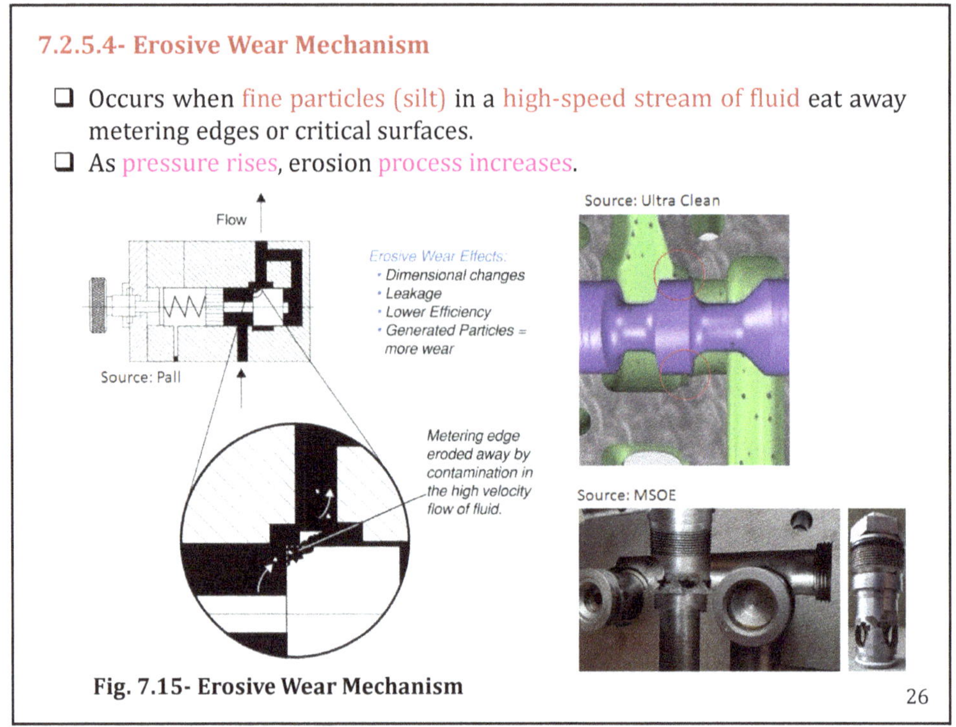

7.2.5.4- Erosive Wear Mechanism
❑ Occurs when fine particles (silt) in a high-speed stream of fluid eat away metering edges or critical surfaces.
❑ As pressure rises, erosion process increases.
Flow
Source: Pall
Erosive Wear Effects:
• Dimensional changes
• Leakage
• Lower Efficiency
• Generated Particles = more wear
Metering edge eroded away by contamination in the high velocity flow of fluid.
Source: Ultra Clean
Source: MSOE
26

Fig. 7.15- Erosive Wear Mechanism

26

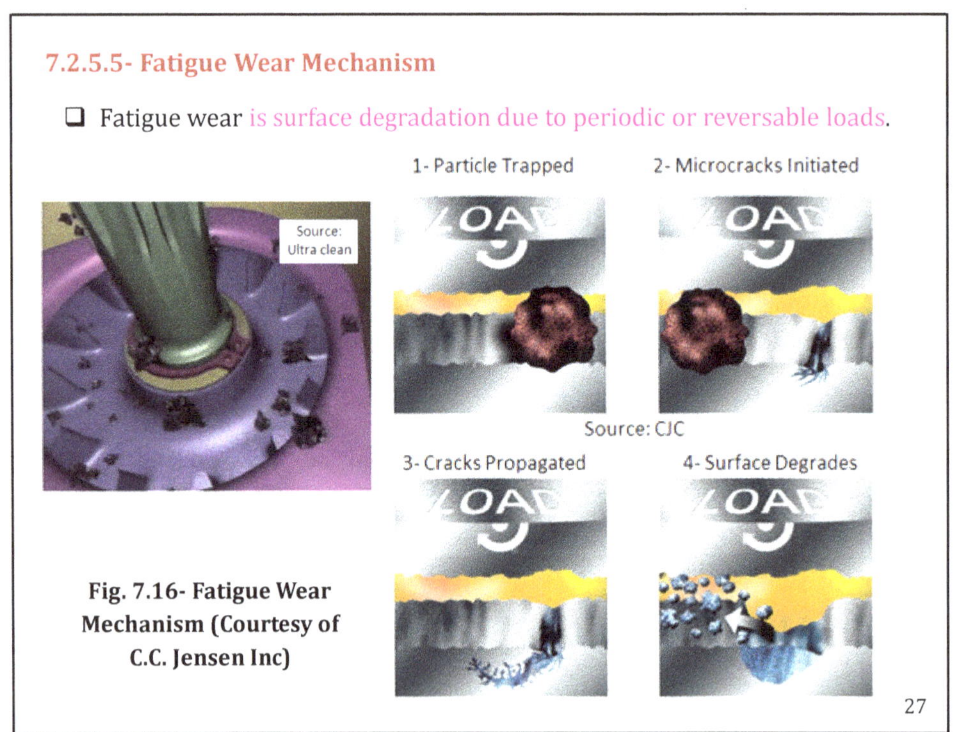

Fig. 7.16- Fatigue Wear Mechanism (Courtesy of C.C. Jensen Inc)

27

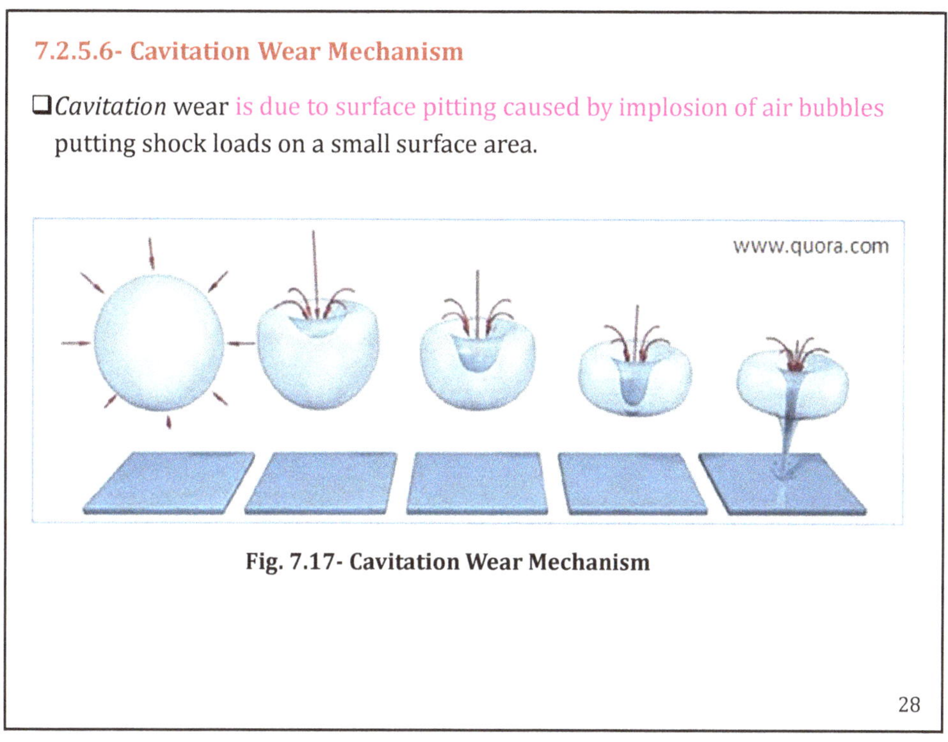
7.2.5.6- Cavitation Wear Mechanism
❑Cavitation wear is due to surface pitting caused by implosion of air bubbles putting shock loads on a small surface area.
www.quora.com
Fig. 7.17- Cavitation Wear Mechanism
28

28

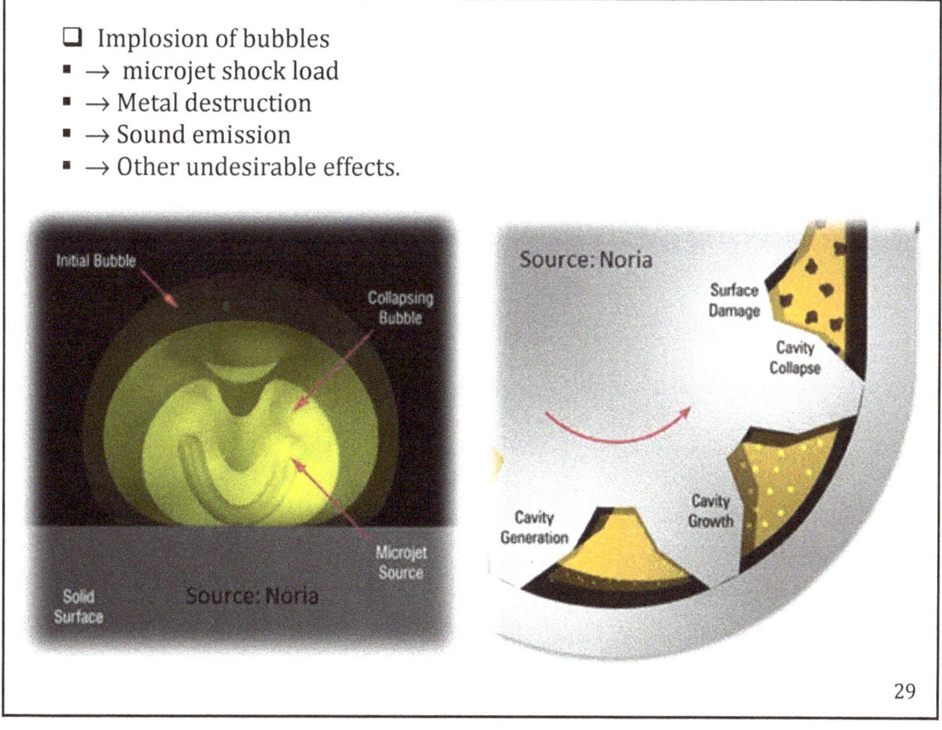
❑ Implosion of bubbles
▪ → microjet shock load
▪ → Metal destruction
▪ → Sound emission
▪ → Other undesirable effects.
Initial Bubble
Collapsing Bubble
Microjet Source
Solid Surface
Source: Noria
Source: Noria
Surface Damage
Cavity Collapse
Cavity Growth
Cavity Generation
29

29

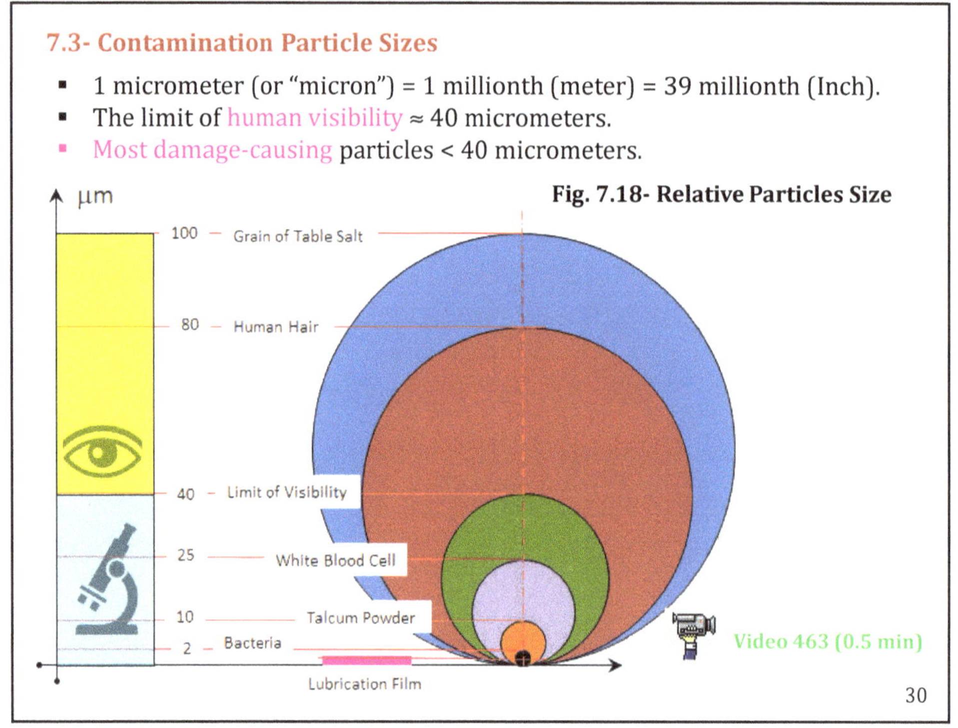

Fig. 7.18- Relative Particles Size

30

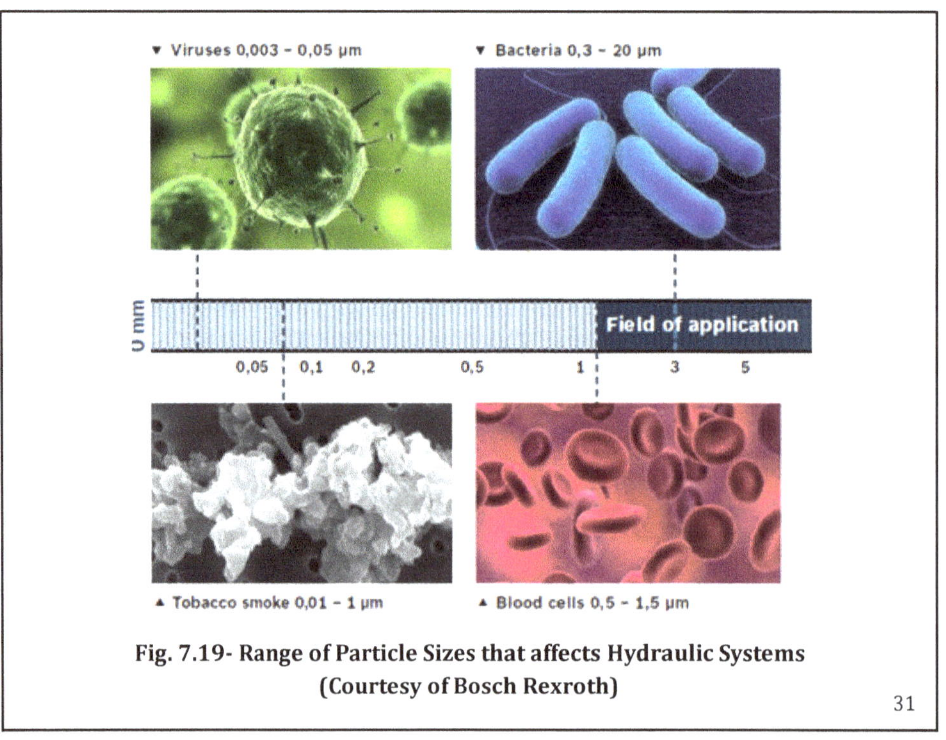

Fig. 7.19- Range of Particle Sizes that affects Hydraulic Systems (Courtesy of Bosch Rexroth)

31

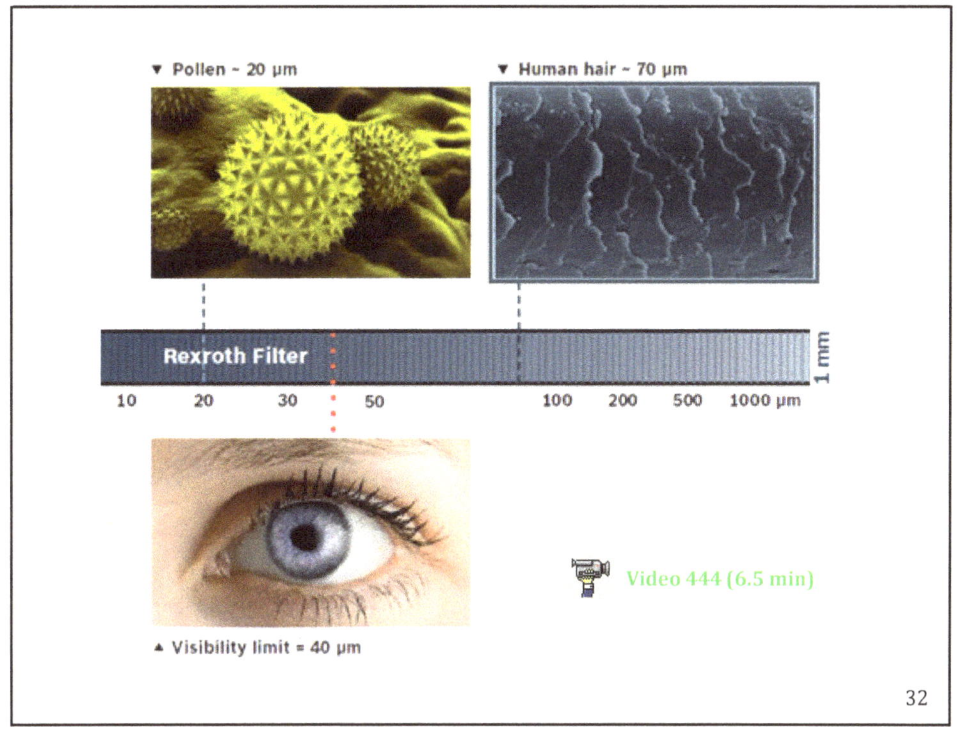

32

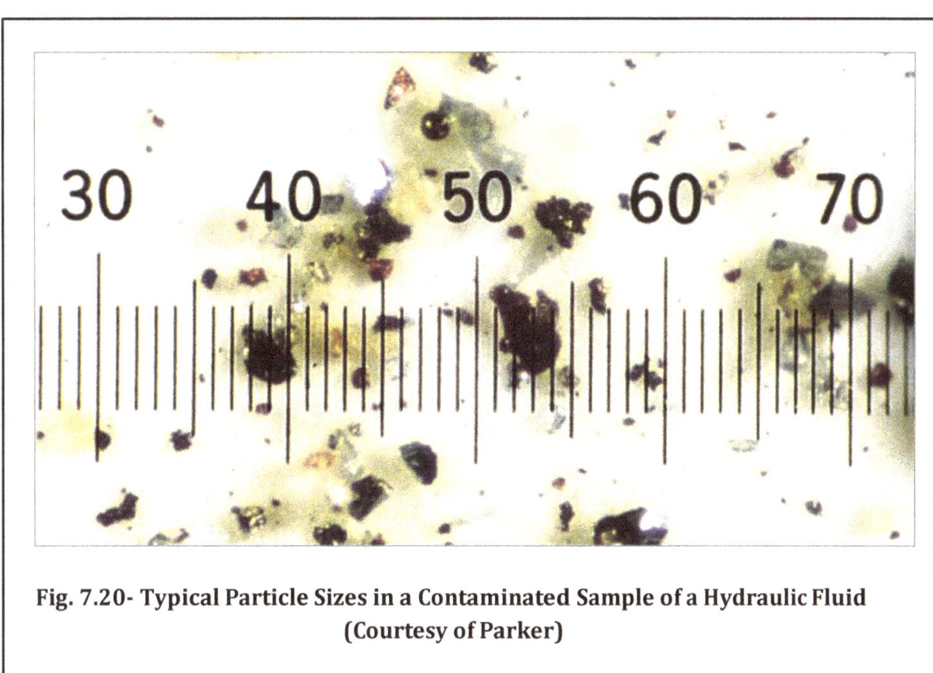

Fig. 7.20- Typical Particle Sizes in a Contaminated Sample of a Hydraulic Fluid (Courtesy of Parker)

33

33

7.4- Critical Clearances in Hydraulic Components

Component	Clearance (μm)
Roller Element Bearings	0.1-1
Journal Bearings	0.5-100
Hydrostatic Bearings	1-25

Table 7.2- Typical Bearing Clearances (Courtesy of Pall)

34

34

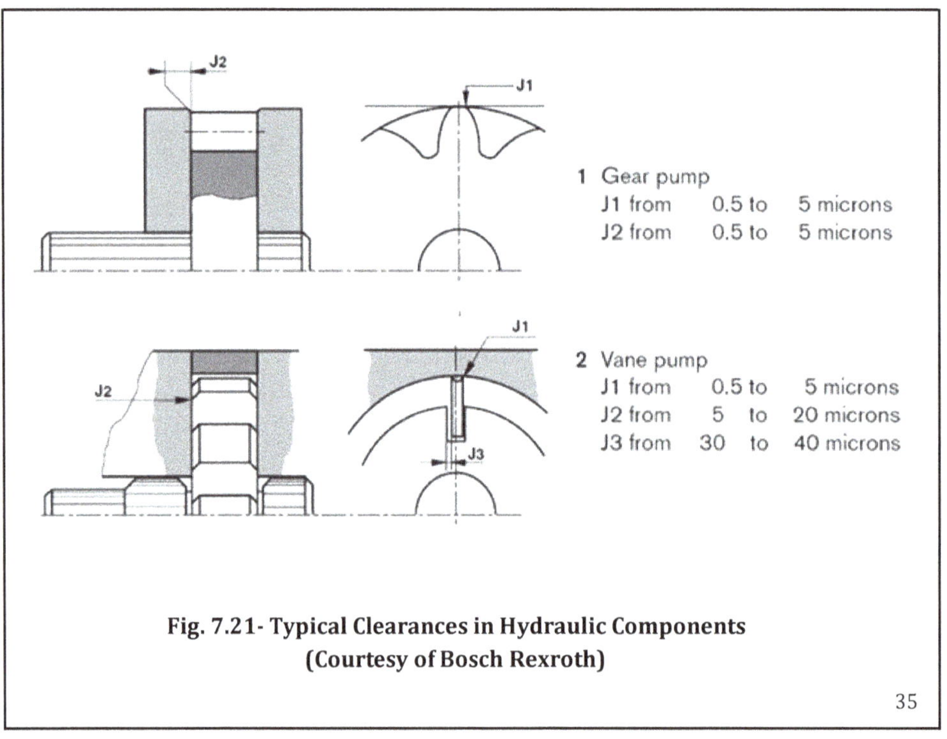

Fig. 7.21- Typical Clearances in Hydraulic Components (Courtesy of Bosch Rexroth)

35

35

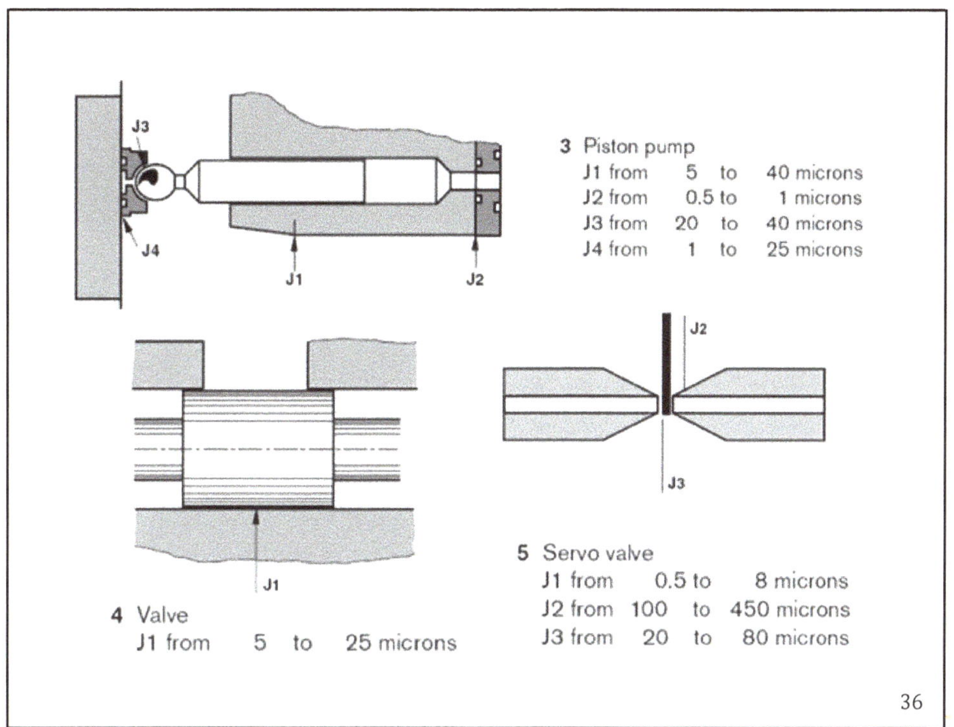

36

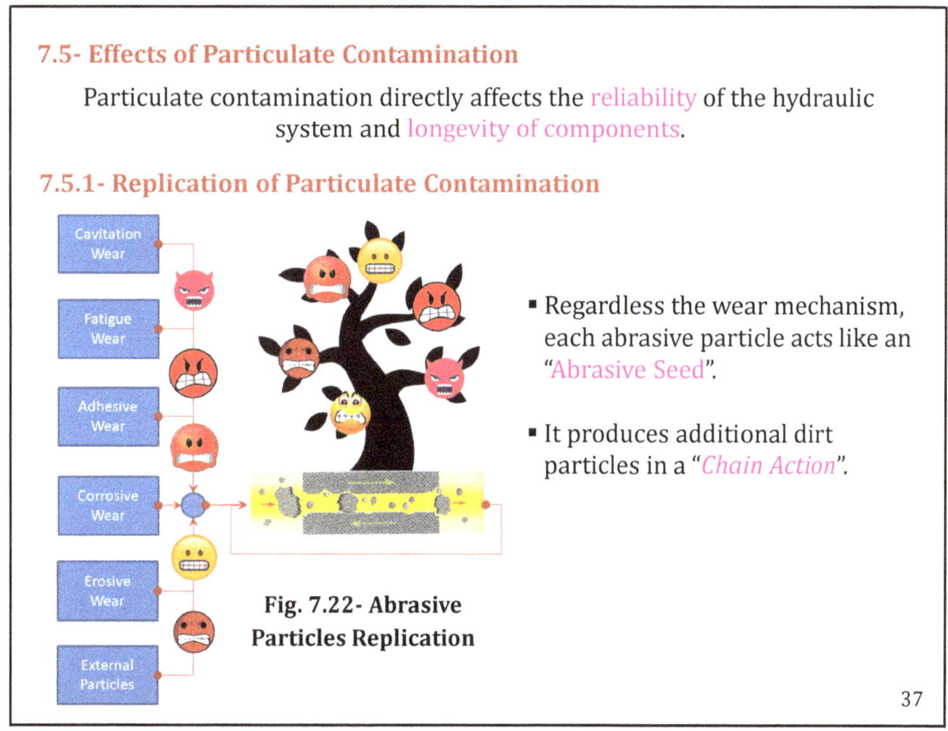

7.5- Effects of Particulate Contamination
Particulate contamination directly affects the reliability of the hydraulic system and longevity of components.
7.5.1- Replication of Particulate Contamination
Cavitation Wear
Fatigue Wear
Adhesive Wear
Corrosive Wear
Erosive Wear
External Particles
▪ Regardless the wear mechanism, each abrasive particle acts like an "Abrasive Seed".
▪ It produces additional dirt particles in a "Chain Action".
37

Fig. 7.22- Abrasive Particles Replication

37

7.5.2- Factors Affecting Level of Damage due to Particulate Contamination

Particle Size, Particle Shape, and Particle Material

7.5.2.1- Effect of Particle Size

- **Very Small Particles (Silt):** → Surface degradation failure
- A long-term failure.

- **Less than Clearance Size Particles:** → Normal Life Failure
- Failure at the end of a component life due to low rate of wear.

- **Clearance Size Particles:** → *Catastrophic Failures*
- Occurs rapidly or suddenly such as seizure or component breakage.

- **Larger than Clearance Size Particles**: → Intermittent Failures.
- Occurs frequently such as blocking a control orifice or a seat of a poppet valve. Usually in a control orifice or a valve seat, when a particle is removed, the component will continue to function normally.

38

38

7.5.2.2- Effect of Particle Shape

Particles with irregular shape and sharp edges (left side) cause deep scratches and are more dangerous than spherical particles (right side).

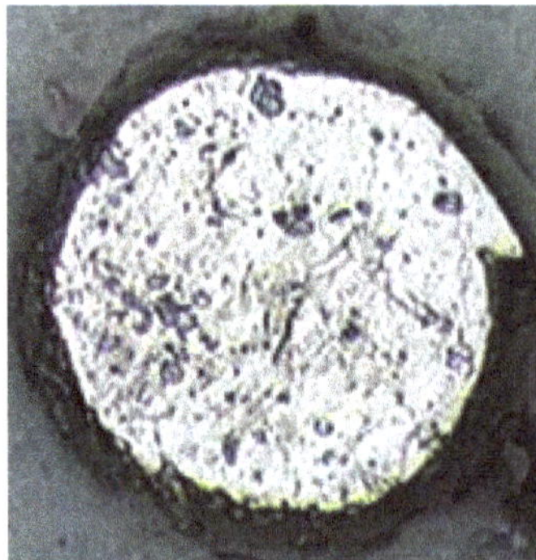

Fig. 7.23- Shapes of Particulate Contamination (Courtesy of Noria Corporation)

39

39

7.5.2.3- Effect of Particle Material

- **Very Severe Damage** results from particles of:
  - Rust.
  - Scale.
  - Carbide steel.
  - Iron.
  - Silica (sand) and other very hard materials.

- **Severe Damage** results from particles of:
  - Brass.
  - Aluminum.
  - Bronze.
  - Calcium and Sulphur products.

- **Slight Damage** results from particles of:
  - Packaging plastics.
  - Laminated fabrics.
  - Elastomeric and rubber particles from seal residues.
  - Paint chips or overspray.
  - Gelatinous particles.

40

40

7.5.3- Typical Failures due to Particulate Contamination

Particulate contamination in a hydraulic system can lead to one or a combination of the following consequences:

Mechanical Efficiency:
Increased friction between surfaces can decrease the efficiency of hydraulic components.

Volumetric Efficiency:
Internal clearances grow larger increasing internal leakage and decreasing pump and motor volumetric efficiency.

Lubrication:
Blocked lubrication passages can cause catastrophic component failure.

Damage to Rotating Components:
Under high friction and temperature, seizure of rotating parts in pumps and motors can occur.

41

41

Damage to Cylinders:
- Cylinder barrel scratching → Internal Leakage Passages → load creeping or drifting.
- Cylinder rod scratching → rod seal failure → external leakage.

Filter Clogging:
- Pressure and return filter clogging → back pressure to the system.
- Suction filter clogging → pump cavitation.
- Frequent filter replacement → increases operating & disposal cost.

System Efficiency:
- Loss of components efficiency → reduced overall system performance.
- Machine efficiency loss is gradual, 20% loss before noticed.

System Performance:
- Increased orifices dimensions → loss of component controllability.
- Stick-Slip spool valve motion → jerkiness of actuator motion.
- Internal leakage → slower system performance.

System Productivity: Increased machine down time → reduces productivity.

42

42

Silt Lock:
- Accumulation of silt → seizure or jamming of components
- → usually occurs in control valves
- → sudden and unpredicted actuator stop → catastrophic failures
- → loss of human life in accident such as (aircraft, spacecraft, passenger cars, elevators, turbine generators, tower cranes, etc).

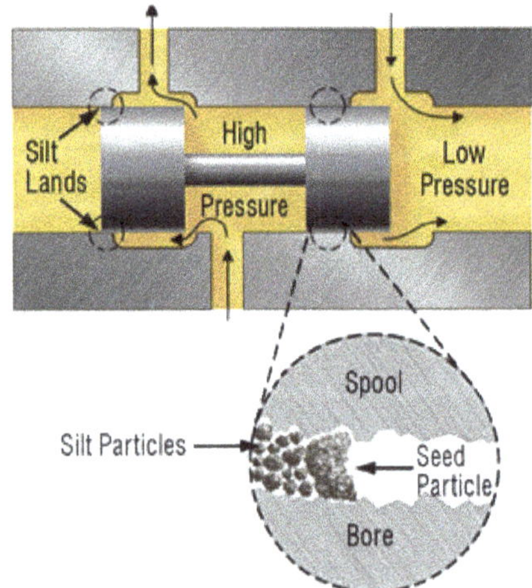

Fig. 7.24- Silt Lock in Spool Valves (Courtesy of Noria Corporation)

43

43

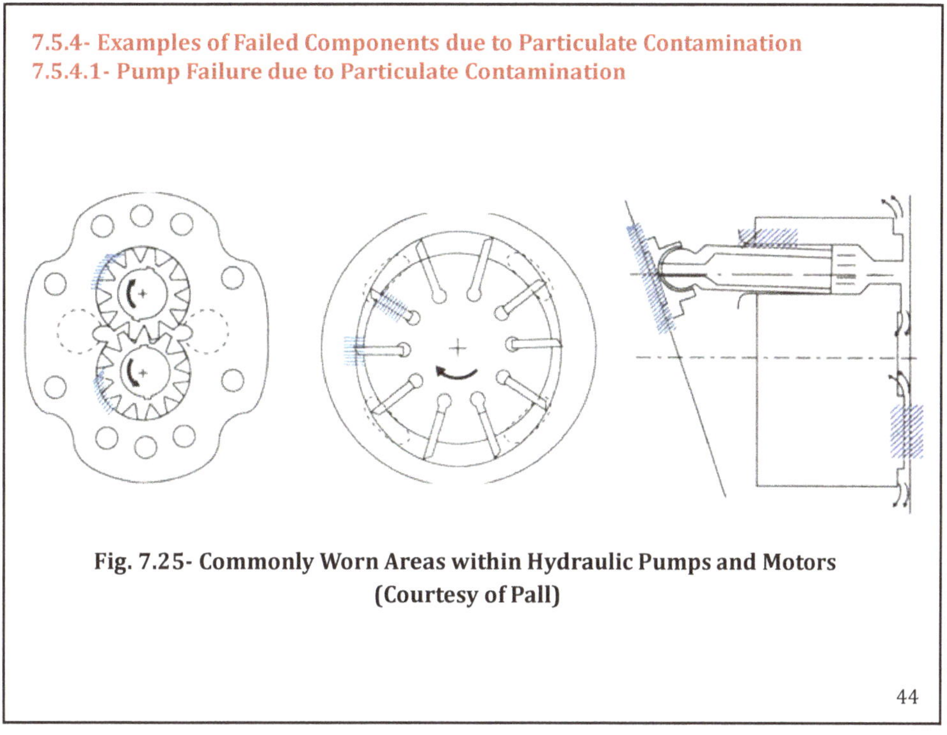

7.5.4- Examples of Failed Components due to Particulate Contamination

7.5.4.1- Pump Failure due to Particulate Contamination

Fig. 7.25- Commonly Worn Areas within Hydraulic Pumps and Motors (Courtesy of Pall)

44

44

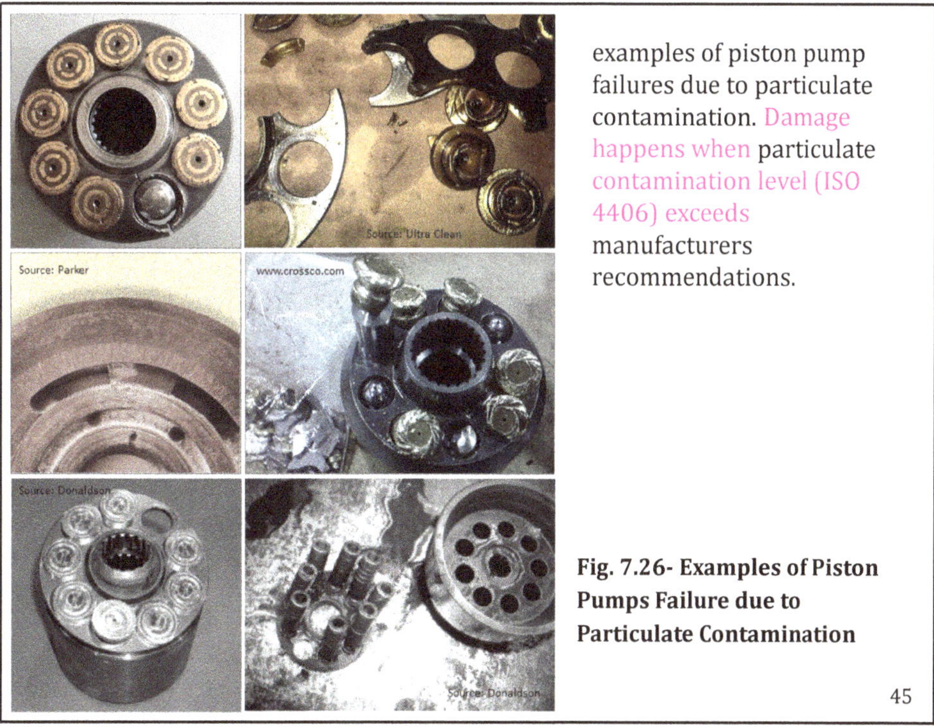

examples of piston pump failures due to particulate contamination. Damage happens when particulate contamination level (ISO 4406) exceeds manufacturers recommendations.

Fig. 7.26- Examples of Piston Pumps Failure due to Particulate Contamination

45

45

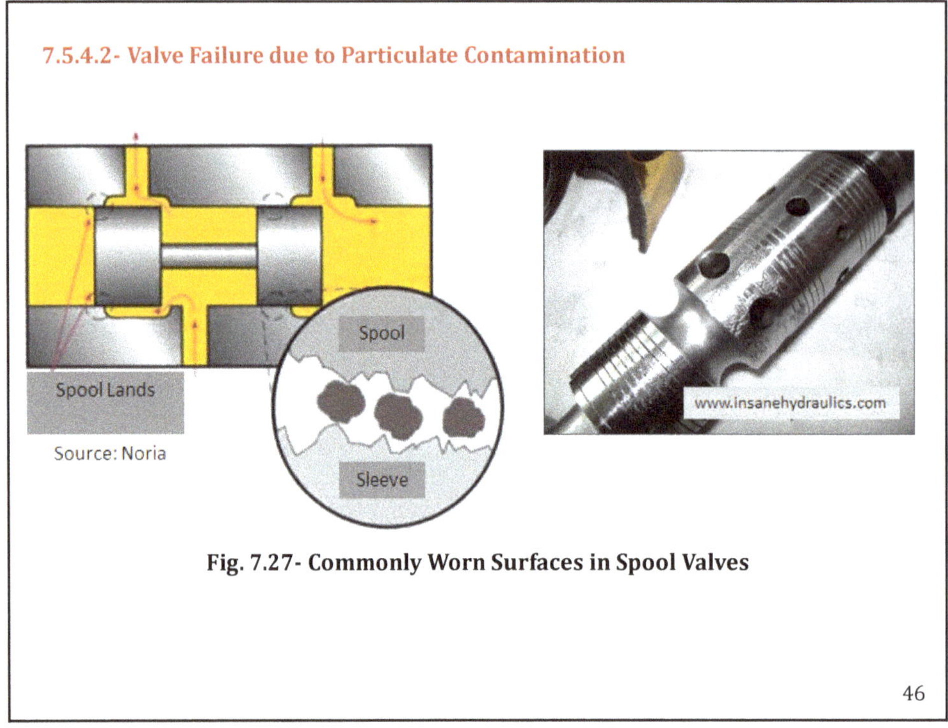

Fig. 7.27- Commonly Worn Surfaces in Spool Valves

46

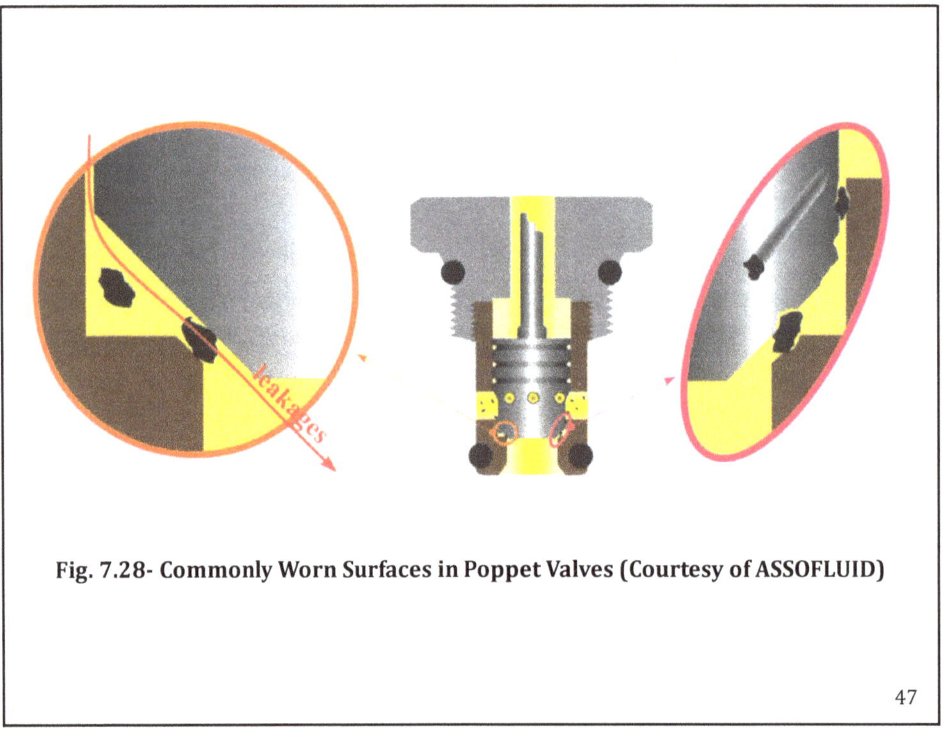

Fig. 7.28- Commonly Worn Surfaces in Poppet Valves (Courtesy of ASSOFLUID)

47

7.5.4.3- Cylinder Failure due to Particulate Contamination

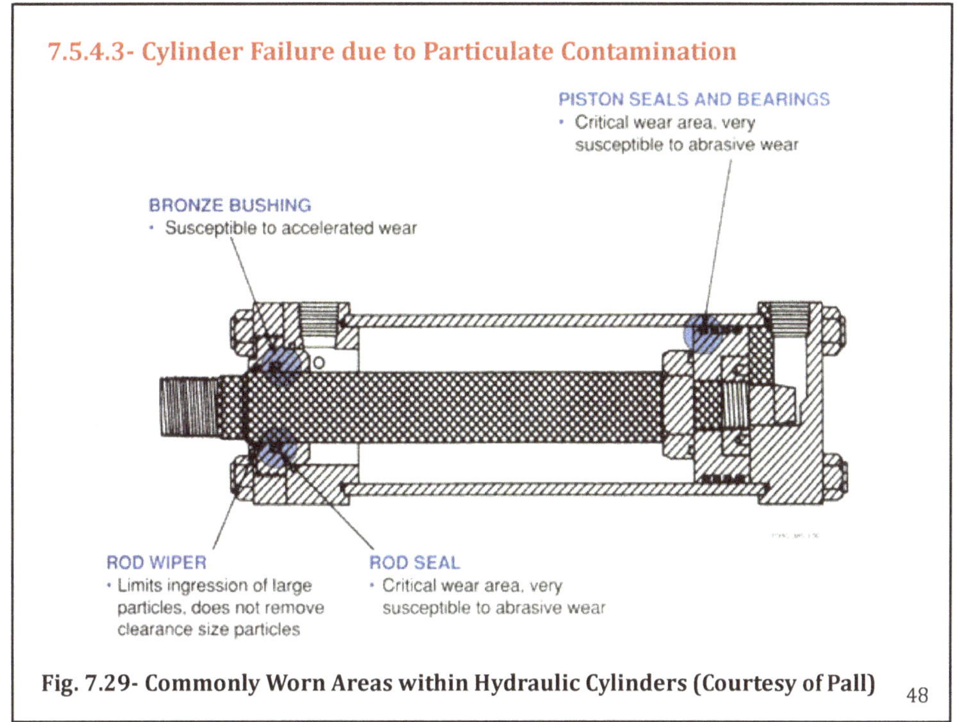
PISTON SEALS AND BEARINGS
• Critical wear area, very susceptible to abrasive wear
BRONZE BUSHING
• Susceptible to accelerated wear
ROD WIPER
• Limits ingression of large particles, does not remove clearance size particles
ROD SEAL
• Critical wear area, very susceptible to abrasive wear

Fig. 7.29- Commonly Worn Areas within Hydraulic Cylinders (Courtesy of Pall) 48

48

axleaddict.com
Source: Donaldson
Source: Parker

- Visible leakage due to seal failure caused by abrasive particulate contamination.
- Piston rings that were eaten away by contaminants.
- Scored cushion plunger resulting in a loss of cushioning effect.

Fig. 7.30- Examples of Hydraulic Cylinder Failures due to Particulate Contamination

49

49

7.5.4.4- Bearing Failure due to Particulate Contamination

- Positive displacement pumps are unbalanced pumps.
- Bearing load is not evenly distributed and concentrated on one side.
- Severe bearing wear occurs in presence of particulate contamination.

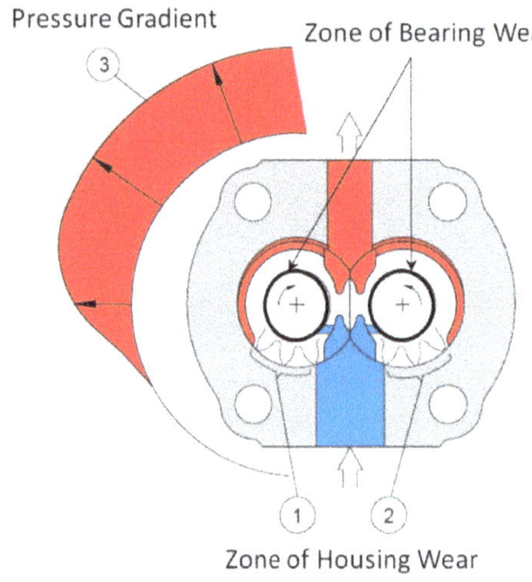

Fig. 7.31- Wear Zones in Gear Pump and Motor Bearings

50

50

- Examples of bearing failure due to particulate contamination:

1. A destroyed raceway of a ball bearing.
2. A chip impeded in a surface of an anti-friction bearing.
3. A destroyed roller bearing in a piston pump.

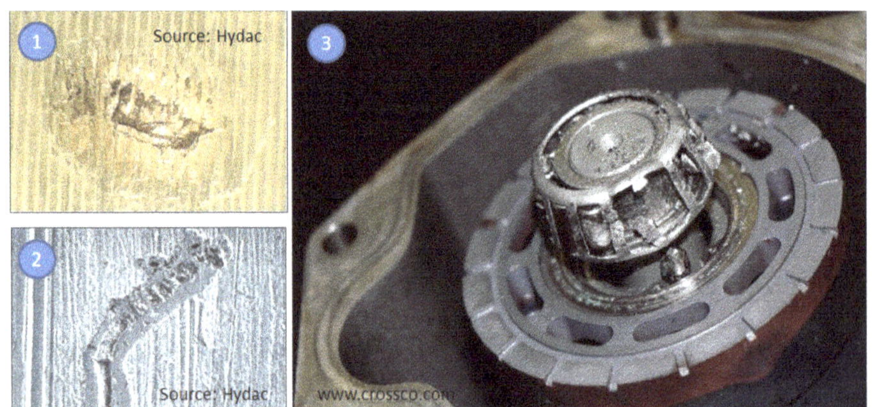

Fig. 7.32- Examples of Bearing Failures due to Particulate Contamination

51

51

7.5.4.5- Filter Clogging due to Particulate Contamination

- The filter appears normal.
- Particles clogging it are smaller than the limit of vision.
- A pressure indicator is needed to detect when the filter is clogged.

Video 286 (1.5 min)

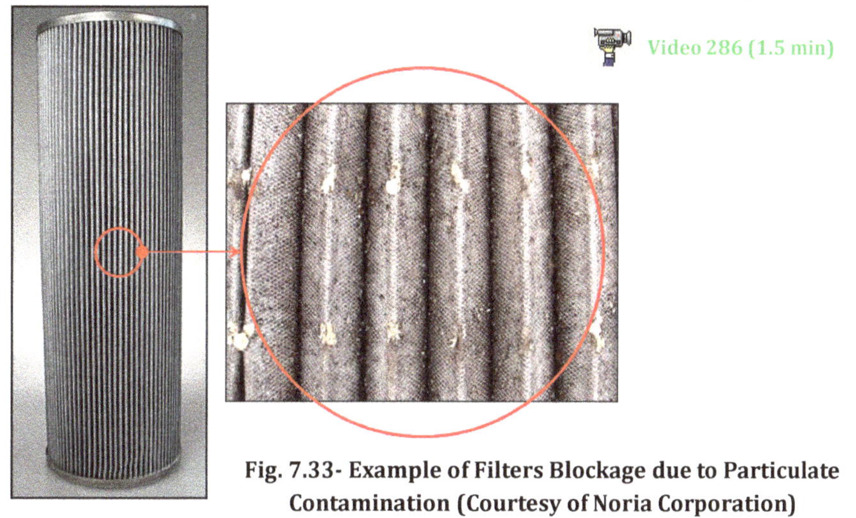

Fig. 7.33- Example of Filters Blockage due to Particulate Contamination (Courtesy of Noria Corporation)

52

52

7.6- Best Practices for Controlling Particulate Contamination
7.6.1- Preventive Practices to Control Particulate Contamination

Form of Particulate Contamination	General Preventive Actions
Built-in	▪ **Contamination Limits** for new components should be verified. ▪ **Hydraulic Transmission Lines** should be cleaned before and after assembly. ▪ **System Flushing** before first use and after major maintenance.
Introduced (Ingested and Induced)	▪ **Service and Maintenance** proper procedures help in minimizing ingested and induced contamination.
Generated	▪ **Filtration System Design** based on the system requirements to maintain recommended fluid cleanliness level. ▪ **Hydraulic Fluid Analysis** in order to predict possible future causes of failure and the required action that should be taken to prevent it. ▪ **Hydraulic Reservoir Design and Maintenance** is an important preventive action for controlling generated contamination.

Table 7.3- General Preventive Actions for Controlling Particulate Contamination

53

53

❑ **Hydraulic Fluid Analysis:**

- Cleanliness level must be checked frequently.
- Cleanliness level must comply with the manufacturer's recommendations.
- Always observe maximum cleanliness and accuracy during sampling.
- Always use independent high quality analysis resources.
- For sensitive systems, use online particle counters or contamination sensors for continuous monitoring of contamination.
- Check the oil after machine malfunctions or major maintenance.
- When replacing seals, compatibility with the oil must be checked.
- Never apply new additives without consulting the oil supplier.

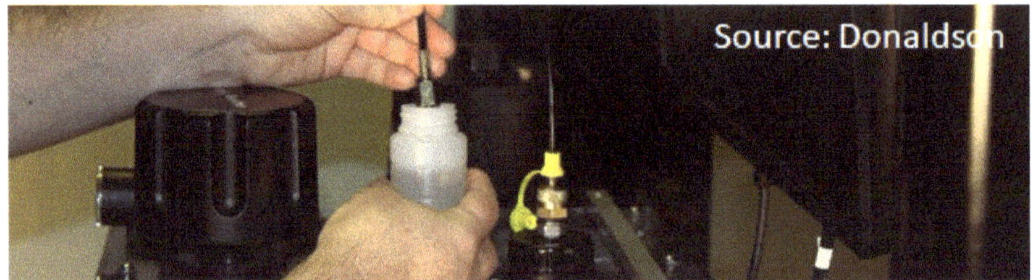

Fig. 7.34- Hydraulic Fluid Analysis (Courtesy of Donaldson) 54

54

❑ **Filtration System Design:**

- Filtration system design depends mainly on the components sensitivity.
- Filter should be equipped with a visual or electronic clogging indicator.
- Clogging indicator should be activated at a differential pressure below the bypass valve cracking pressure.
- This gives time to service the element, before the bypass valve open.

Fig. 7.35- Hydraulic Filter Differential Pressure Indicator 55

55

❑ New Oil is not Clean!!

- New fluid added to the system can be a source of contamination.
- New oil considered contaminated until a sample has been analyzed.
- New oil should always be introduced to the system through an appropriate filter.
- Never transfer fluid using buckets, containers, funnels, etc.

❑ New Transmission Lines are not Clean!!:

- New transmission lines contain built-in contaminants.
- Particulate contaminants ingress into lines during packaging, shipping, and storage.
- Particulate contaminants induce into lines during assembly.
- Transmission lines must be cleaned before and after assembly.

56

56

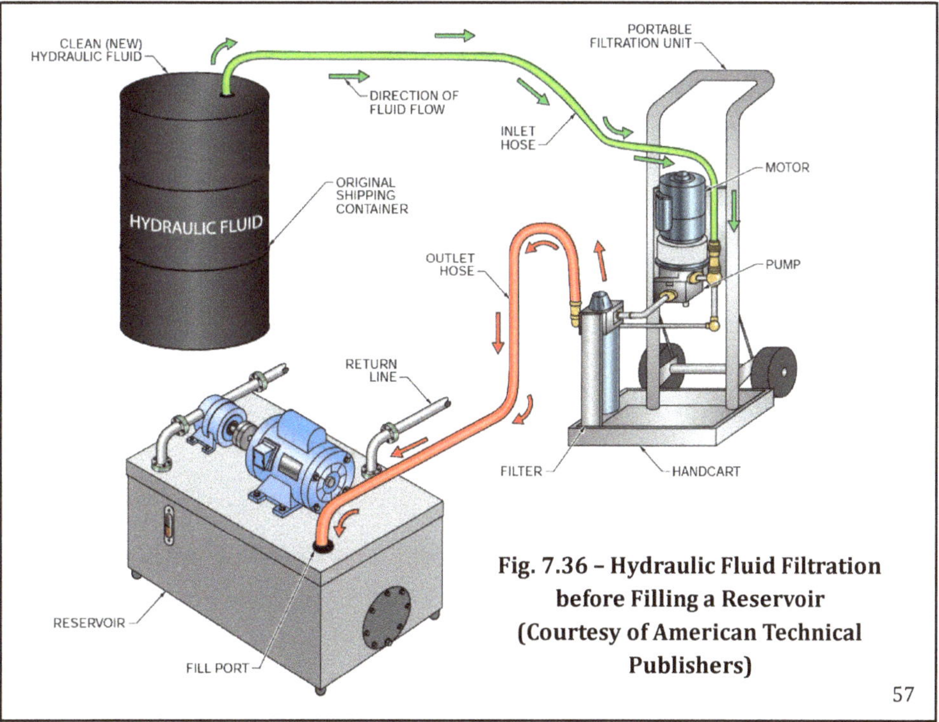

Fig. 7.36 – Hydraulic Fluid Filtration before Filling a Reservoir (Courtesy of American Technical Publishers)

57

57

- **New Hydraulic Components are not Clean!!:**
  - Do not assume that the new components are 100% clean.
  - It is wise to pre-clean all hydraulic system components prior to assembly.
  - MSOE evaluated more than 100 new hydraulic components.
  - →1/3 of new components has particulate contaminants exceeded 8 mg.

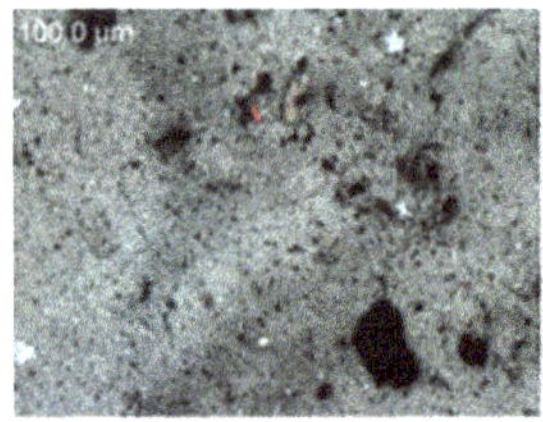

Debris from a new Hose

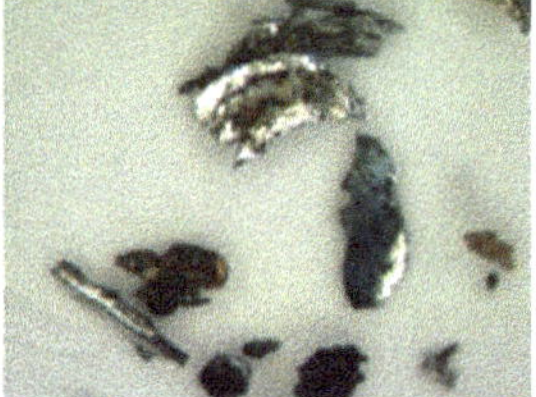
Debris from a new Valve

Debris from a new Reservoir

Debris from a new Cylinder

Fig. 7.37- Debris from New Components (Courtesy of MSOE)

58

58

- Manufacturer is responsible for providing components that meet the requirements agreed upon with the purchaser.
- For more information about cleanliness requirement of new components, refer to the following ISO Standards:
  - **ISO 18413:** Hydraulic fluid power - Cleanliness of components - Inspection document and principles related to contaminant extraction, analysis, and data reporting.
  - **ISO 12669:** Hydraulic fluid power - Method for determining the required cleanliness level (RCL) of a system.
  - **ISO/TR 10949:** Hydraulic fluid power – Component cleaning – Guidelines for achieving and controlling cleanliness of components from manufacture to installation

59

59

- Method to set a max. allowable *Contamination Limit* in new components.
- Component manufactures should comply with this method.
- The method involves a cost/benefit analysis.
- Cost to achieve a given level of cleanliness / cost of possible damage.
- The method is based on the *Volume-to-Area Ratio*
- In this case, the area is the wet surfaces that are in direct contact with the hydraulic fluid.

COMPONENT	VOLUME-TO-AREA RATIO
Reservoirs	1 to 5
Hoses and Tubes	0.2
Cylinders	0.5 to 0.6
Pumps and Motors	0.001 to 0.05
Valves	0.001
Complete Systems	0.2 to 4

Table 7.4- Volume-to-Area Ratio of Hydraulic Components (Courtesy of MSOE) 60

60

- Particulate contamination limits are specified in milligrams (mg).
- Different units of measure are used to define contamination levels:
  - Mass per unit volume (mg/liter) - for components that have a high volume-to-area ratio.
  - Mass per unit weight (mg/kg) OR Mass per unit area (mg/m^2) for components that have a low volume-to-area ratio.
  - Mass per unit length (mg/m). for hydraulic transmission lines.

UNIT OF MEASURE	TYPICAL RANGE
Mass per unit volume (mg/liter)	3 to 10
Mass per unit weight (mg/Kg)	0.5 to 5
Mass per unit area (mg/M^2)	25 to 1,000
Mass per unit length (mg/M)	6 to 12

Table 7.5- Contamination Limits for New Hydraulic Components (Courtesy of MSOE) 61

61

The following are examples on how to use the previous data:

- Example 1: A hydraulic reservoir that has a volume-to-area ratio of 5 has a contamination limit range from 3-10 mg/liter.
- Example 2: A hydraulic cylinder that has a volume-to-area ratio of 0.5 has a contamination limit range from 0.5-5 mg/kg or 25-1000 mg/m^2.
- Example 3: A hydraulic pipe has a contamination limit range from 6-12 mg/m.

62

62

❑ **Service and Maintenance:**

- Maintain organized and clean housekeeping.
- Repair work should be performed in a dust-free environment.
- Parts and seals should be kept in sealed plastic bags until needed.
- Only pre-filtered solvents should be used.
- Make sure solvents are compatible with seals and other parts.
- When replacing or cleaning filters, consult the service manual.

Fig. 7.38- Organized, Dry and Clean Housekeeping

63

63

- Maintaining clean outside surfaces and their surrounding areas.
- When cleaning, only lint-free wipes that contain no fibers should be used. Ordinary shop towels and waste rags should not be used.

Fig. 7.39- Keeping the Hydraulic System Clean is an Important Practice 64

64

- Covering cylinder rods minimizes ingression of particulate contaminants through the rod wipers.

Video 488 (1.5 min)

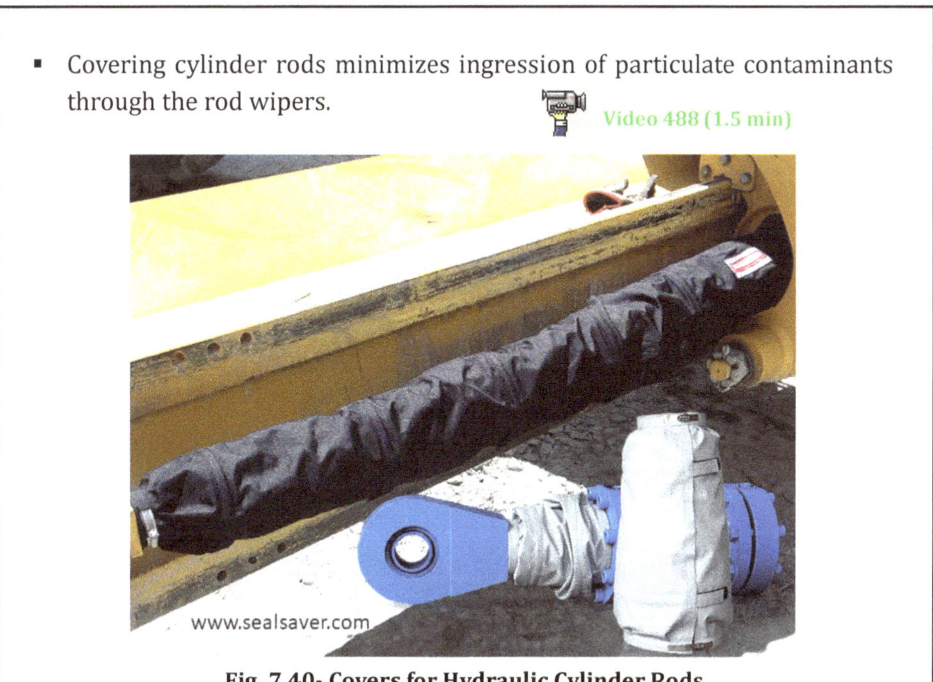

Fig. 7.40- Covers for Hydraulic Cylinder Rods 65

65

- Use pipe plugs, tube caps, etc. during disassembly, assembly, shipping, and storage.
- Make sure installation and removal of caps and plugs does not generate contaminants in the threaded area of the component.

Fig. 7.41- Covers for Hydraulic Components and Parts (www.capsnplugs.com)

66

66

Hydraulic Reservoir Design and Maintenance:

- A well-designed reservoir helps hydraulic fluid getting rid of all contaminants (particulate, fluidic, gaseous, and thermal).
- Drain-plug or strainer magnets capture ferrous particulates and rust.
- Reservoir breathes through a breather filter.
- When changing the oil, the tank and the system should be emptied completely, and the tank should be cleaned with an appropriate compatible solvent.

Hydraulic System Flushing:

- Before first commissioning.
- After major maintenance.

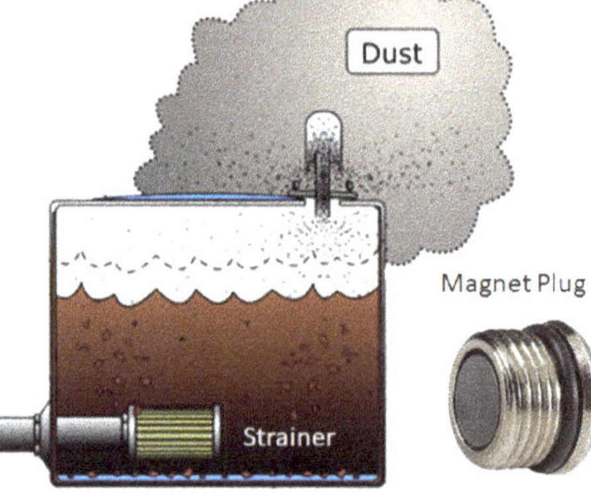

Fig. 7.42- Reservoir Design and Maintenance for Controlling Generated Particulate Contamination

67

67

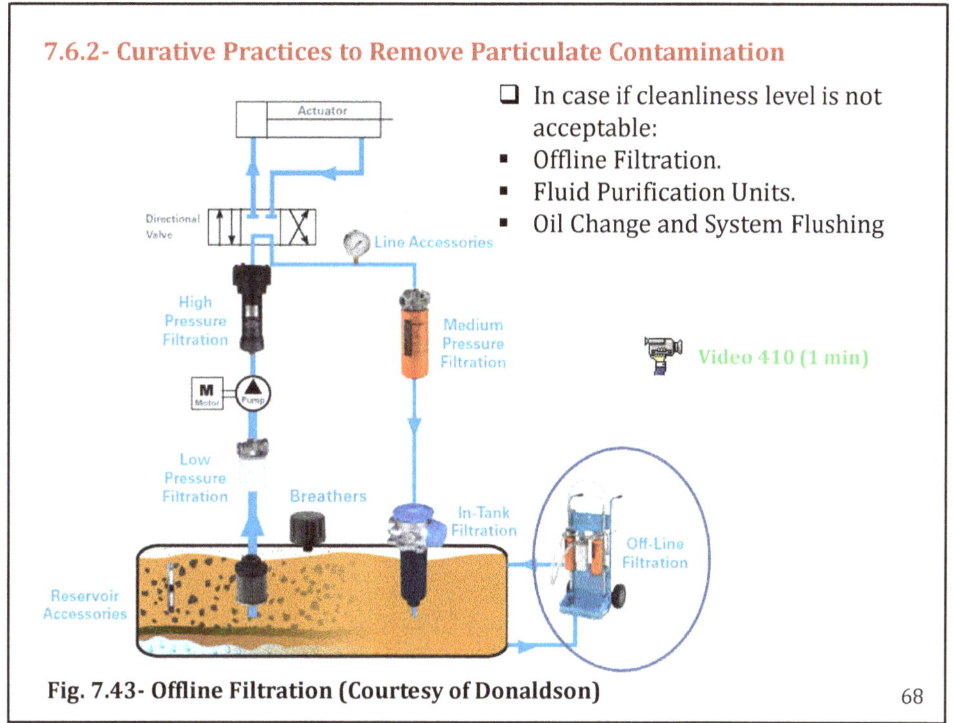

Fig. 7.43- Offline Filtration (Courtesy of Donaldson)

68

Chapter 7 Reviews

1. Particulate contaminants can be in which one of the following forms?
   A. Silt.
   B. Nonabrasive particles such as fibers.
   C. Gelatinous particles or microorganisms.
   D. All of the above

2. Residual sand from a foundry during casting a pump housing is considered?
   A. Built in contaminants.
   B. Introduced (ingested) contaminants.
   C. Introduced (induced) contaminants.
   D. Generated contaminants.

3. Humidity condensed into a reservoir is considered?
   A. Built in contaminants.
   B. Introduced (ingested) contaminants.
   C. Introduced (induced) contaminants.
   D. Generated contaminants.

4. Debris resulted from tube flaring is considered?
   A. Built in contaminants.
   B. Introduced (ingested) contaminants.
   C. Introduced (induced) contaminants.
   D. Generated contaminants.

5. Bearing wear is considered?
   A. Built in contaminants.
   B. Introduced (ingested) contaminants.
   C. Introduced (induced) contaminants.
   D. Generated contaminants.

Chapter 7 Assignment

Student Name: --- Student ID: ------------------

Date: --- Score: -----------------------

A: Explain the difference between the ingested and the induced contaminants.

B: List 4 examples of sources for built in particulate contaminates.

Chapter 8
Hydraulic Fluid Analysis

Objectives:

This chapter discusses standard methods for hydraulic fluid analysis including methods for particle and material analysis. The chapter covers the various standard cleanliness classes used to evaluate the contamination level in hydraulic fluids. The chapter also provides examples for interpretation of hydraulic fluid analysis reports.

0

0

Brief Contents:

1

1

8.1- Introduction to Hydraulic Fluid Analysis

Why is hydraulic *fluid analysis* important?

- It identifies:
- Tells what is happening inside an equipment.
- Contamination level.
- Type of contaminants.
- Potential component wear.
- Requirements for optimizing filtration performance.
- Comply with the warranty support programs.
- Minor problems before they become major.
- → minimizes downtime.
- → extends equipment life.
- → maximizes asset reliability and extends equipment life.

2

2

Equipment warranty support programs require routine hydraulic fluid analysis to maintain coverage just like medical service providers require periodic checkup to maintain one's health

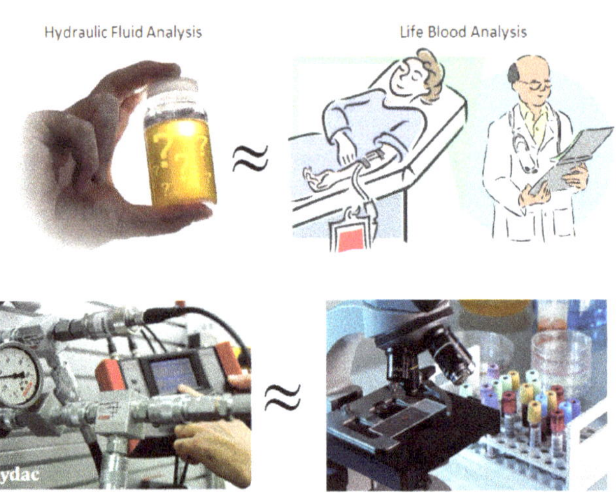

Hydraulic Fluid Analysis and System Reliability

3

3

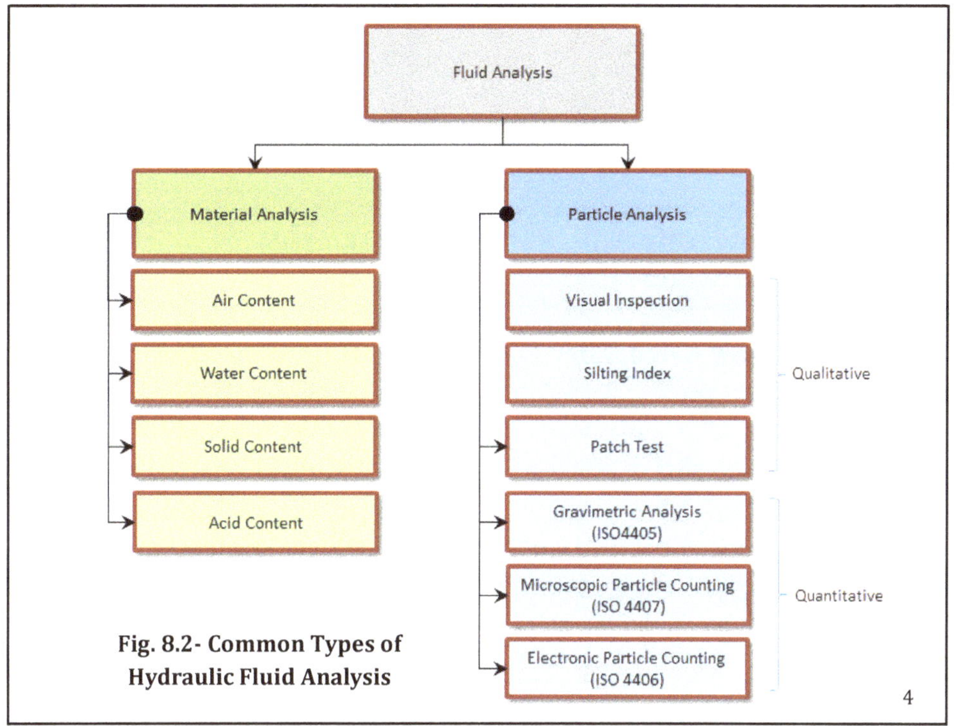

Fig. 8.2- Common Types of Hydraulic Fluid Analysis

4

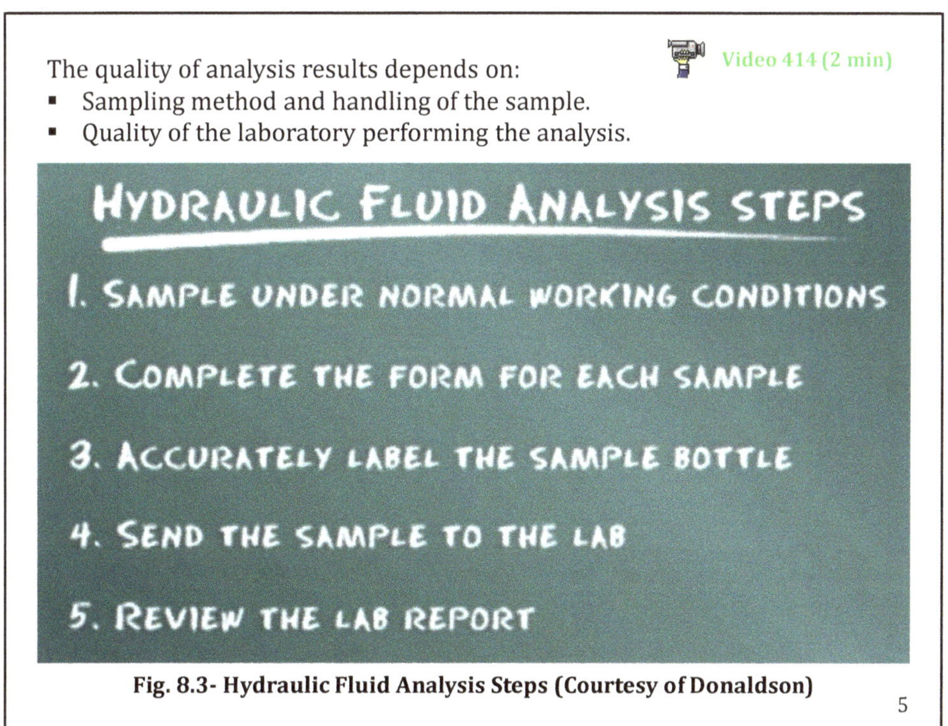

The quality of analysis results depends on:
▪ Sampling method and handling of the sample.
▪ Quality of the laboratory performing the analysis.
Video 414 (2 min)
HYDRAULIC FLUID ANALYSIS STEPS
1. SAMPLE UNDER NORMAL WORKING CONDITIONS
2. COMPLETE THE FORM FOR EACH SAMPLE
3. ACCURATELY LABEL THE SAMPLE BOTTLE
4. SEND THE SAMPLE TO THE LAB
5. REVIEW THE LAB REPORT
5

Fig. 8.3- Hydraulic Fluid Analysis Steps (Courtesy of Donaldson)

5

8.2- Hydraulic Fluid Sampling

For a representative hydraulic fluid *sample*, the following should be considered:

- **Sampling Interval:** Always take samples at regularly scheduled intervals.
- **Sampling Location:** Points of withdrawing the fluid should be defined.
- **Sampling Kit:** Standard sampling kit should be used.
- **Sampling Procedure:** Fluid sampling should follow prescribed procedure.

6

6

Recommended Sampling Interval for industrial hydraulic equipment?

A. 500 hours.
B. Quarterly.
C. Semi-annually.
D. Annually.

7

7

8.2.1- Hydraulic Fluid Sampling Intervals

Video 413 (0.5 min)

- Machinery manufacturers will often suggest a sampling interval.
- Otherwise, use the following as a guide.

Industrial and Marine			
Equipment Type	*Normal Use Sampling Frequency (Hours)	(Calender)	Occasional Use Sampling Frequency (Calendar)
Steam Turbines	500	Monthly	Quarterly
Hydro turbines	500	Monthly	Quarterly
Gas Turbines	500	Monthly	Quarterly
Diesel Engines-Stationary	500	Monthly	Quarterly
Natural Gas Engines	500	Monthly	Quarterly
Air/Gas Compressors	500	Monthly	Quarterly
Refrigeration Compressors	500	Monthly	Quarterly
Gearboxes-Heavy Duty	500	Monthly	Quarterly
Gearboxes-Medium Duty		Quarterly	Semi-Annually
Gearboxes-Low Duty		Semi-Annually	Annually
Motors-2500 hp and higher	500	Monthly	Quarterly
Motors-200 to 2500 hp		Quarterly	Semi-Annually
Hydraulics		Quarterly	Semi-Annually
Diesel Engines-On and Off Highway	150 hours/10,000 miles	Monthly	Quarterly

Table 8.1- Fluid Analysis Intervals for Common Industrial Machines (Courtesy of Spectro Scientific) 8

8

Because mobile machines work outdoor where the contamination is more than industrial applications, sampling intervals are reduced to 300 hours instead of 500.

Off-Highway/Mobile Equipment	
Equipment Type	Normal Use Sampling Frequency (Hours/Miles)
Gasoline Engines	5,000 miles
Differentials	300 hours/20,000 miles
Fina Drives	300 hours/2,000 miles
Transmissions	300 hours/20,000 miles
Hydraulic Systems	1,000 hours/Annually

Table 8.2- Fluid Analysis Intervals for Mobile Equipment (Courtesy of Spectro Scientific)

9

9

For applications that are associated with safety of human life, such as aerospace industry, sampling intervals are reduced to 50-100 hours.

Aviation	
Equipment Type	*Normal Use Sampling Frequency in hours
Reciprocating Engines	50 hours
Gas Turbines	100 hours
Gearboxes	100 hours
Hydraulics	100 hours

Table 8.3- Fluid Analysis Intervals for Aerospace Industry (Courtesy of Spectro Scientific)

10

10

Among the places shown below, the correct sampling location is?

A. After the return filter.

B. After the heat exchanger.

C. At the inlet of the pump.

D. On the return line at a point of a turbulent flow.

11

11

8.2.2- Hydraulic Fluid Sampling Locations

- **Do NOT take samples from places where:**
  - Oil flow is restricted.
  - Contaminants or component wear products tend to settle.
  - Oil is cold after oil coolers.
  - Bottom of reservoirs.

- **Sampling location should be:**
  - Low pressure lines with turbulent flow such as elbows and tees.
  - Easily accessible for operators to quickly and easily take the sample.
  - Does not require disassembly of other parts.
  - Equipped with sampling valve.
  - Labeled sampling location.

Fig. 8.4- Labeling of Sampling Points (Spectro Scientific)

12

12

8.2.2.1- Sampling from Low Pressure Return Line (ISO 4021)

- Withdraw the sample before any filtration.
- Per **ISO 4021**, preferably from an upwards pointing sampling point at an elbow with turbulent flow.
- Sampling points fitted on the lower or side perimeter of a pipe tend to allow depositing of particles in the sampling valve.

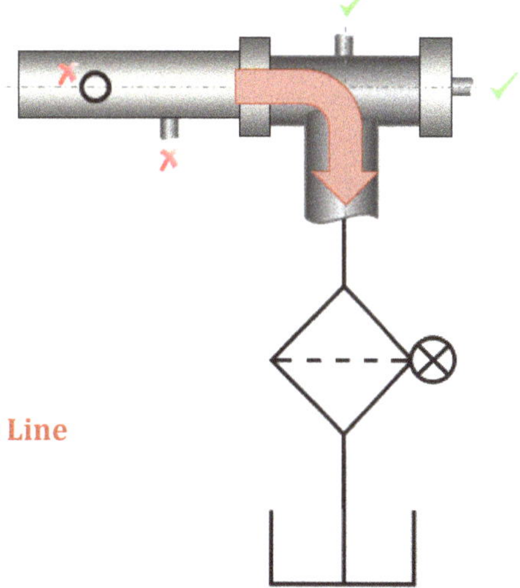

Fig. 8.5- Sampling from a Return Line

8.2.2.2- Sampling from High Pressure Line

The following precautions must be considered:

- A warning label of a high-pressure jet hazard shall be posted at the sampling location.
- Sampling valve shall be shielded and equipped by a check valve.

13

13

8.2.2.3- Sampling from Reservoir

- Fluid in the reservoir does not represent the fluid flowing in the system.
- **DO NOT** Sample from a reservoir.
- Sampling from the reservoir **IS NOT RECOMMENDED** by today's standards.
- The reason this part is presented here is to sample fluids from non-hydraulic driven machines such as engine oil sumps and transmission gear boxes.
- Sampling fluids from these machines is required for material and properties analysis, not for particle analysis.

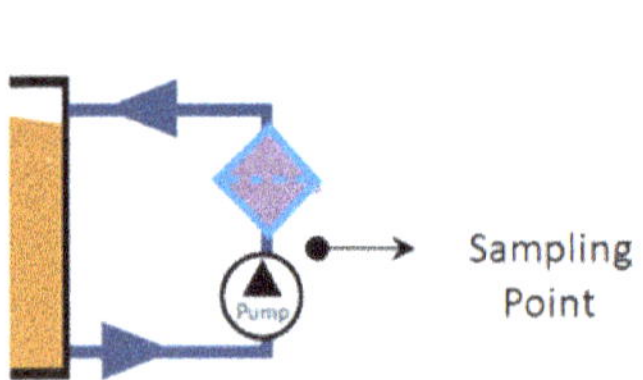

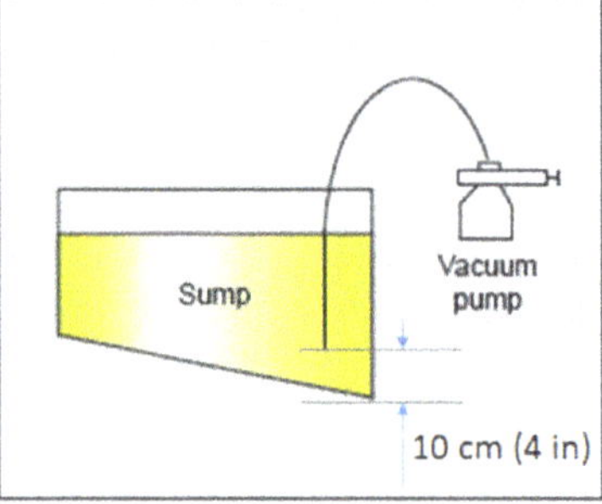

Fig. 8.6- Sampling from a Reservoir

14

14

8.2.3- Hydraulic Fluid Sampling Kit

❑ **Vacuum Pump:**

- Is a tool for extracting an oil sample from the sample port.
- Used with a sample port adapter, flexible tubing, and a sample bottle.
- Manual and electric pumps are available.

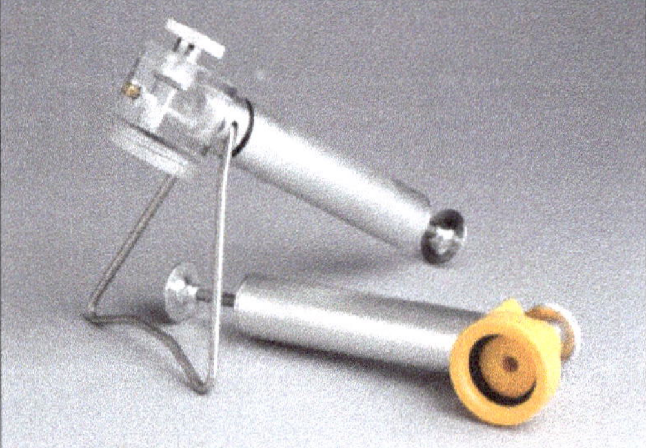

Fig. 8.7- Hydraulic Fluid Sampling Vacuum Pump (www.tricocorp.com)

15

15

- **Sampling Bottle:** a qualified sampling container must satisfy the following conditions:
  - Container: Glass or plastic bottle with wide mouth.
  - Size: Approximately 250 mL (4 ounce) size.
  - Cover: screw-on cap with plastic film between the cap and the bottle.
  - Label: to record the sampling data.
  - Cleanliness: Cleaned by filtered air, designated as *"Super Clean"*, and qualified in accordance with ISO 3722.

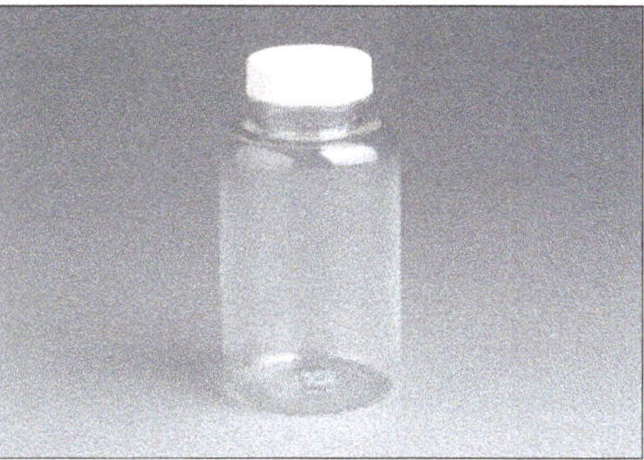

Fig. 8.8- Super Clean Sampling Bottle (www.tricocorp.com)

16

16

V489 (2 min)

Example from Industry:

- Ultra Clean Vacuum Device bottles (UCVD).
- Cleaned to an ISO code of 11/9/4 and sealed.
- Unlike other "super clean bottles" there is no need to open the cap.
- The bottle may be used in conjunction with a sampling probe, and it also avoids the need for a sampling vacuum pump.
- The operator simply connects the tubing from the port to the bottle and opens the valve.
- When finished, bottle valve is just closed

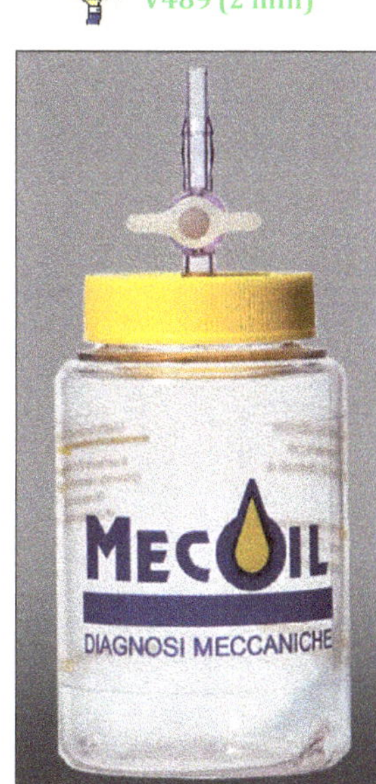

Fig. 8.9- Ultra Clean Sampling Bottle (www. mecoil.net)

17

17

- **Sampling Port:**
  - Designed to draw samples from the sampling points
  - Should provide superior leak protection.
  - Available in several types and sizes to match the varying requirements of manufacturers.

Fig. 8.10- Hydraulic Fluid Sampling Ports (www.tricocorp.com)

- **Sampling Tubing:**
  - Special sampling tubing must be as clean as the bottle, so it does not add contaminants.
  - A good idea is to be pre-flush with system fluid of 10 times tube volume prior to using it to take samples

Fig. 8.11- Hydraulic Fluid Sampling Tubing (www.tricocorp.com)

18

18

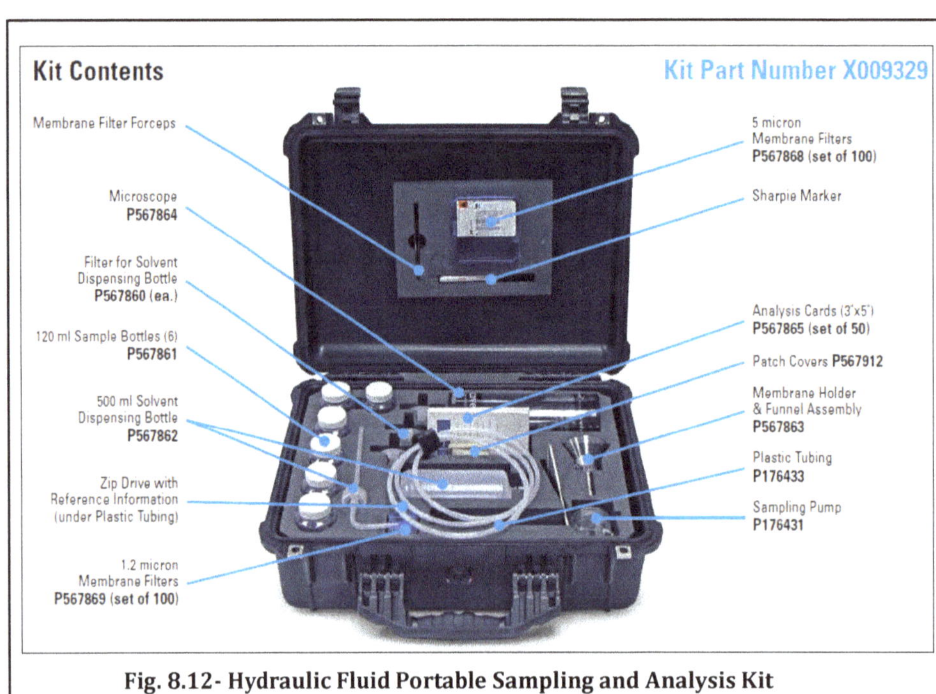

Fig. 8.12- Hydraulic Fluid Portable Sampling and Analysis Kit (Courtesy of Donaldson)

19

19

8.2.4- Hydraulic Fluid Sampling Procedure

- Detailed sampling procedure is prescribed in ISO 4021.
- However, in brief, to get a representative sample:
  - DO NOT sample immediately after oil change or addition of makeup fluids.
  - Run the system for at least 30 minutes or until it is warmed-up.
  - Shift directional valves several times to ensure that the fluid has been well circulated and is well mixed.

20

20

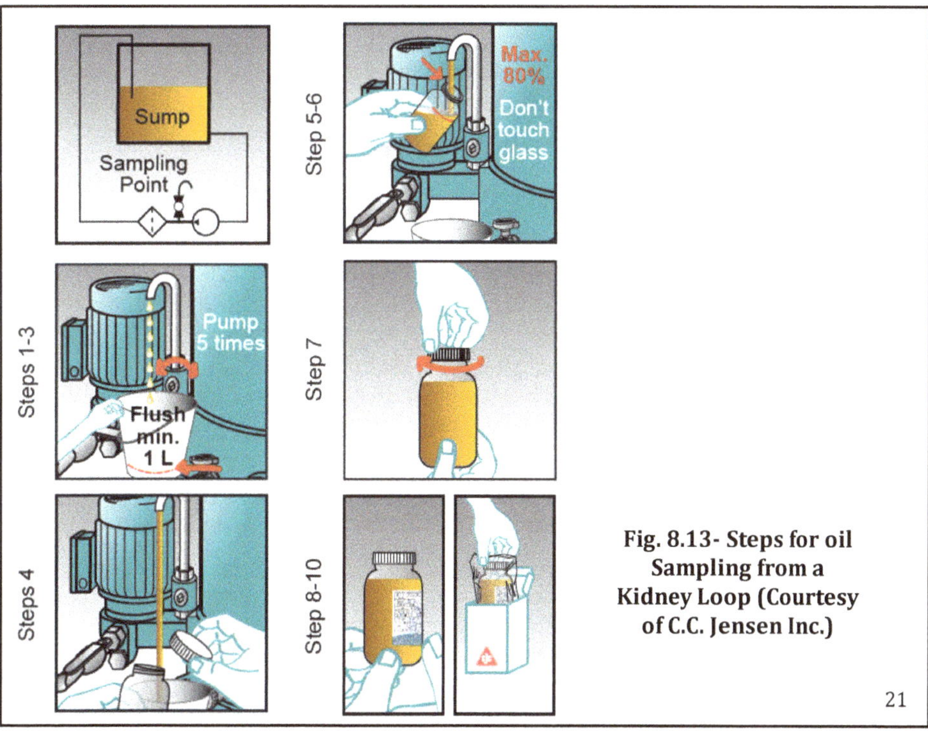

Fig. 8.13- Steps for oil Sampling from a Kidney Loop (Courtesy of C.C. Jensen Inc.)

21

21

❑ Notes:

- Suitable length of tube off the roll.
- Use new tube every time.
- Always flush tube before taking the sample.
- Be careful not to let the tube touch the walls or the bottom of the reservoir.
- Fill the bottle to approximately 80%.

Video 291 (1 min)

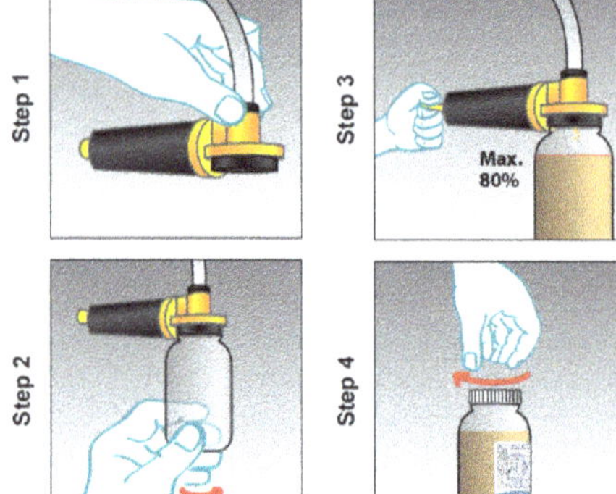

Fig. 8.14- Oil Sampling from a Reservoir using a Vacuum Pump (Courtesy of C.C. Jensen Inc.)

8.3- Hydraulic Fluid Material Analysis

8.3.1- Air Content

Standard methods for measuring air content were explained in Chapter 4.

8.3.2- Water Content

Standard method for measuring water content was explained in Chapter 5.

8.3.3- Solids Content

- Knowing the wear metal content of the fluid, helps predict which component is about to fail and possible catastrophic failure.
- This information is used as input for proactive maintenance plans.

❑ **Wear Metal Analysis (ASTM D5185):**

- Atomic Absorption Spectrograph is performed to determine wear metal content in ppm found in a sample.
- Wear metals could be iron, copper, lead, zinc, silicone, aluminum, tin, nickel or chromium,.
- The sample is vaporized over an extremely hot flame
- A light of fixed characteristic wavelength, for the metallic element being tested for, is passed through the sample.
- The amount of light absorbed by the sample indicates the quantity of that metallic element present in the sample.
- Note: This method only looks for particles 5-6 μm or smaller. It is not a substitute for particle counting.

❑ **Visual Inspection:**

- Content analysis for solid particles can be performed by observing the fluid sample under a microscope.
- Requires a trained operator.

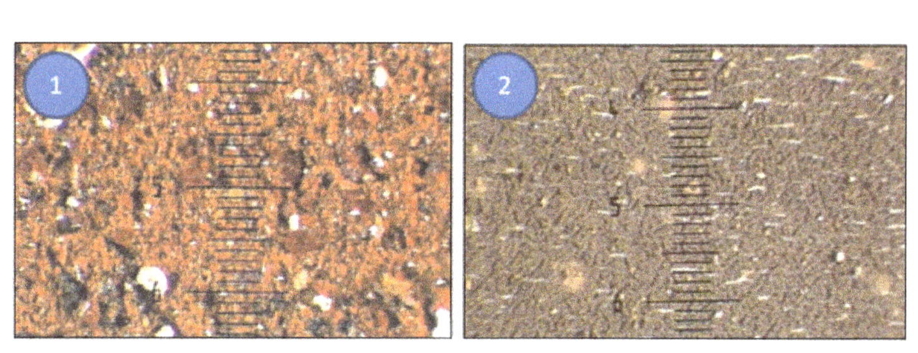

Fig. 8.15- Particulate Content Observation (Courtesy of Hydac)

Sample #	Particle Type	Effect
1	▪ Mainly rust. ▪ White particles. ▪ Additives.	▪ Rapid oil aging. ▪ Pumps and valves breakdown.
2	▪ Oil aging products.	▪ Blocking filters. ▪ Silting-of systems.

Table 8.4- Particulate Content Analysis (Courtesy of Hydac)

24

24

Continued

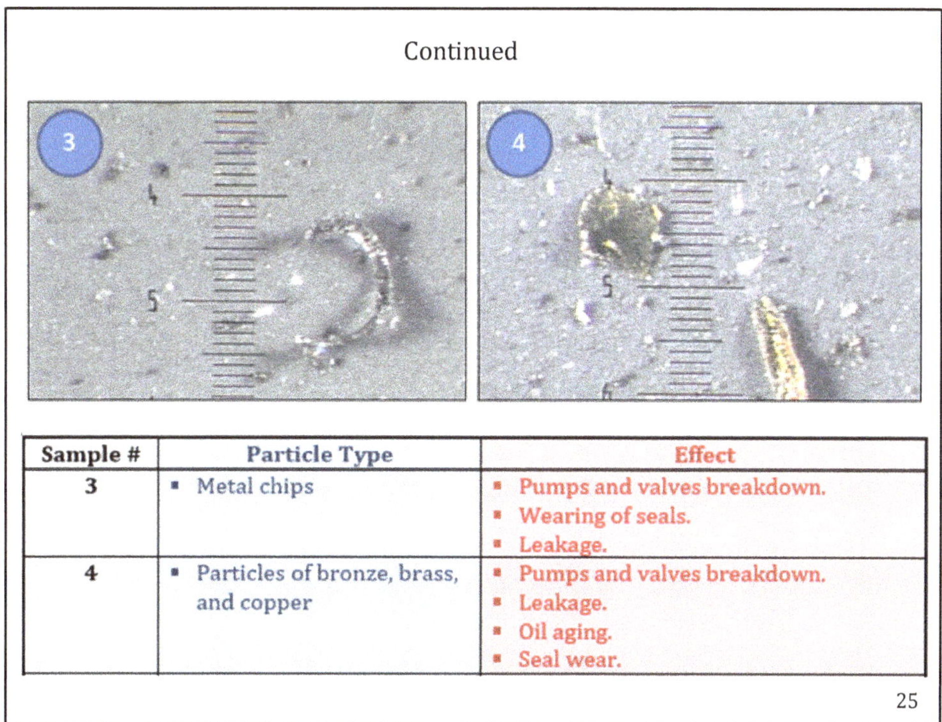

Sample #	Particle Type	Effect
3	▪ Metal chips	▪ Pumps and valves breakdown. ▪ Wearing of seals. ▪ Leakage.
4	▪ Particles of bronze, brass, and copper	▪ Pumps and valves breakdown. ▪ Leakage. ▪ Oil aging. ▪ Seal wear.

25

25

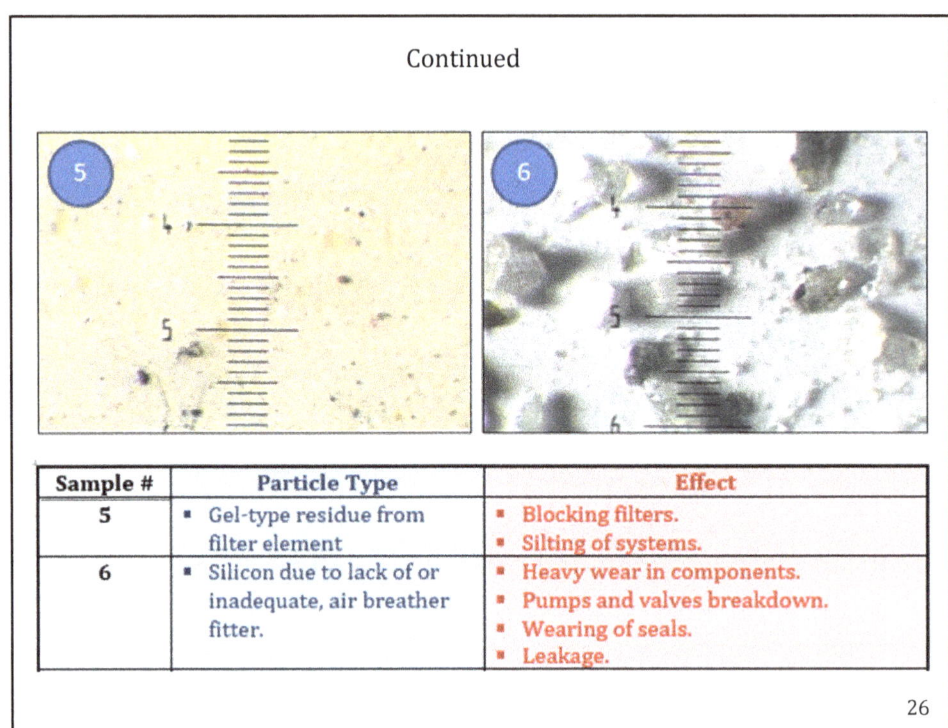

Sample #	Particle Type	Effect
5	▪ Gel-type residue from filter element	▪ Blocking filters. ▪ Silting of systems.
6	▪ Silicon due to lack of or inadequate, air breather fitter.	▪ Heavy wear in components. ▪ Pumps and valves breakdown. ▪ Wearing of seals. ▪ Leakage.

26

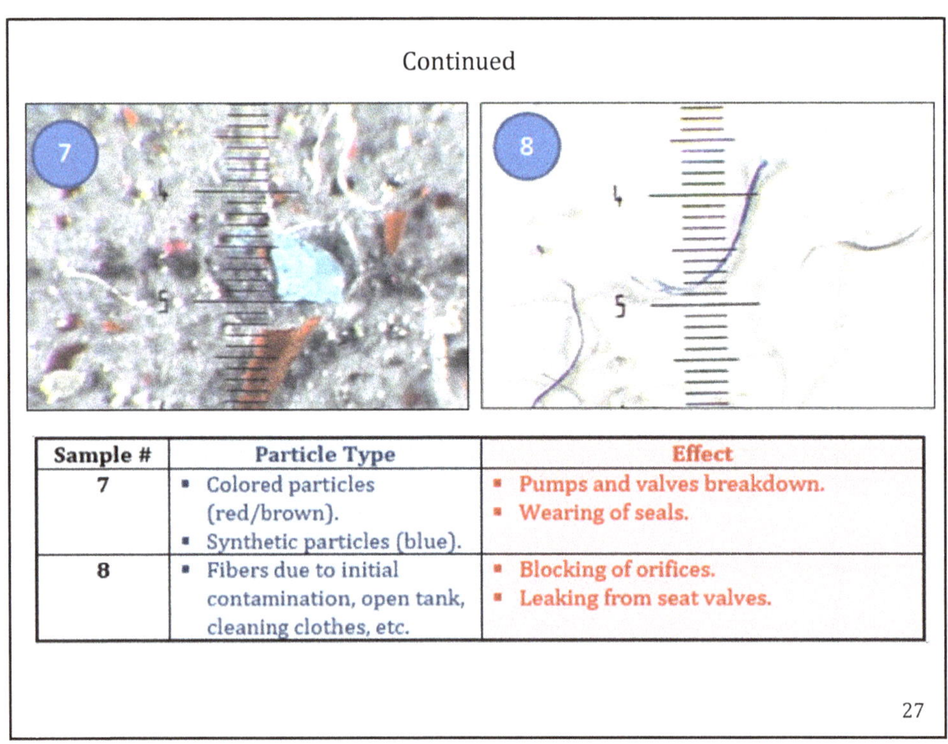

Sample #	Particle Type	Effect
7	▪ Colored particles (red/brown). ▪ Synthetic particles (blue).	▪ Pumps and valves breakdown. ▪ Wearing of seals.
8	▪ Fibers due to initial contamination, open tank, cleaning clothes, etc.	▪ Blocking of orifices. ▪ Leaking from seat valves.

27

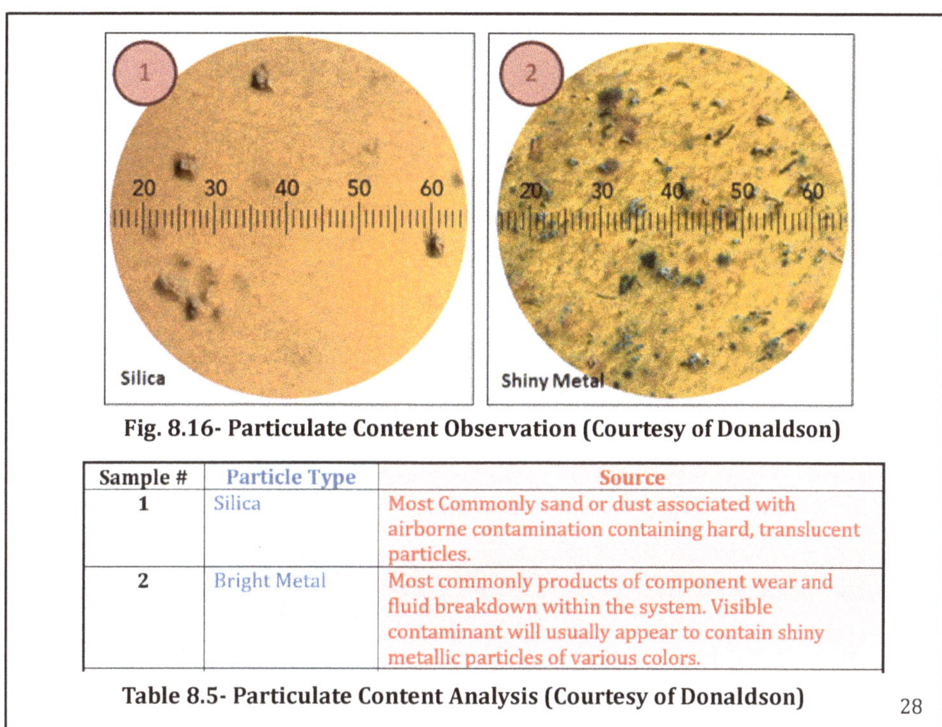

Fig. 8.16- Particulate Content Observation (Courtesy of Donaldson)

Sample #	Particle Type	Source
1	Silica	Most Commonly sand or dust associated with airborne contamination containing hard, translucent particles.
2	Bright Metal	Most commonly products of component wear and fluid breakdown within the system. Visible contaminant will usually appear to contain shiny metallic particles of various colors.

Table 8.5- Particulate Content Analysis (Courtesy of Donaldson)

28

28

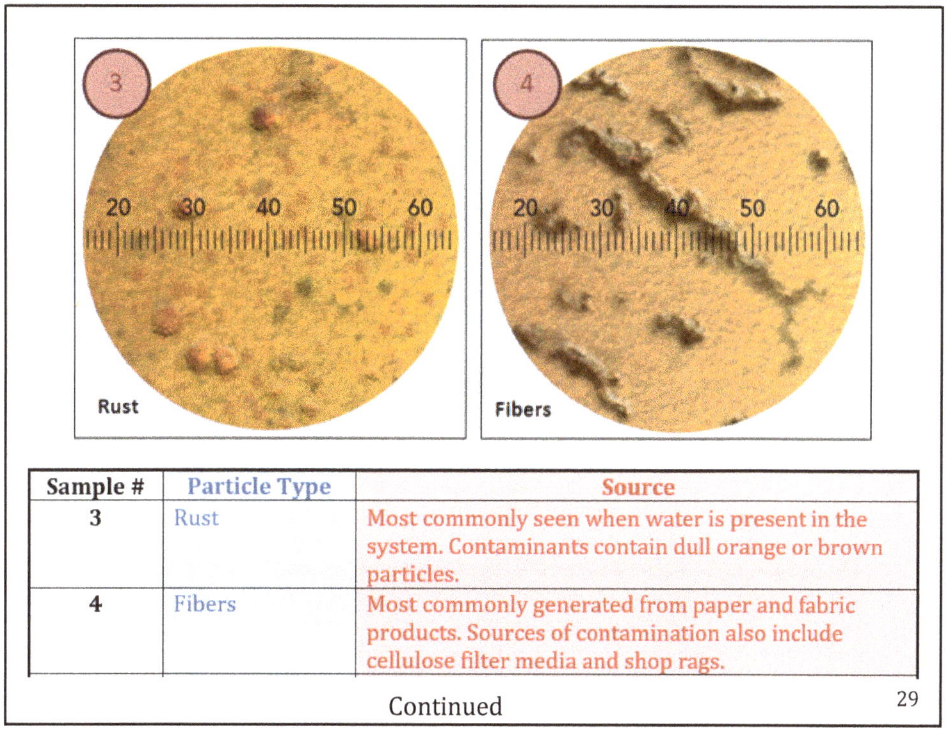

Sample #	Particle Type	Source
3	Rust	Most commonly seen when water is present in the system. Contaminants contain dull orange or brown particles.
4	Fibers	Most commonly generated from paper and fabric products. Sources of contamination also include cellulose filter media and shop rags.

Continued

29

29

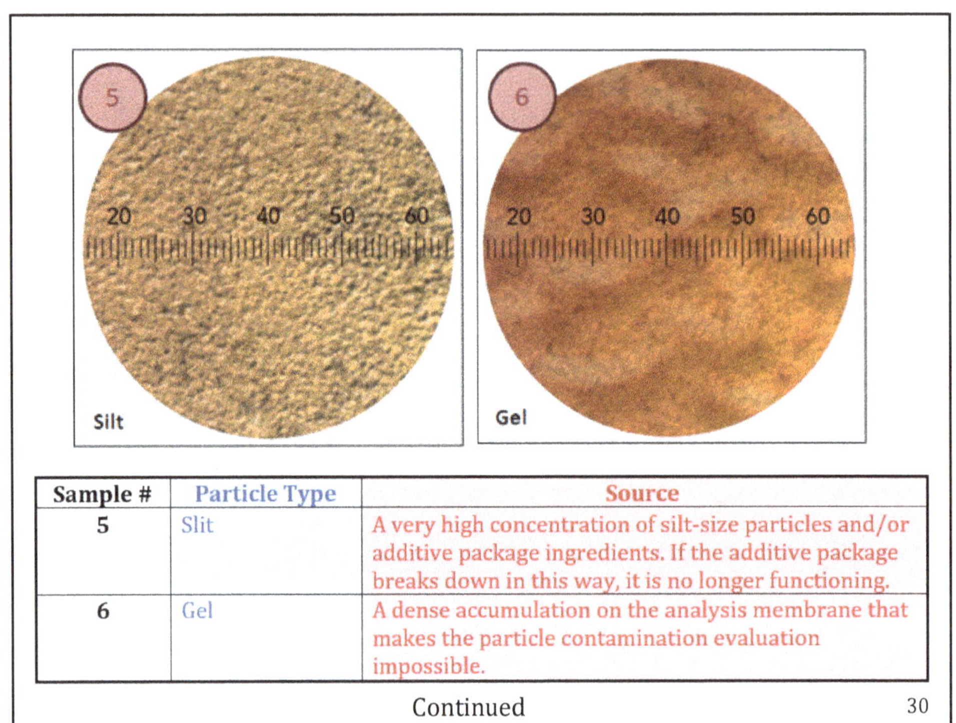

Sample #	Particle Type	Source
5	Slit	A very high concentration of silt-size particles and/or additive package ingredients. If the additive package breaks down in this way, it is no longer functioning.
6	Gel	A dense accumulation on the analysis membrane that makes the particle contamination evaluation impossible.

Continued 30

30

Fig. 8.17- Particulate Content Observation (Courtesy of Bosch Rexroth)

31

31

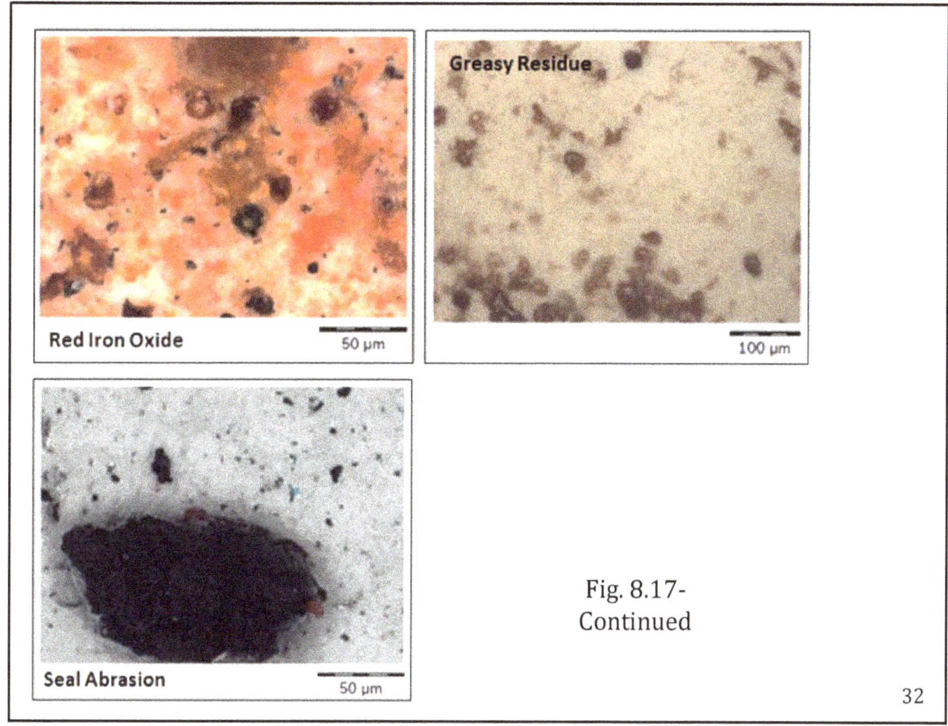

Fig. 8.17- Continued

32

32

8.4- Hydraulic Fluid Cleanliness Standards

- ❑ Based on:
- Several statistical and experimental investigations.
- Critical clearances in the hydraulic components,
- specific particle sizes that are most harmful to hydraulic components.

8.4.1- Two-Code ISO Standard 4406-1987

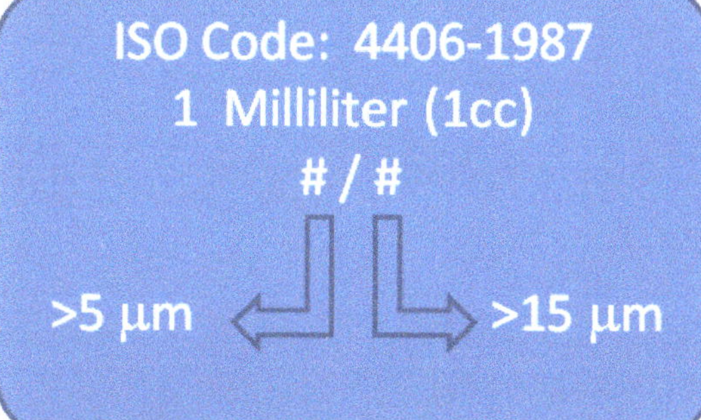

discussed here only as background information as it has been updated in 1999 to be three-code standard.

Fig. 8.18- Structure of ISO Code 4406-1987

33

33

Particle Concentration (Particles per milliliter)	Range Number
10,000,000	30
5,000,000	29
2,500,000	28
1,300,000	27
640,000	26
320,000	25
160,000	24
80,000	23
40,000	22
20,000	21
10,000	20
5,000	19
2,500	18
1,300	17
640	16
320	15
160	14
80	13
40	12
20	11
10	10
5	9
2.5	8
1.3	7
0.64	6
0.32	5
0.16	4
0.08	3
0.04	2
0.02	1
0.01	0.9
0.005	0.8
0.0025	0.7

For Example:

- 19/14 cleanliness code indicates
- 2501-5000 > particles > 5µm
- 81-160 particles at 15µm per ml of fluid.

Table 8.6- Particle Concentration per ISO Code 4406-1987

34

8.4.2- Three-Code ISO Standard 4406-1999

- In some cases, the code may appear as */18/13. This code means that the particle size less than 4 µm has no considerable effect on the system.
- In some cases, also it appears as 12/08/*. This indicates that the particle greater than 14 µm was too few to statically provide an accurate value.

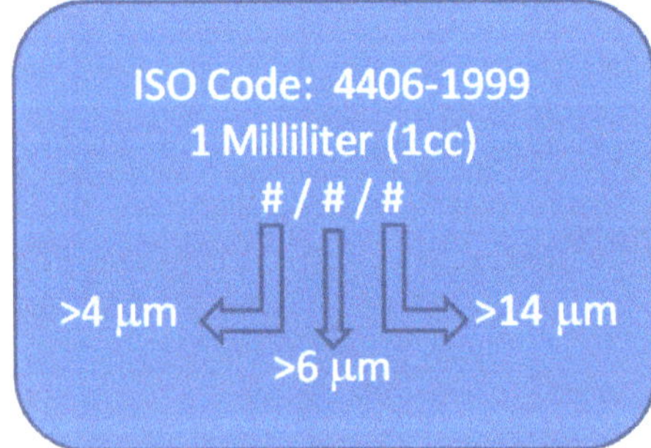

Fig. 8.19- Structure of ISO Code 4406-1999

Video 462 (1 min)

35

ISO 4406-1999 Range Numbers		
	Number of Particles per Millimeter	
Range Number	More Than	Up to and Including
28	1,300,000	2,500,000
27	640,000	1,300,000
26	320,000	640,000
25	160,000	32,000
24	80,000	160,000
23	40,000	80,000
22	20,000	40,000
21	10,000	20,000
20	5,000	10,000
19	2,500	5,000
18	1,300	2,500
17	640	1,300
16	320	640
15	160	320
14	80	160
13	40	80
12	20	40
11	10	20
10	5	10
9	2.5	5
8	1.3	2.5
7	0.64	1.3
6	0.32	0.64
5	0.16	0.32
4	0.08	0.16
3	0.04	0.08
2	0.02	0.04
1	0.01	0.02
0	0	0.01

For Example:

- 20/18/14cleanliness code means:
-]5,000 - 10,000] > 4 µm.
-] 1,300 - 2,500] > 6 µm.
-]80 - 160] > 14 µm.

Table 8.7- Particle Concentration per ISO Code 4406-1999

36

36

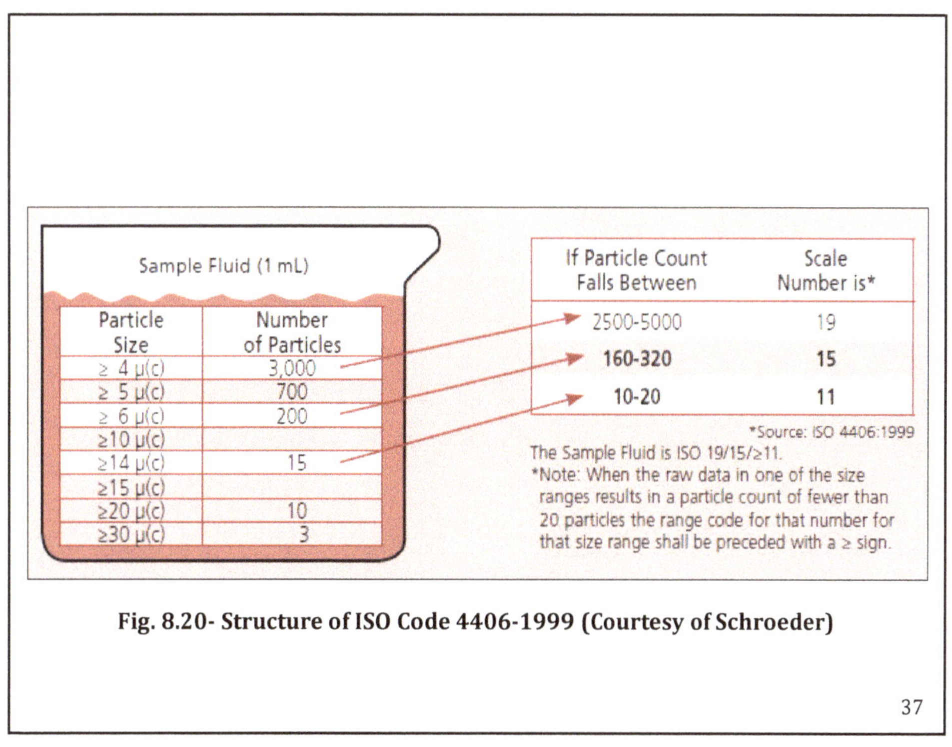

Sample Fluid (1 mL)
Particle Size
Number of Particles
≥ 4 µ(c) 3,000
≥ 5 µ(c) 700
≥ 6 µ(c) 200
≥10 µ(c)
≥14 µ(c) 15
≥15 µ(c)
≥20 µ(c) 10
≥30 µ(c) 3
If Particle Count Falls Between
Scale Number is*
2500-5000 19
160-320 15
10-20 11
*Source: ISO 4406:1999
The Sample Fluid is ISO 19/15/≥11.
*Note: When the raw data in one of the size ranges results in a particle count of fewer than 20 particles the range code for that number for that size range shall be preceded with a ≥ sign.

Fig. 8.20- Structure of ISO Code 4406-1999 (Courtesy of Schroeder)

37

37

- It is mistakenly understood that the new supplied oil is cleaner than what you have in the system!
- This is usually incorrect because the fluid in the system is continuously filtered.
- Typically, new fluid as delivered from the drum, has a cleanliness level of ISO Code 23/21/19.

Video 412 (0.5 min)

Fig. 8.21- Typical Cleanliness ISO Code for New Hydraulic Fluid

38

38

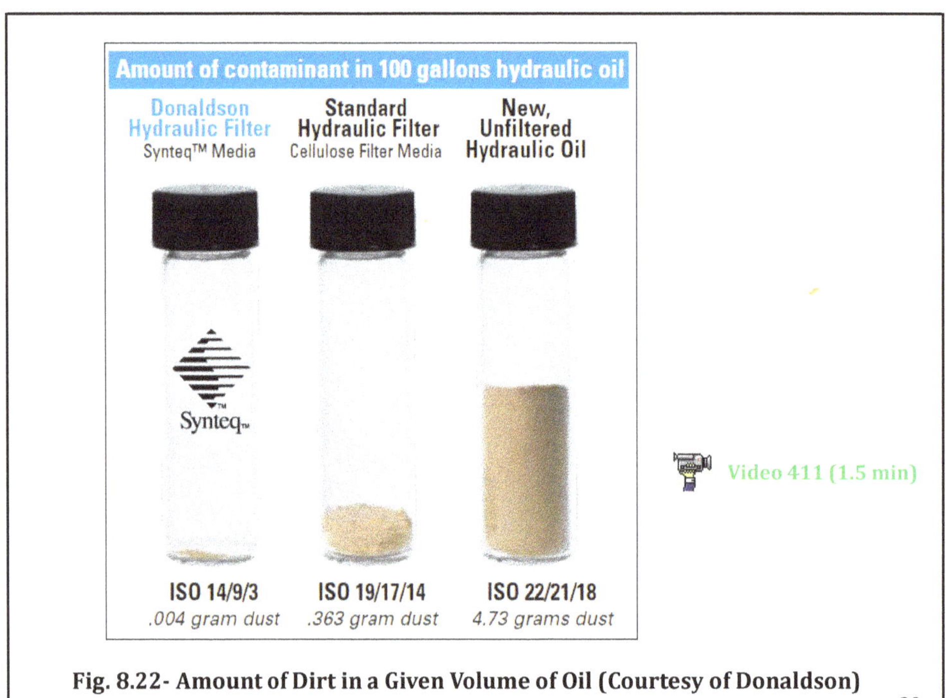

Video 411 (1.5 min)

Fig. 8.22- Amount of Dirt in a Given Volume of Oil (Courtesy of Donaldson)

39

39

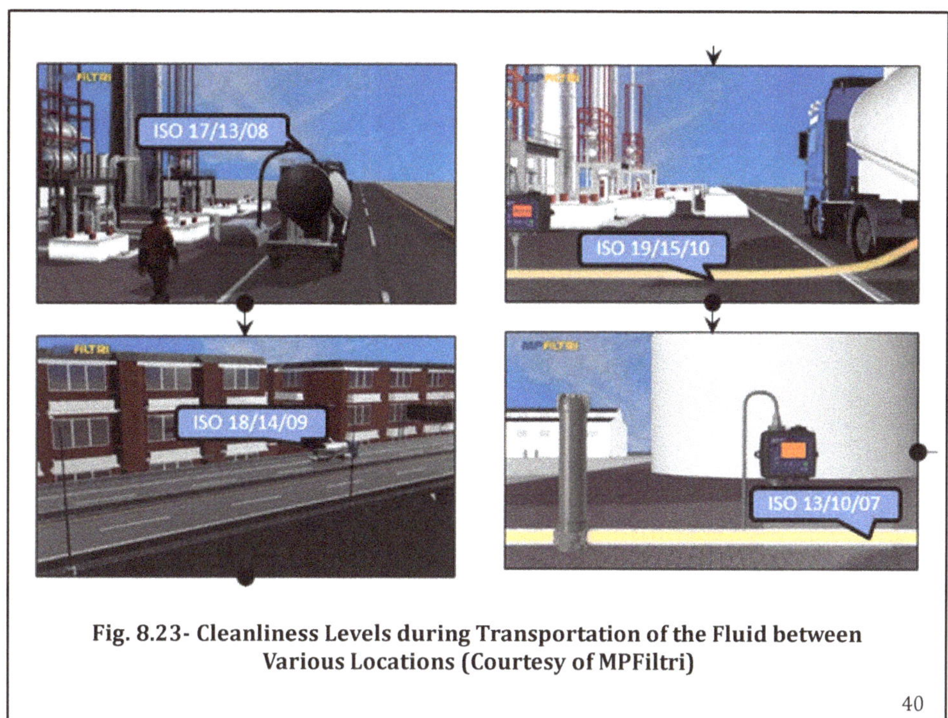

Fig. 8.23- Cleanliness Levels during Transportation of the Fluid between Various Locations (Courtesy of MPFiltri)

40

40

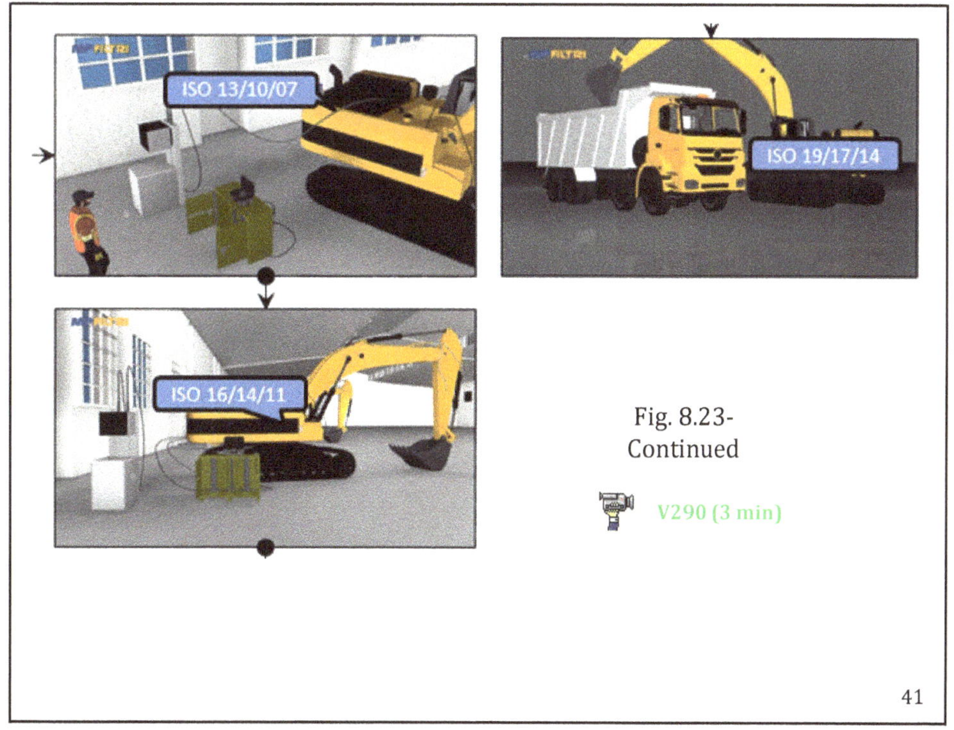

Fig. 8.23- Continued

V290 (3 min)

41

41

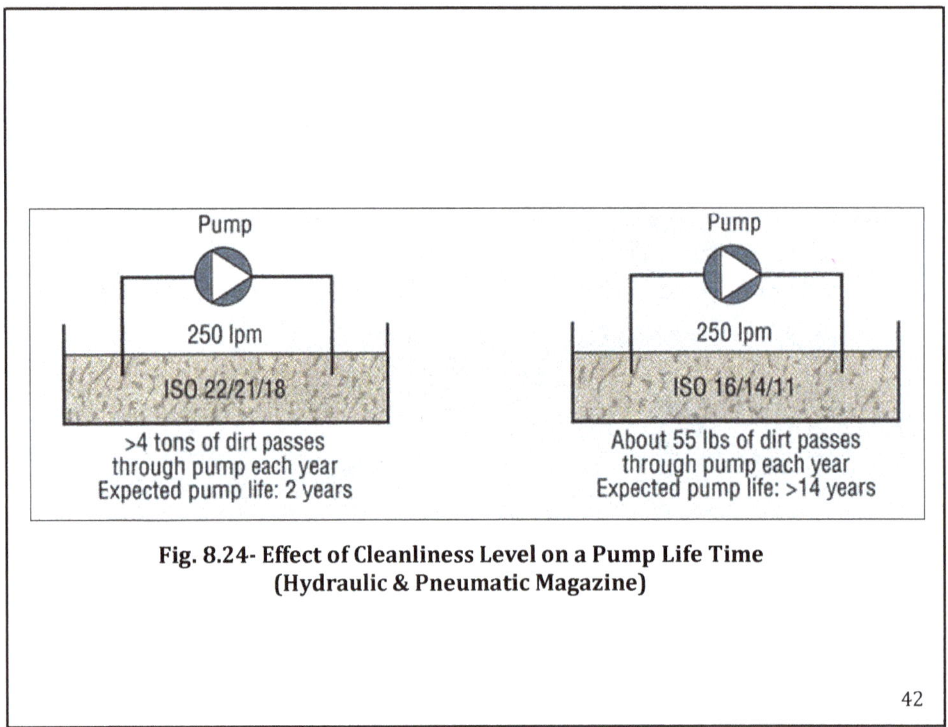

Fig. 8.24- Effect of Cleanliness Level on a Pump Life Time (Hydraulic & Pneumatic Magazine)

42

42

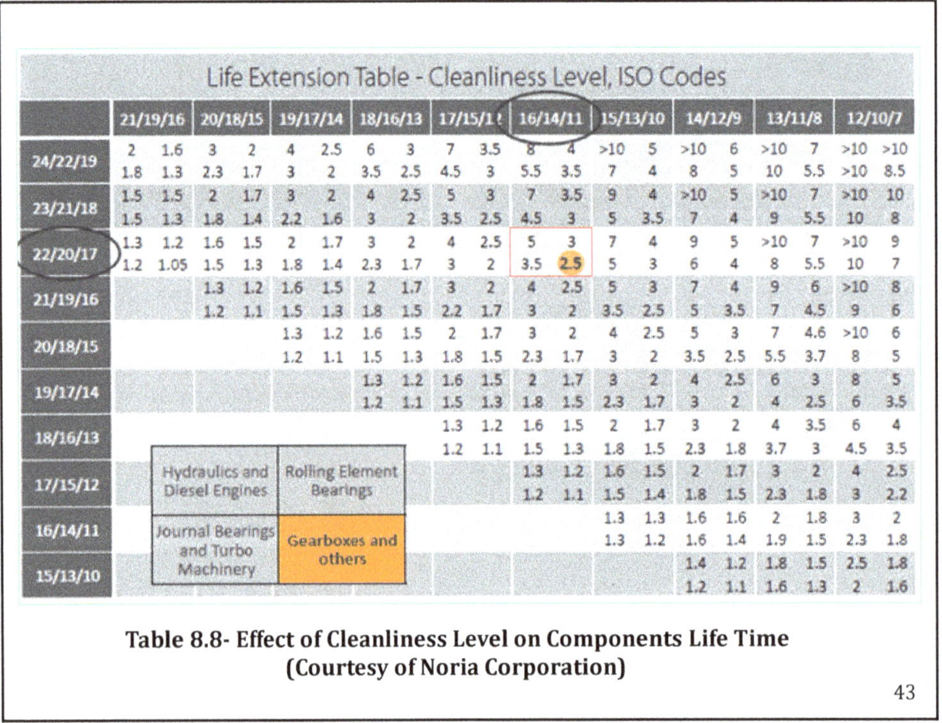

Life Extension Table - Cleanliness Level, ISO Codes

	21/19/16		20/18/15		19/17/14		18/16/13		17/15/12		16/14/11		15/13/10		14/12/9		13/11/8		12/10/7	
24/22/19	2	1.6	3	2	4	2.5	6	3	7	3.5	8	4	>10	5	>10	6	>10	7	>10	>10
	1.8	1.3	2.3	1.7	3	2	3.5	2.5	4.5	3	5.5	3.5	7	4	8	5	10	5.5	>10	8.5
23/21/18	1.5	1.5	2	1.7	3	2	4	2.5	5	3	7	3.5	9	4	>10	5	>10	7	>10	10
	1.5	1.3	1.8	1.4	2.2	1.6	3	2	3.5	2.5	4.5	3	5	3.5	7	4	9	5.5	10	8
22/20/17	1.3	1.2	1.6	1.5	2	1.7	3	2	4	2.5	5	3	7	4	9	5	>10	7	>10	9
	1.2	1.05	1.5	1.3	1.8	1.4	2.3	1.7	3	2	3.5	2.5	5	3	6	4	8	5.5	10	7
21/19/16			1.3	1.2	1.6	1.5	2	1.7	3	2	4	2.5	5	3	7	4	9	6	>10	8
			1.2	1.1	1.5	1.3	1.8	1.5	2.2	1.7	3	2	3.5	2.5	5	3.5	7	4.5	9	6
20/18/15					1.3	1.2	1.6	1.5	2	1.7	3	2	4	2.5	5	3	7	4.6	>10	6
					1.2	1.1	1.5	1.3	1.8	1.5	2.3	1.7	3	2	3.5	2.5	5.5	3.7	8	5
19/17/14							1.3	1.2	1.6	1.5	2	1.7	3	2	4	2.5	6	3	8	5
							1.2	1.1	1.5	1.3	1.8	1.5	2.3	1.7	3	2	4	2.5	6	3.5
18/16/13									1.3	1.2	1.6	1.5	2	1.7	3	2	4	3.5	6	4
									1.2	1.1	1.5	1.3	1.8	1.5	2.3	1.8	3.7	3	4.5	3.5
17/15/12											1.3	1.2	1.6	1.5	2	1.7	3	2	4	2.5
											1.2	1.1	1.5	1.4	1.8	1.5	2.3	1.8	3	2.2
16/14/11													1.3	1.3	1.6	1.6	2	1.8	3	2
													1.3	1.2	1.6	1.4	1.9	1.5	2.3	1.8
15/13/10															1.4	1.2	1.8	1.5	2.5	1.8
															1.2	1.1	1.6	1.3	2	1.6

Hydraulics and Diesel Engines	Rolling Element Bearings
Journal Bearings and Turbo Machinery	Gearboxes and others

Table 8.8- Effect of Cleanliness Level on Components Life Time (Courtesy of Noria Corporation)

43

43

Pump/Motors	Target Cleanliness Class
Fixed Gear or Vane	20/18/15
Fixed Piston	19/17/14
Variable Vane	18/16/13
Variable Piston	18/16/13
Drives	
Cylinders	20/18/15
Hydrostatic Drives	16/15/12
Test Rigs	15/13/10
Valves	
Check Valve	20/18/15
Directional Valve	20/18/15
Standard Flow Control Valve	20/18/15
Poppet Valve	19/17/14
Proportional Valve	18/16/13
Servo valve	15/13/10
Bearings	
Anti-Friction Bearing	18/15/12
Transmission	17/15/12
Ball Bearing	15/13/10
Roller Bearing	16/14/11

Table 8.9- Guideline for Cleanliness Levels per ISO 4406-1999

44

44

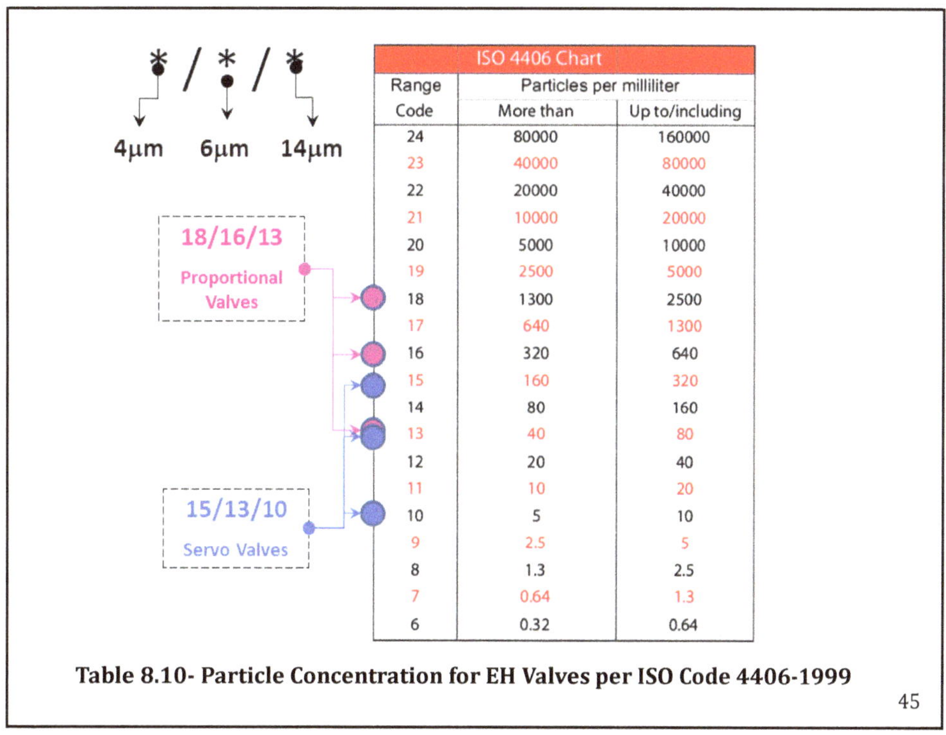

ISO 4406 Chart

Range Code	Particles per milliliter More than	Up to/including
24	80000	160000
23	40000	80000
22	20000	40000
21	10000	20000
20	5000	10000
19	2500	5000
18	1300	2500
17	640	1300
16	320	640
15	160	320
14	80	160
13	40	80
12	20	40
11	10	20
10	5	10
9	2.5	5
8	1.3	2.5
7	0.64	1.3
6	0.32	0.64

Table 8.10- Particle Concentration for EH Valves per ISO Code 4406-1999

45

45

8.4.3- NAS Standard 1638

❑ **National Aerospace Standard (NAS 1638)**

- It became the American National Aerospace Standard in 1964.
- Used to identify contamination in aircraft hydraulic components.
- Today, use of NAS-1638 is very limited for the sake of ISO 4406.
- Like ISO 4406, contamination level is based on particle size and concentration in a given volume of oil.
- Unlike ISO 4406:
  - Fluid Sample volume (100 ml = 100 cc).
  - Particle counts at various size ranges → broad-base classes (0 – 12).
  - Cleanest possible fluid (class 00) to the dirtiest oil (class 12).
  - NAS class is assigned based on **heights** particle number among the individual range of particle sizes.

- For example, if in a 100 mi sample of fluid:
  - For particle size 5-15 μm, 1000 particles were found.
  - For particle size 15-25 μm, 356 particles were found.
  - For particle size 25-50 μm, 126 particles were found.
  - For particle size 50-100 μm, 45 particles were found.
  - For particle size > 100 μm, 16 particles were found.
  - Contamination class is "Class 6"

46

46

Contamination Class	Particle Size in μm (in 100 ml)				
	5-15	15-25	25-50	50-100	>100
00	125	22	4	1	0
0	250	44	8	2	0
1	500	89	16	3	1
2	1,000	178	32	6	1
3	2,000	356	63	11	2
4	4,000	712	126	22	4
5	8,000	1,425	253	45	8
6	16,000	2,850	506	90	16
7	32,000	5,700	1,012	180	32
8	64,000	11,400	2,025	360	64
9	128,000	22,800	4,045	720	128
10	256,000	45,600	8,100	1,440	256
11	512,000	91,200	16,200	2,880	512
12	1,024,000	182,400	32,400	5,760	1,024

Table 8.11- Particle Concentration per NAS 1638

47

47

8.4.4- SAE Standard AS 4059(E)

❑ **Society of Automotive Engineering SAE AS 4059(E).**

- Like NAS 1638, SAE 4059 (E) cleanliness classes are based on:
- Particle size → labeled with letters (A - F).
- Particle concentration in 100 mil. → broad-band classes (1 – 12).
- Unlike NAS 1638 (Method of evaluating the particle size).

Example 1: Cleanliness Class is (AS 4059:6) means:

All particle sizes should not be > the number indicated for class 6.

Example 2: Cleanliness Class is (AS 4059 :6 B) means:

Particles of size B should not be > Class 6 (19,500)

Example 3: Cleanliness Class is (AS 4059 :7 B / 6 C) means:

- Max # of particles Size B (5 µm or 6 µm(c)) = 38,900 / 100 ml.
- Max # of particles Size C (15 µm or 14 µm(c)) = 3,460 / 100 ml.

Example 4: Cleanliness Class is (AS 4059:6 B-F) means:

Maximum # of particles size range B-F should not exceed the number for class 6 for each relevant size range. Other size ranges are not in interest.

48

48

Maximum Particle Concentration* (particles/100ml)						
ISO 4402 *	> 1 µm	> 5 µm	> 15 µm	> 25 µm	> 50 µm	> 100 µm
ISO 11171**	> 4 µm(c)	> 6 µm(c)	> 14 µm(c)	> 21 µm(c)	> 38 µm(c)	> 70 µm(c)
Size Coding	A	B	C	D	E	F
000	195	76	14	3	1	0
00	390	152	27	5	1	0
0	780	304	54	10	2	0
1	1,560	609	109	20	4	1
2	3,120	1,220	217	39	7	1
3	6,250	2,430	432	76	13	2
4	12,500	4,860	864	152	26	4
5	25,000	9,730	1,730	306	53	8
6	50,000	19,500	3,460	612	106	16
7	100,000	38,900	6,920	1,220	212	32
8	200,000	77,900	13,900	2,450	424	64
9	400,000	156,000	27,700	4,900	848	128
10	800,000	311,000	55,400	9,800	1,700	256
11	1,600,000	623,000	111,000	19,600	3,390	1,020
12	3,200,000	1,250,000	222,000	39,200	6,780	

* ISO 4402 or Optical Microscope. Particle size is based on longest dimension

** ISO 11171 or Electron Microscope. Particle size is based on projected area equivalent diameter

Table 8.12- Particle Concentration per SAE AS4059(E)

49

49

8.4.5- Contamination Standards Cross-Reference

Sample #	NAS 1638	ISO 4406: 1999	SAE AS 4059
1	Class 3	14/12/9	Class 4
2	Class 4	15/13/10	Class 5
3	Class 5	16/14/11	Class 6
4	Class 6	17/15/12	Class 7
5	Class 7	18/16/13	Class 8
6	Class 8	19/17/14	Class 9
7	Class 9	20/18/15	Class 10
8	Class 10	21/19/16	Class 11
9	Class 11	22/20/17	Class 12
10	Class 12	23/21/18	Class 13

Table 8.13- Approximate Cross-Reference for Contamination Classes (Courtesy of Hydac)

50

50

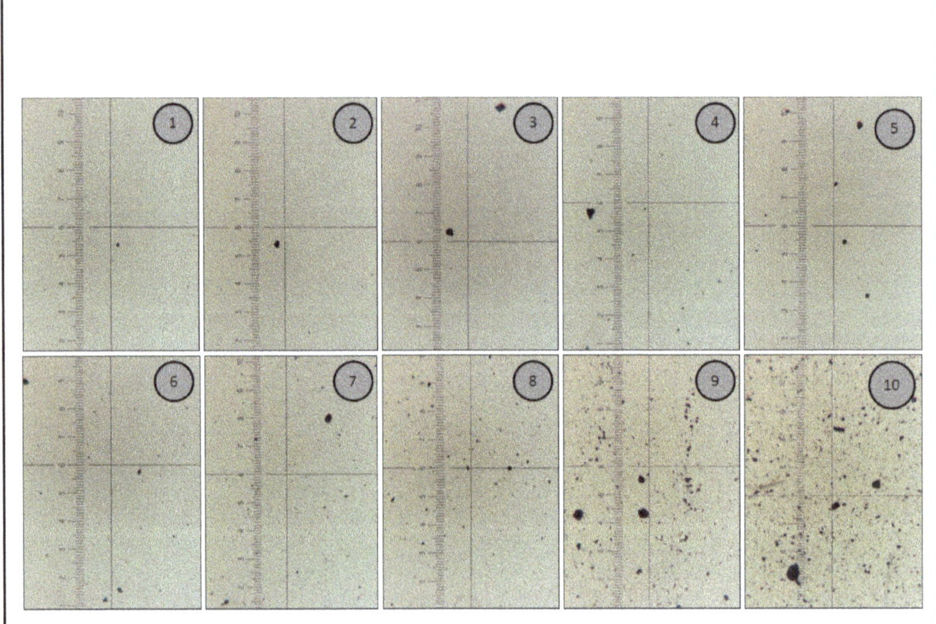

Fig. 8.25- Samples of Fluid Contamination Levels (Courtesy of Hydac)

51

51

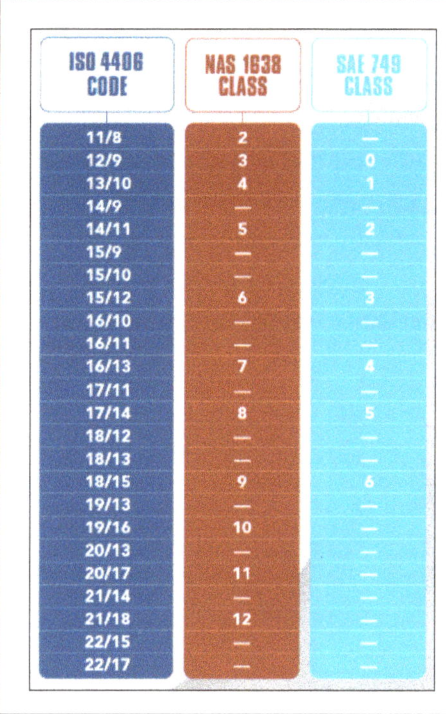

ISO 4406 CODE	NAS 1638 CLASS	SAE 749 CLASS
11/8	2	—
12/9	3	0
13/10	4	1
14/9	—	—
14/11	5	2
15/9	—	—
15/10	—	—
15/12	6	3
16/10	—	—
16/11	—	—
16/13	7	4
17/11	—	—
17/14	8	5
18/12	—	—
18/13	—	—
18/15	9	6
19/13	—	—
19/16	10	—
20/13	—	—
20/17	11	—
21/14	—	—
21/18	12	—
22/15	—	—
22/17	—	—

It is to be mentioned that the 2-codes ISO standard and the SAE 749 standard are no longer in use and they are stated here for historical purposes.

Table 8.14- Approximate Cross-Reference for Contamination Classes (Courtesy of Donaldson)

52

52

ISO 4406:1999	SAE AS 4059	NAS 1638-01/196	MIL-STD 1246A 1967	ACFTD Gravimetric Level-mg/L
24				
23/20/18		12		
22/19/17	12	11		
21/18/16	11	10		
20/17/15	10	9	300	
19/16/14	9	8		
18/15/13	8	7	200	1
17/14/12	7	6		
16/13/11	6	5		
15/12/10	5	4		0.1
14/11/9	4	3	100	
13/10/8	3	2		
12/9/7	2	1		0.01
11/8/6	1	0		
10/7/5	0	00		
8/7/4	00		50	
5/3/01			25	
2/0/0			5	

Table 8.15- Approximate Cross-Reference for Contamination Classes (Courtesy of Schroeder)

53

53

The following counts of particles of sizes were found in a 1 milliliter of a hydraulic sample (2500 particles of > 4 microns, 640 particles of size > 6 microns, and 80 particles of size > 14 microns. What is the corresponding cleanliness level?

A. 17/15/12.
B. 18/16/13.
C. 19/17/14.
D. 20/18/15.

54

54

The following counts of particles of sizes were found in a 1 milliliter of a hydraulic sample, 1000 particles of size > 4 microns, 200 particles of size > 6 micros, and 30 particles of size > 14 micron. What is the corresponding cleanliness level?

A. 17/15/12 per ISO 4406/1999.
B. Class 6 per NAS 1638.
C. Class 7 per SAE AS 4059.
D. All of the above.

55

55

8.5- Hydraulic Fluid Particle Analysis
8.5.1-Visual Inspection

- Visual inspection:
  - Is the first and easiest qualitative contamination test.
  - No equipment is required and can be done in field.
  - Comparison between new and used fluid samples,
  - Holding the sample up to the light:
    - If particles can be seen, the fluid is very dirty.
    - If particles can't be seen, then further analysis is required.

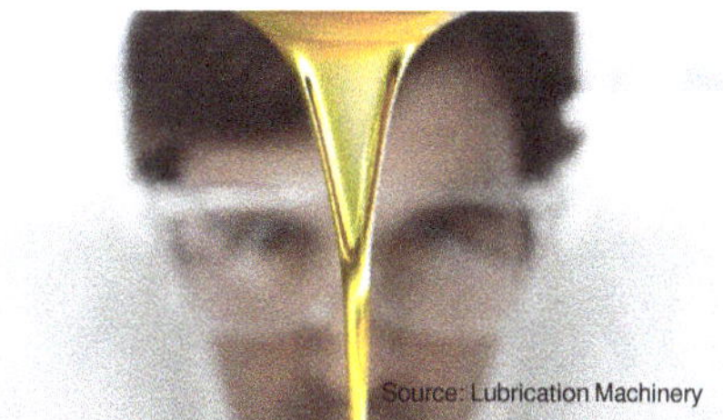

56

- Oil color and odor are good signs for a quick field test.

1. Signs of oxidation and sludge the reservoirs and around filters.
2. Signs of thermal degradation or darkened oil due to varnish formation.
3. Signs of foaming where oil color tends to be milky or cloudy.
4. Signs of Stable oil-water emulsion and fluid could also be milky or cloudy.
5. Signs of oil change is overdue where oil is very dark.

Fig. 8.26- Hydraulic Fluid Visual Inspection V287 (1.5 min)

57

8.5.2- Silt Index Test

- Silt is very fine particles generated from continuous erosion of metal parts.
- Silt Index Test is not a very famous test method and was phased out for the sake of other advanced and more accurate methods.
- The way it works is a sample of fluid is forced through a porous filter.
- The silting index is calculated based on difference in the pressure during passing the first and the second half of the sample.

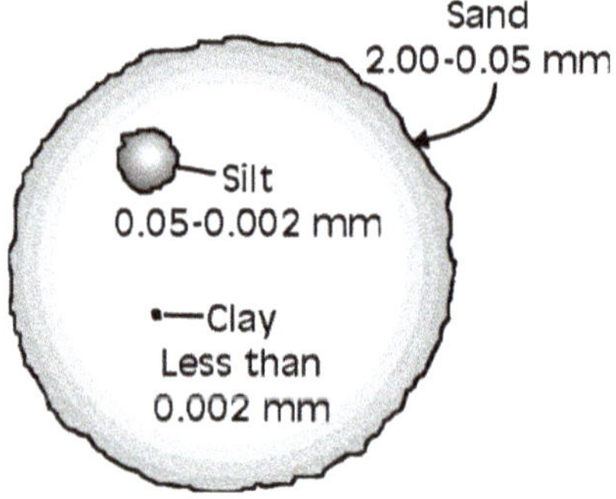

Fig. 8.27- Relative Size of Silt

58

58

8.5.3-Patch Test

- Patch Test is another qualitative contamination test.

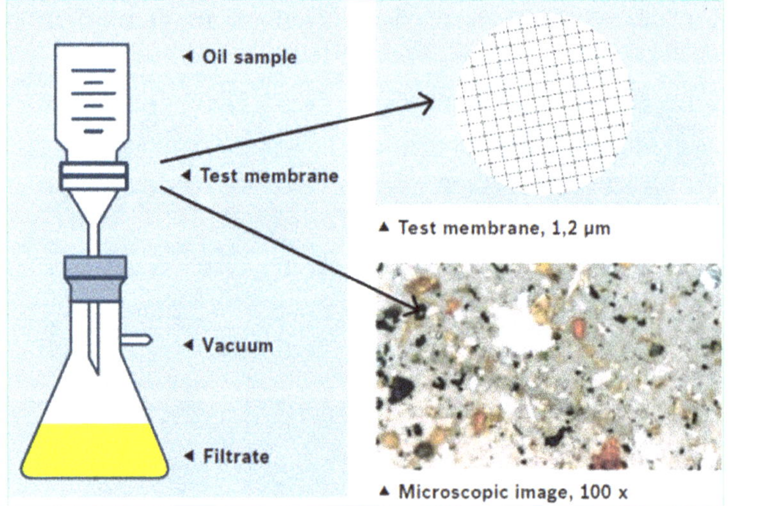

Fig. 8.28- Patch Test Device (Courtesy of Bosch Rexroth)

59

59

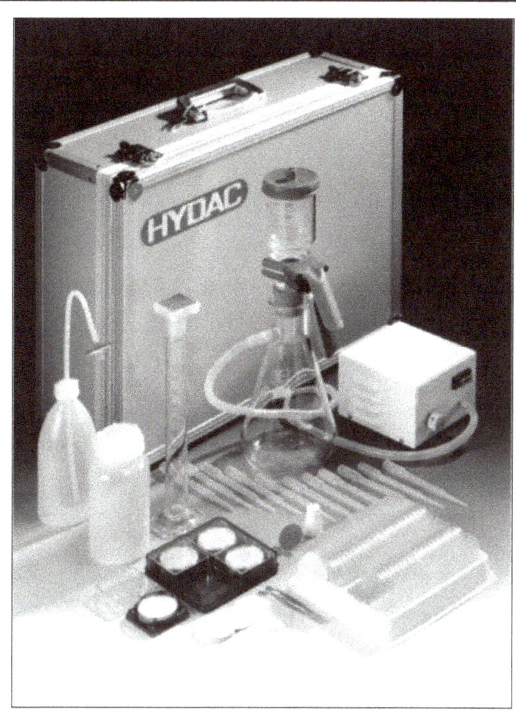

Portable patch test kit includes *Fluid Cleanliness Comparison Guide.*

Fig. 8.29- Hydraulic Fluid Portable Patch Test Kit (Courtesy of Hydac)

60

60

Basic Steps for Patch Test:

- Assemble the pump, the funnel, and clamp on empty lower flask.
- Flush the fitter assembly with pre-filtered solvent.
- Place a patch membrane on filter holder.
- Dilute the oil sample with filtered solvent and mix vigorously.
- Turn on the vacuum pump.
- Pour the fluid sample into the funnel and fill to the 25 ml level.

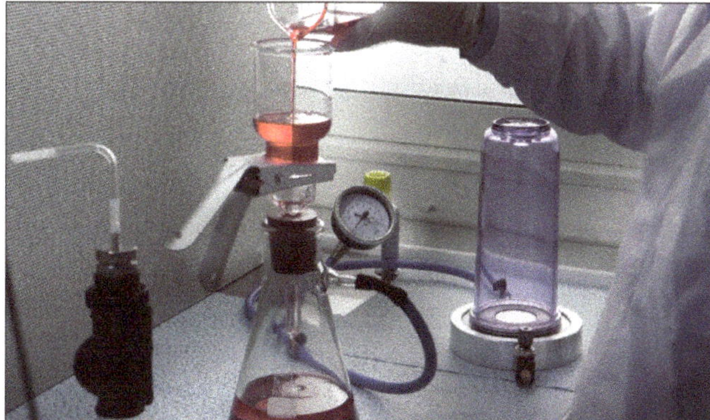

Fig. 8.30- Performing Patch Test (Courtesy of MPFiltri)

61

61

- When sample passes completely through the patch membrane, remove membrane with forceps, and air dry it.
- Place on clean index card and immediately cover with adhesive analysis laminated cover.
- Inspect for debris.
- Cleanliness level is based on color and shade of the membrane patch indicates the category of contamination such as Normal (1), Abnormal (2), and Critical (3).

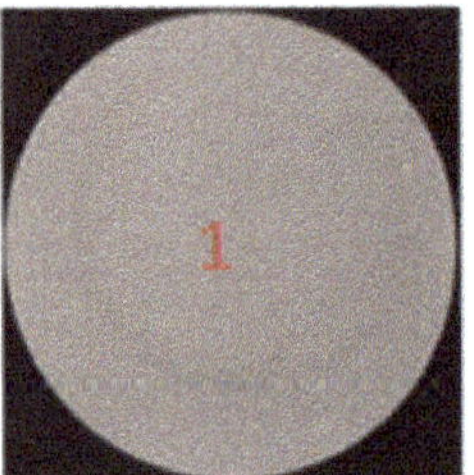

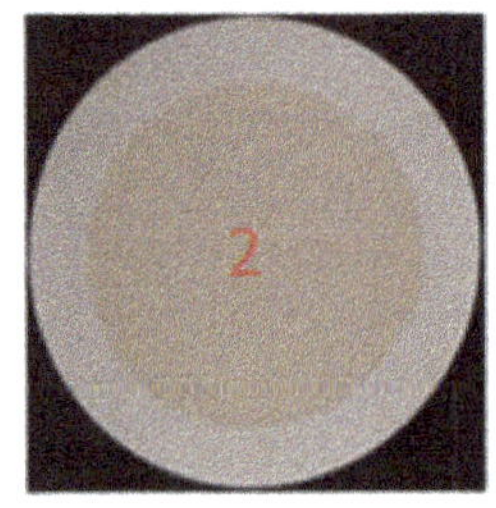

Fig. 8.31- Observations from Patch Test (Courtesy of MPFiltri)

62

62

8.5.4- Gravimetric Analysis (ISO 4405)

- It uses same apparatus used for the patch test,
- + sensitive weight scale
- + filter membrane whose weight has been previously defined.

- <u>The way it works is:</u>
- Passing (100 ml) of oil sample through the filter membrane.
- The cleanliness level is based on the difference between the weight of membrane before and after passing the sample of fluid through it.
- Results are given as mg/l.

 Video 32 (0.5 min)

63

63

8.5.5- Microscopic Particle Counting (ISO 4407)

- It is also referred to as *Optical* or *Visual* particle counting.
- Like patch and the gravimetric tests, a sample passes through filter.
- Unlike patch and gravimetric tests:
  - A special membrane filter is used that has average pore size <1 μm and grid markings.
  - Filter membrane is dried, viewed under microscope, and compared to reference library.
  - The experience of the operator is important in obtaining accurate results.

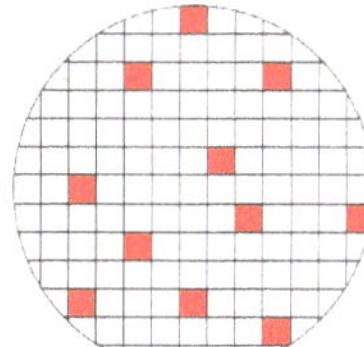

Fig. 8.32- Scaled Paper Membrane for Microscopic Particle Counting (Courtesy of Hydac)

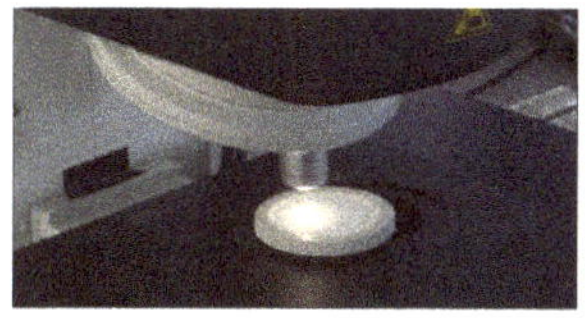

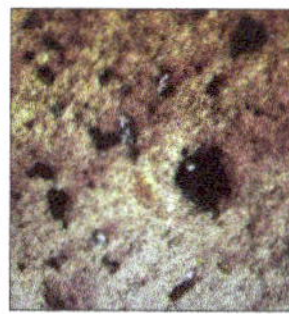

Fig. 8.33- Microscopic Particle Counting (Courtesy of MPFiltri)

64

64

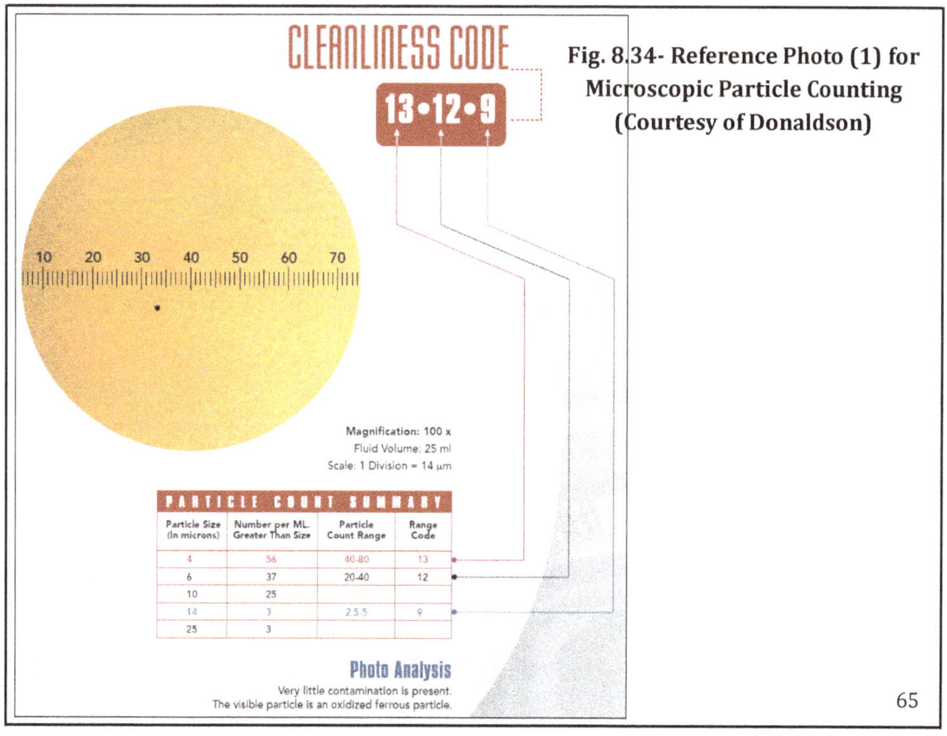

Particle Size (in microns)	Number per ML. Greater Than Size	Particle Count Range	Range Code
4	56	40-80	13
6	37	20-40	12
10	25		
14	3	2.5-5	9
25	3		

Fig. 8.34- Reference Photo (1) for Microscopic Particle Counting (Courtesy of Donaldson)

65

65

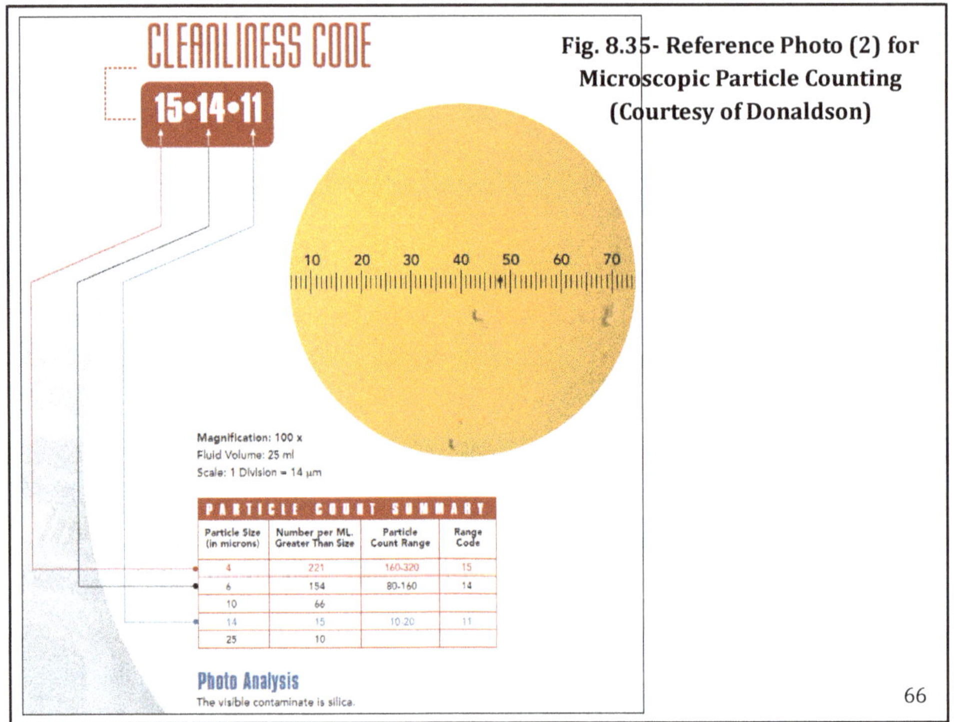

Particle Size (in microns)	Number per ML. Greater Than Size	Particle Count Range	Range Code
4	221	160-320	15
6	154	80-160	14
10	66		
14	15	10-20	11
25	10		

Fig. 8.35- Reference Photo (2) for Microscopic Particle Counting (Courtesy of Donaldson)

66

66

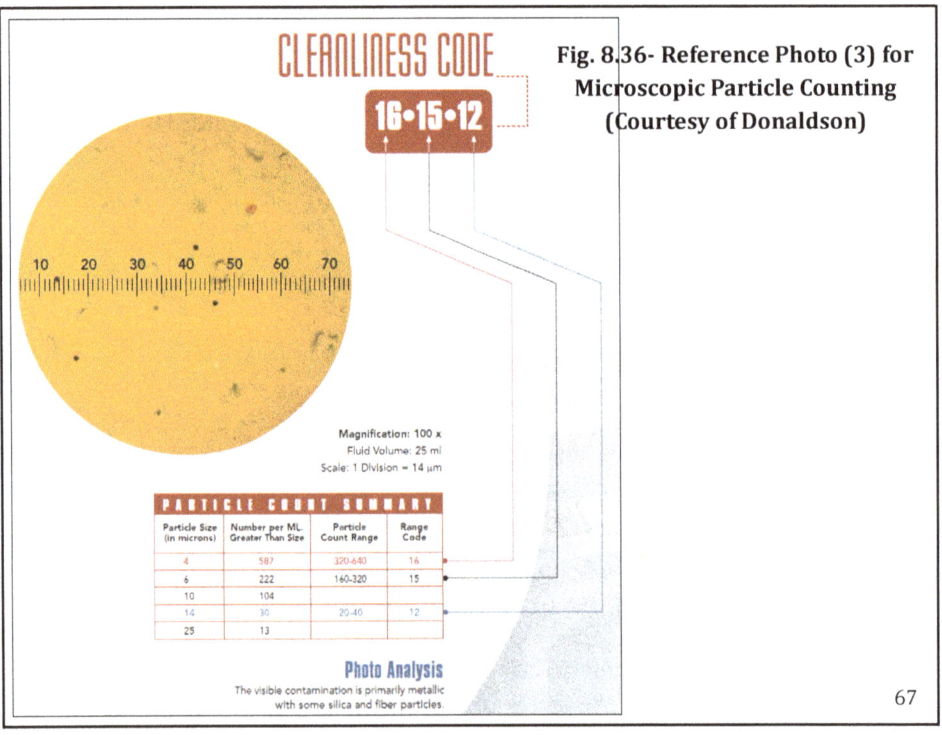

Particle Size (in microns)	Number per ML. Greater Than Size	Particle Count Range	Range Code
4	587	320-640	16
6	222	160-320	15
10	104		
14	30	20-40	12
25	13		

Fig. 8.36- Reference Photo (3) for Microscopic Particle Counting (Courtesy of Donaldson)

67

67

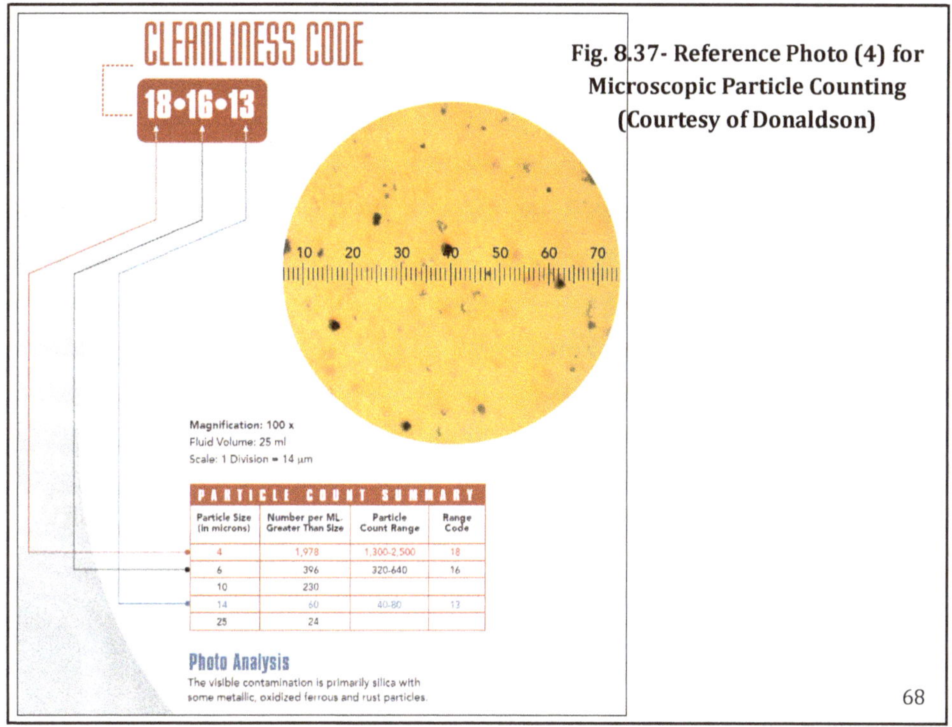

Particle Size (in microns)	Number per ML. Greater Than Size	Particle Count Range	Range Code
4	1,978	1,300-2,500	18
6	396	320-640	16
10	230		
14	60	40-80	13
25	24		

Fig. 8.37- Reference Photo (4) for Microscopic Particle Counting (Courtesy of Donaldson)

68

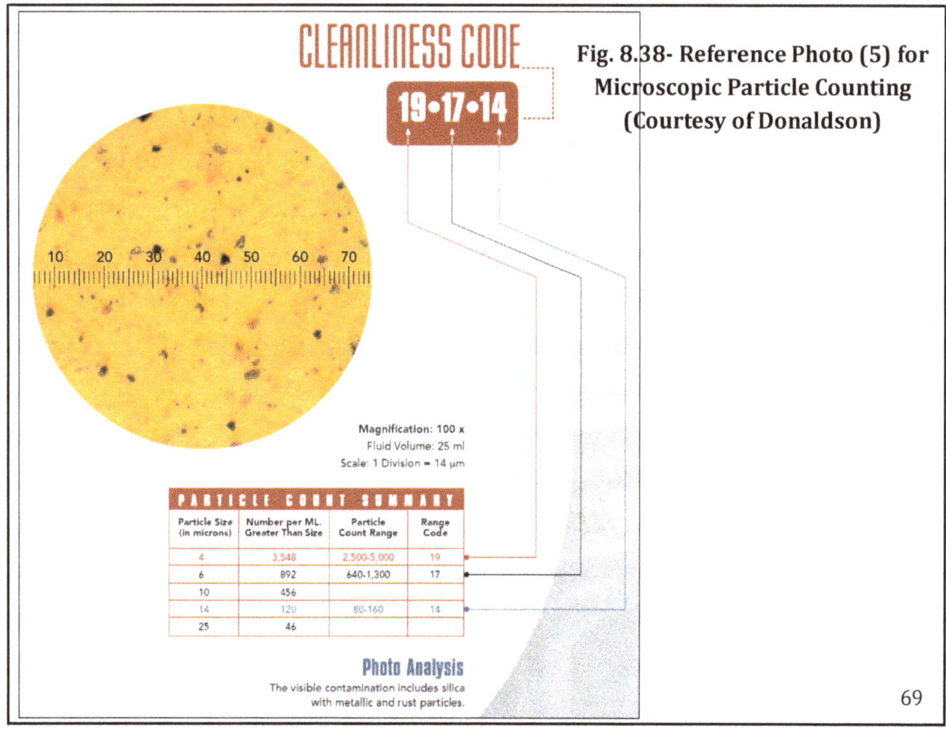

Particle Size (in microns)	Number per ML. Greater Than Size	Particle Count Range	Range Code
4	3,548	2,500-5,000	19
6	892	640-1,300	17
10	456		
14	120	80-160	14
25	46		

Fig. 8.38- Reference Photo (5) for Microscopic Particle Counting (Courtesy of Donaldson)

69

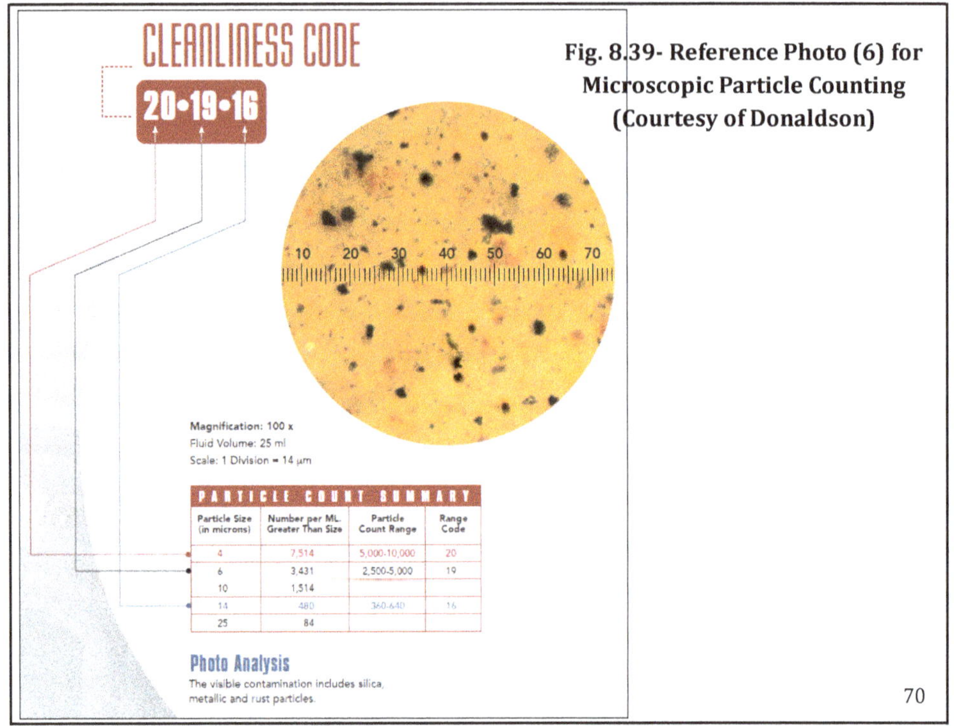

Particle Size (in microns)	Number per ML. Greater Than Size	Particle Count Range	Range Code
4	7,514	5,000-10,000	20
6	3,431	2,500-5,000	19
10	1,514		
14	480	360-640	16
25	84		

Fig. 8.39- Reference Photo (6) for Microscopic Particle Counting (Courtesy of Donaldson)

70

70

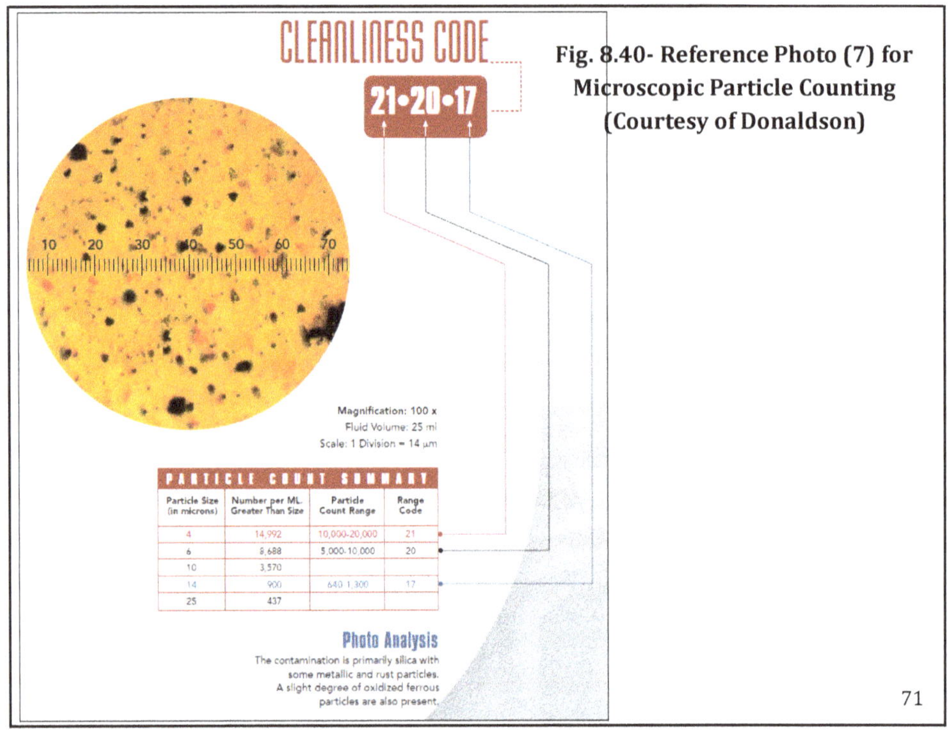

Particle Size (in microns)	Number per ML. Greater Than Size	Particle Count Range	Range Code
4	14,992	10,000-20,000	21
6	8,688	5,000-10,000	20
10	3,570		
14	900	640-1,300	17
25	437		

Fig. 8.40- Reference Photo (7) for Microscopic Particle Counting (Courtesy of Donaldson)

71

71

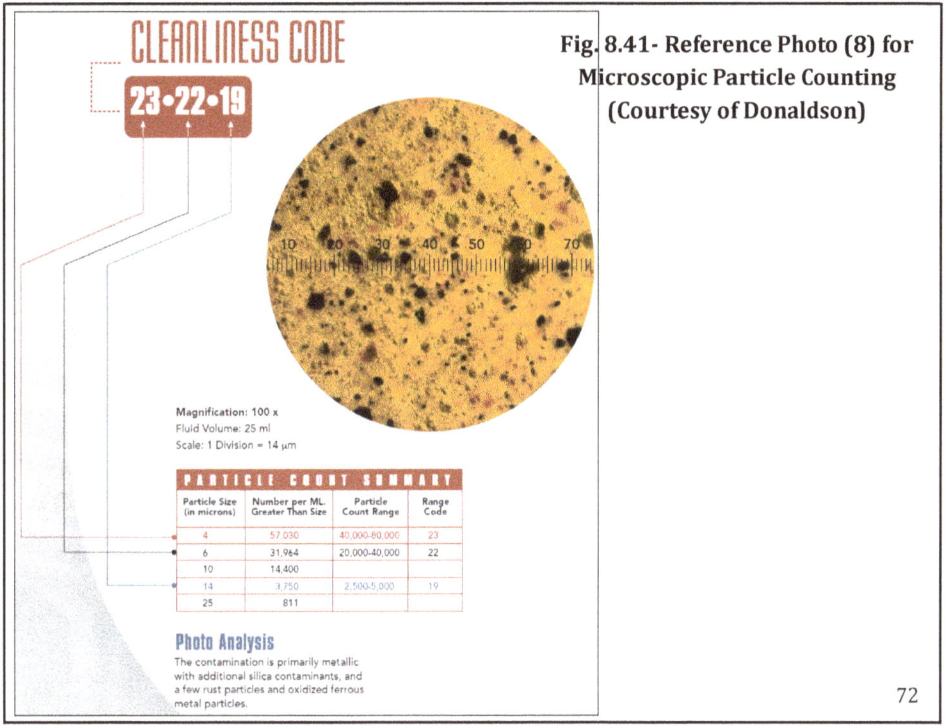

Particle Size (in microns)	Number per ML. Greater Than Size	Particle Count Range	Range Code
4	57,030	40,000-80,000	23
6	31,964	20,000-40,000	22
10	14,400		
14	3,750	2,500-5,000	19
25	811		

Fig. 8.41- Reference Photo (8) for Microscopic Particle Counting (Courtesy of Donaldson)

72

72

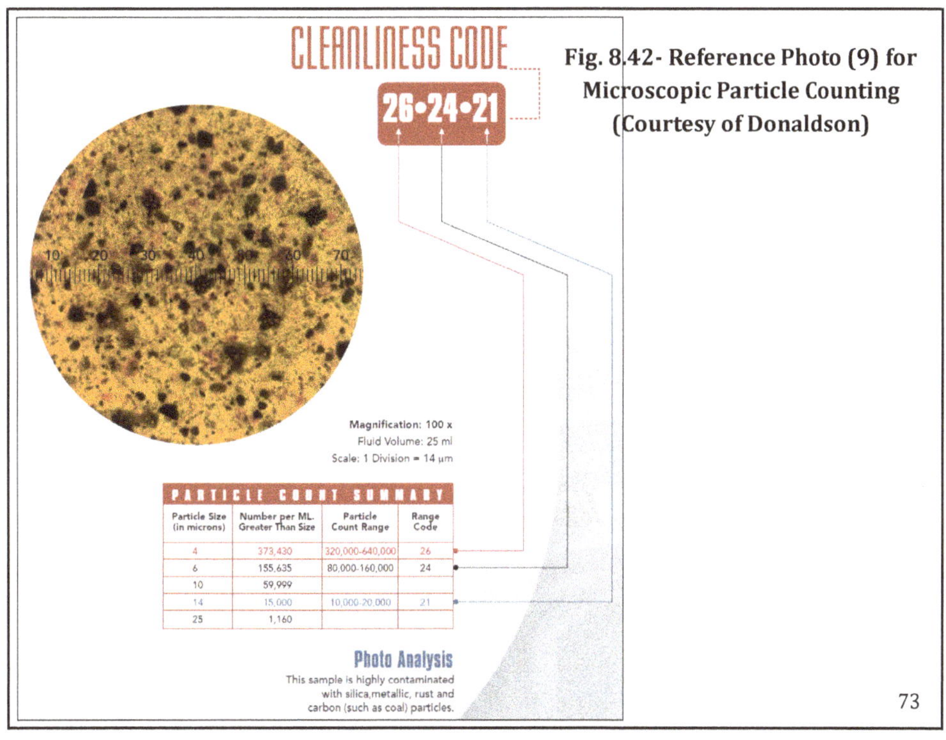

Particle Size (in microns)	Number per ML. Greater Than Size	Particle Count Range	Range Code
4	373,430	320,000-640,000	26
6	155,635	80,000-160,000	24
10	59,999		
14	15,000	10,000-20,000	21
25	1,160		

Fig. 8.42- Reference Photo (9) for Microscopic Particle Counting (Courtesy of Donaldson)

73

73

8.5.6- Automatic Particle Counting (ISO 11500:2008)

Automatic particle counters, based on its ability and design sophistication, are classified as follows:

- Particle Counters.
- Particle Monitors.
- Particle Classifiers.

8.5.6.1- Automatic Particle Counters (APC)

❑ **ISO 11500:2008**

- Specifies the method for determining the cleanliness level in hydraulic-fluid samples of clear, homogeneous, single-phase liquids.
- The method defined in this standard is the most common and accurate fluid analysis in use today.
- This standard is applicable to monitor:
  - Cleanliness level of fluids circulating in hydraulic systems.
  - Progress of a flushing operation.
  - Cleanliness level of support equipment, test rigs and packaged stock.

74

74

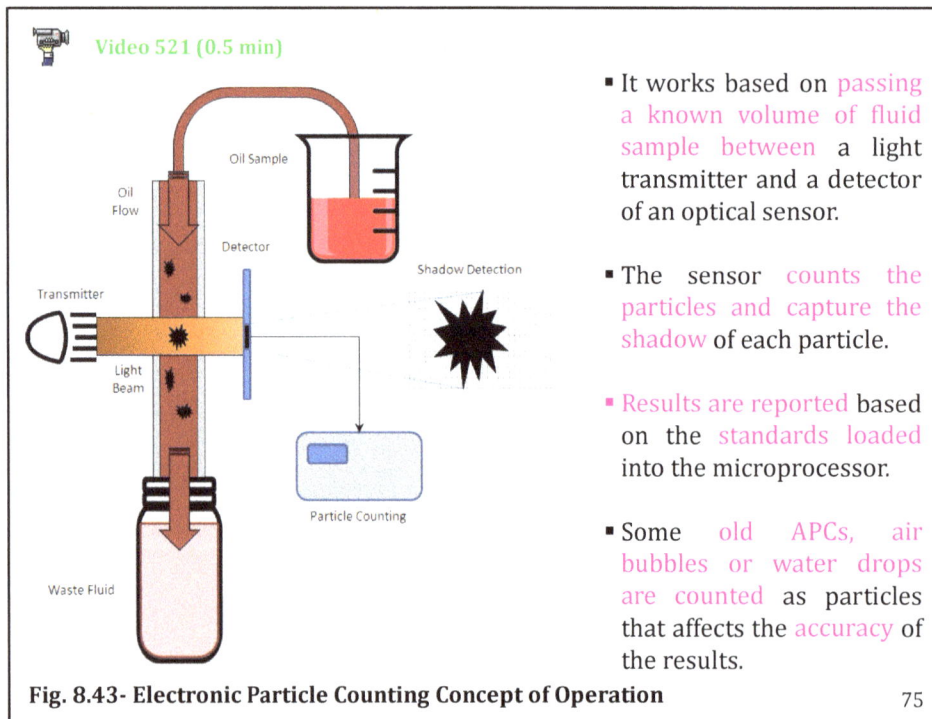

Video 521 (0.5 min)

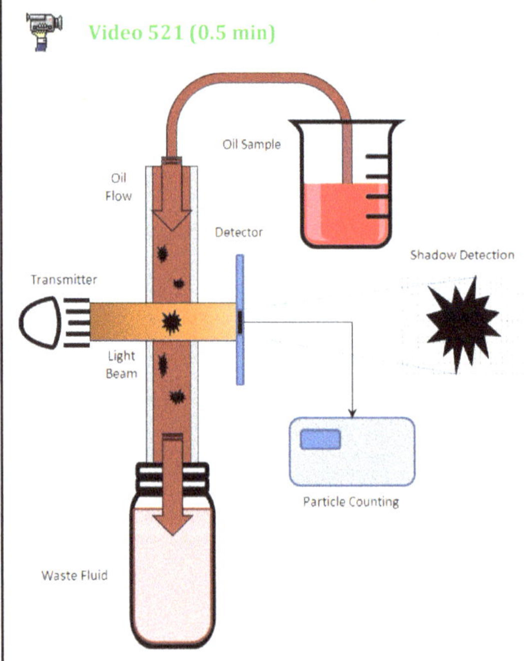

- It works based on passing a known volume of fluid sample between a light transmitter and a detector of an optical sensor.
- The sensor counts the particles and capture the shadow of each particle.
- Results are reported based on the standards loaded into the microprocessor.
- Some old APCs, air bubbles or water drops are counted as particles that affects the accuracy of the results.

Fig. 8.43- Electronic Particle Counting Concept of Operation

75

75

- Two standards for identifying a particle size as follows:
  - **ISO 4402:1991.** based on the longest cord within the body of the particle. It is **no longer in use** and replaced by the ISO 11171.
  - **ISO 11171:1999.** based on diameter of the equivalent circular area.

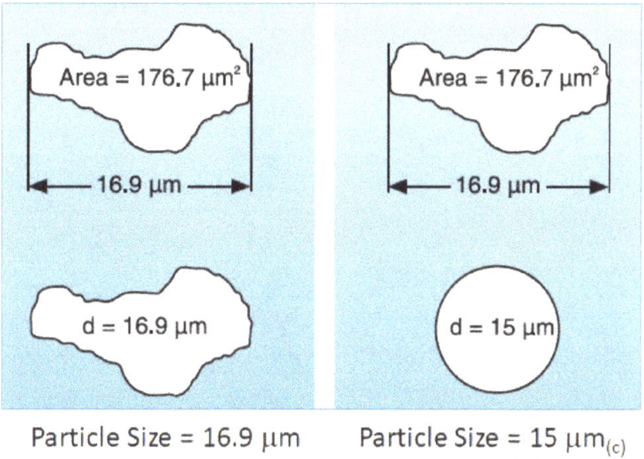

Particle Size = 16.9 µm
ISO 4402

Particle Size = 15 $\mu m_{(c)}$
ISO 11171

Fig. 8.44- Methods of Identifying the Particle Size (Courtesy of Hydac) 76

76

- An APC can be:
  - Desktop, work offline. Video 524 (1.5 min)
  - Portable, used in field. Video 523 (1.5 min)

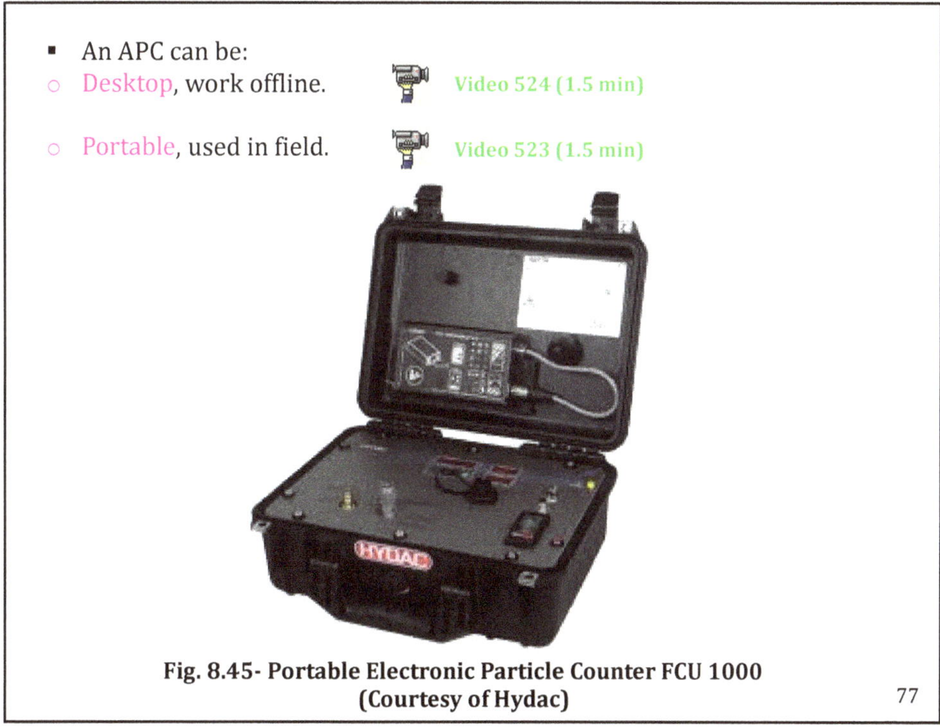

Fig. 8.45- Portable Electronic Particle Counter FCU 1000 (Courtesy of Hydac) 77

77

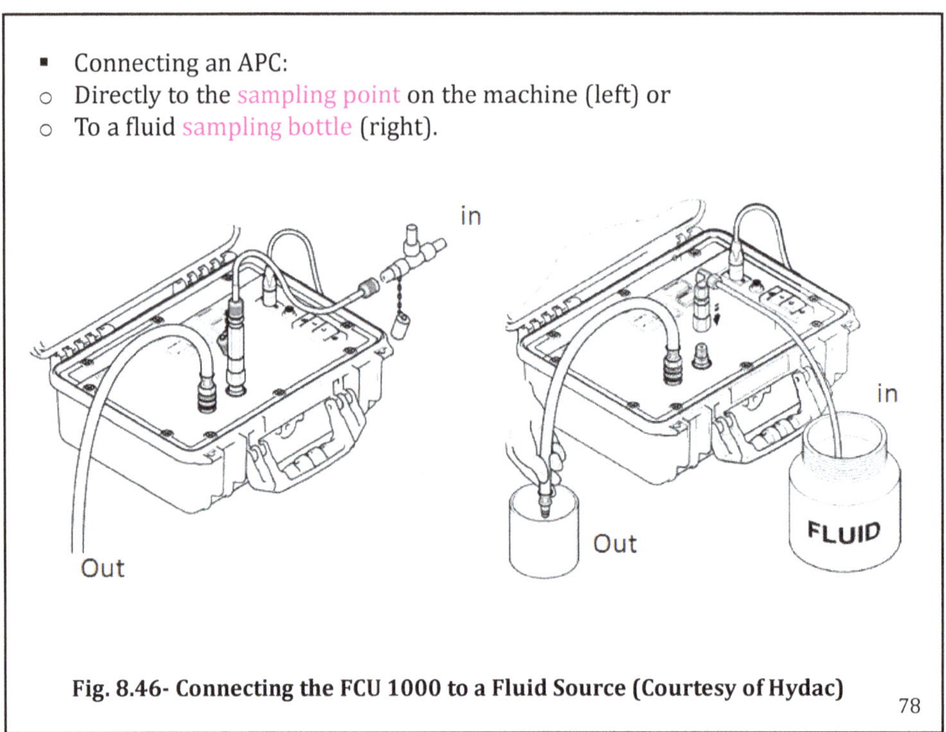

▪ Connecting an APC:
○ Directly to the sampling point on the machine (left) or
○ To a fluid sampling bottle (right).
in
in
Out
Out
FLUID

Fig. 8.46- Connecting the FCU 1000 to a Fluid Source (Courtesy of Hydac) 78

78

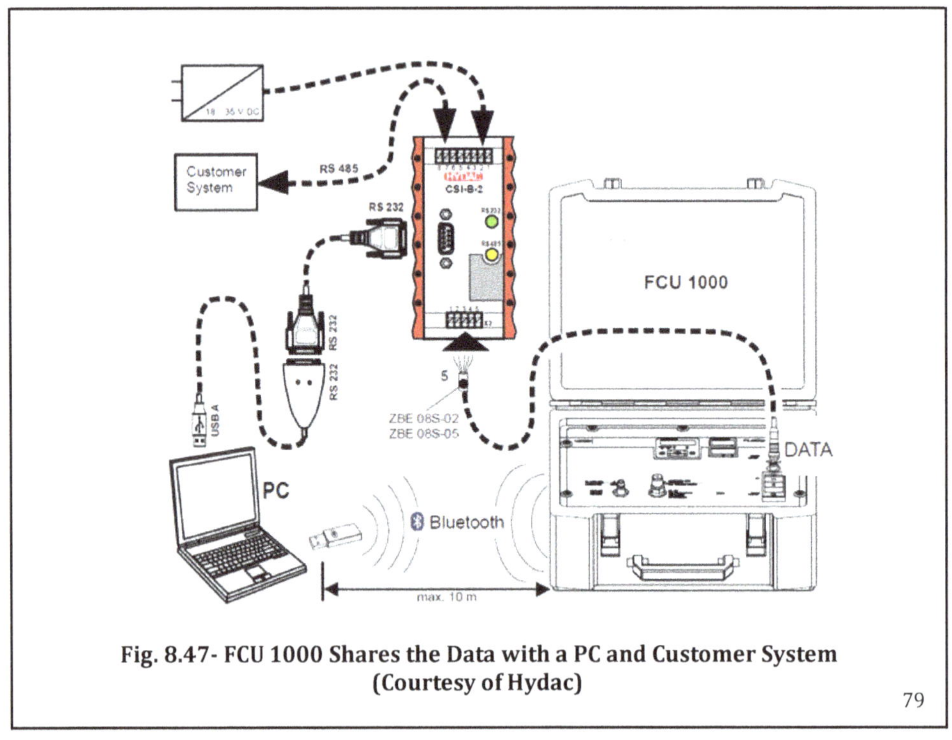

Fig. 8.47- FCU 1000 Shares the Data with a PC and Customer System (Courtesy of Hydac) 79

79

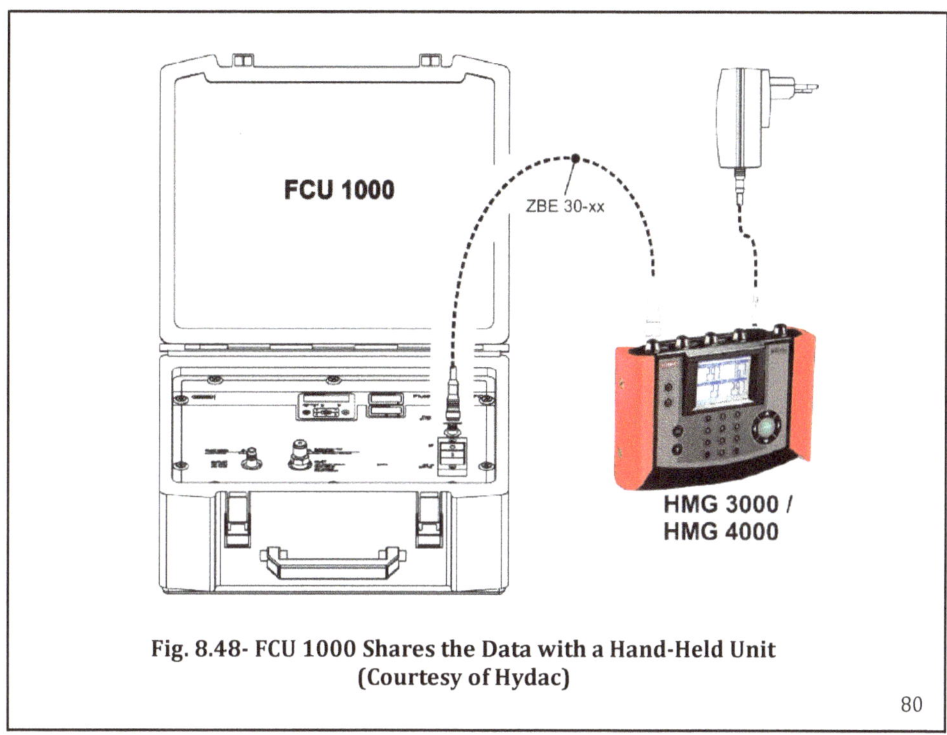

Fig. 8.48- FCU 1000 Shares the Data with a Hand-Held Unit (Courtesy of Hydac)

80

80

8.5.6.2- Particle Monitors

- Inline Contamination Monitor (ICM).
- Automatically measures and displays particulate contamination, moisture and temperature levels in various hydraulic fluids.
- It can be used as a standalone device or controlled by external PC

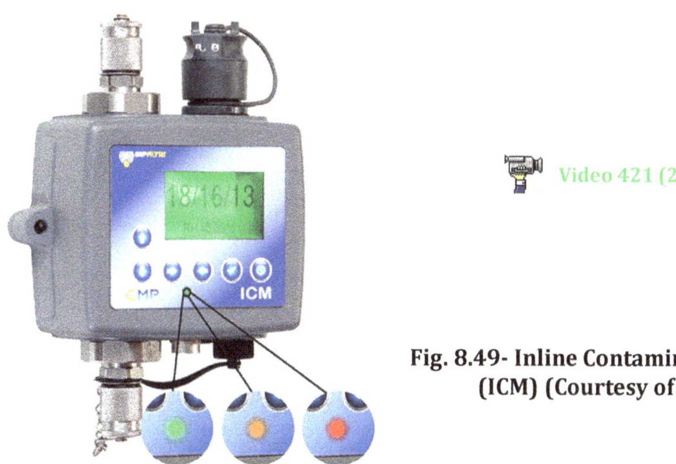

Video 421 (2 min)

Fig. 8.49- Inline Contamination Monitor (ICM) (Courtesy of MPFiltri

81

81

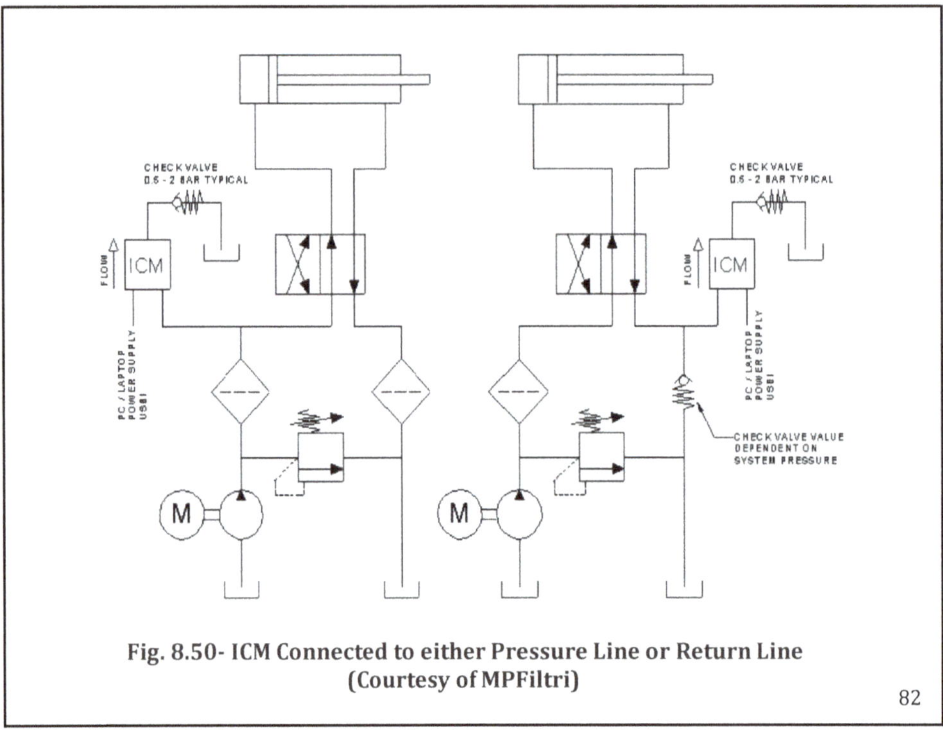

Fig. 8.50- ICM Connected to either Pressure Line or Return Line (Courtesy of MPFiltri)

82

82

8.5.6.3- Particle Classifiers

- *Particle Classifiers* are the high end of particle analysis devices.
- Obtain the cleanliness level.
- Capture images of the particles.
- Provide wear analysis based on the shadow of wear particles.
- Actual photos of various particles for more complete wear analysis.

83

83

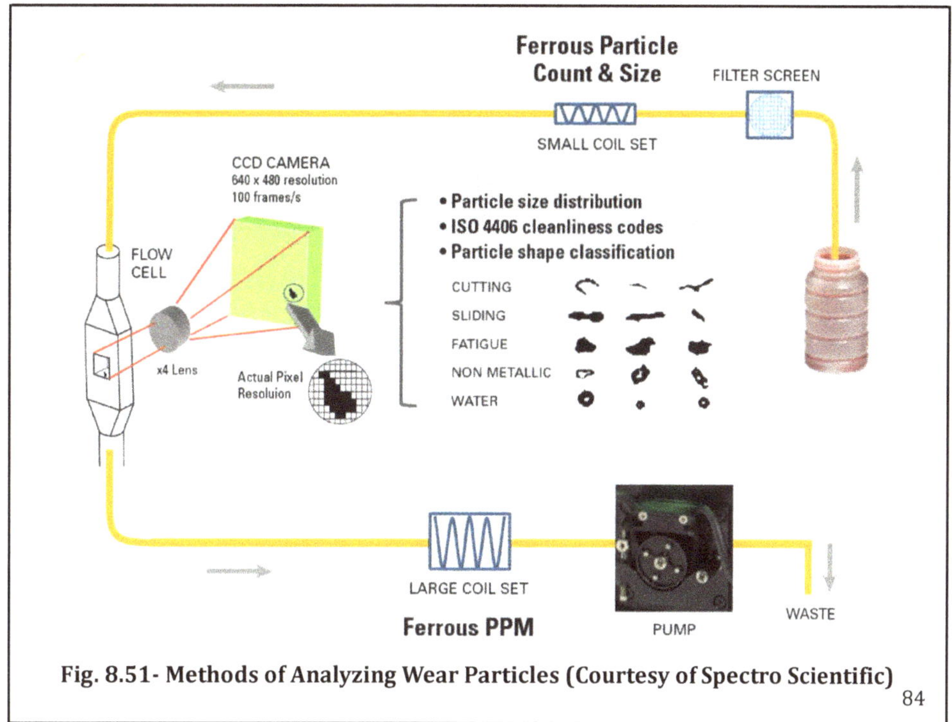

Fig. 8.51- Methods of Analyzing Wear Particles (Courtesy of Spectro Scientific) 84

84

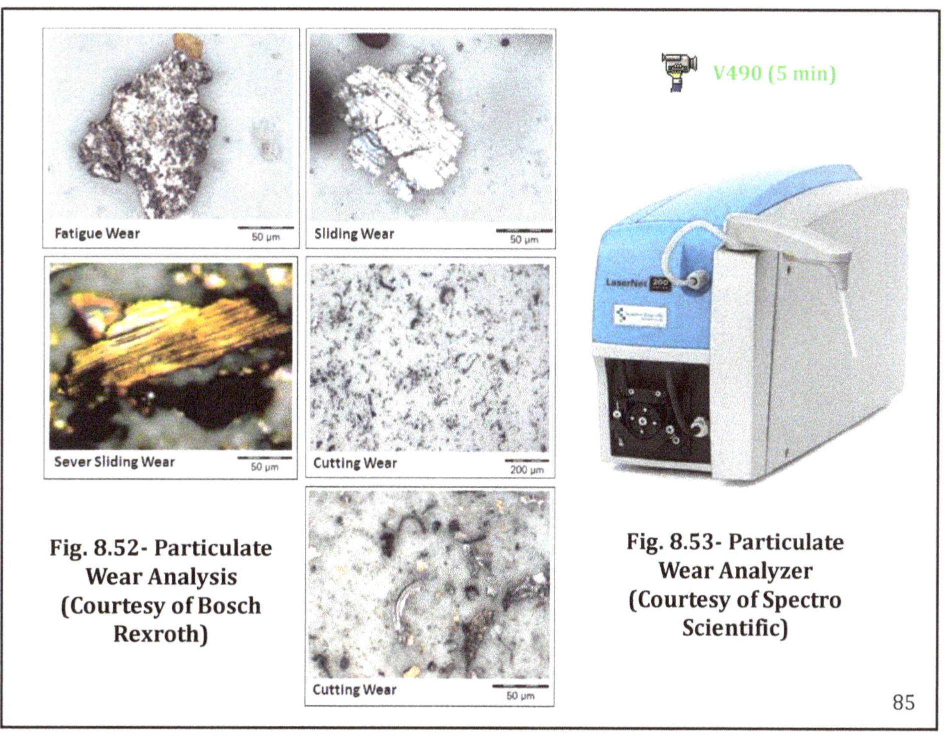

Fig. 8.52- Particulate Wear Analysis (Courtesy of Bosch Rexroth)

Fig. 8.53- Particulate Wear Analyzer (Courtesy of Spectro Scientific)

85

85

Sample of fluid is filtered through a paper filter. Color and shade determine the contamination level.

This test is known as?

A. Silt Index.
B. Microscopic (Visual) Particle Counting.
C. Gravimetric Analysis.
D. Patch Test.

86

86

Sample of fluid is pushed through a porous filter under constant pressure. The difference in pressure during passing the 1st and 2nd half of the fluid sample determines the contamination level.

This test is known as?

A. Silt Index.
B. Microscopic (Visual) Particle Counting.
C. Gravimetric Analysis.
D. Patch Test.

87

87

After a sample of fluid is passed through a paper filter of known weight, the paper is dried. Difference in the weight of the paper before and after passing the sample determines the contamination level.

This test is known as?

A. Silt Index.

B. Microscopic (Visual) Particle Counting.

C. Gravimetric Analysis.

D. Patch Test.

88

88

Sample of fluid is passed through a special filter membrane and then seen under an microscope. The counts and size of the particles determines the contamination level.

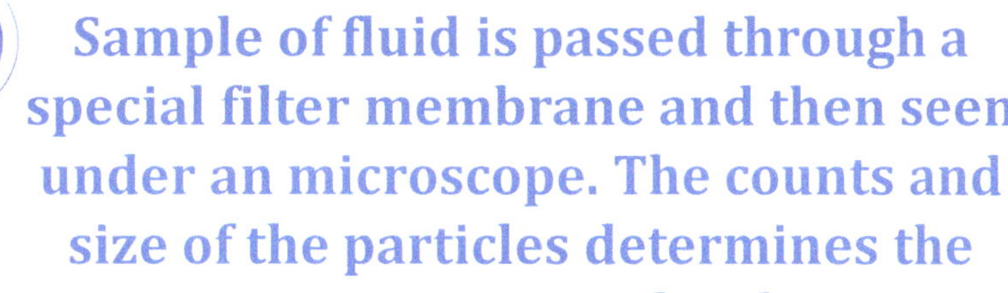

This test is known as?

A. Silt Index.

B. Microscopic (Visual) Particle Counting.

C. Gravimetric Analysis.

D. Patch Test.

89

89

8.5.6.4- Calibration of Automatic Particle Counters

- To ensure the accuracy of measurements, a particle counter should be frequently calibrated according to ISO 11171.
- The calibration method is to pass through the counter a hydraulic fluid sample of known volume and contamination class.
- Results from the counters under calibration are compared to the readings from a reference counter

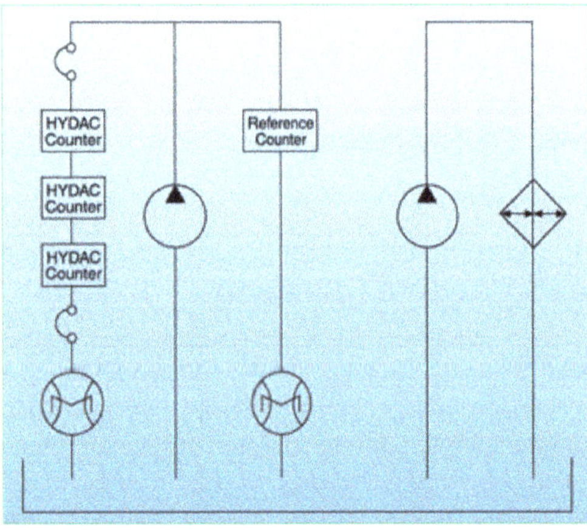

Fig. 8.54- Particle Counter Calibration (Courtesy of Hydac)

90

90

❑ **A Special Test Dust**

- **ACFTD (Air Cleaner Fine Test Dust):**
  - Was the first test dust from Arizona (USA).
  - Made of ground silica granules ranging from 0 to 100 µm.
  - Marketed in the 1960s and is used until 1990 under ISO 4402.
  - Stopped being used by 1990s and replaced by MTD.

- **MTD (ISO Medium Test Dust):**
  - Starting 1997, MTD test dust was considered for use in ISO 11171:1999
  - ISO 12103 led to the following four different categories of test dust:
    - ❖ ISO 12103-A1 UFT (ISO Ultra Fine Test Dust).
    - ❖ ISO 12103-A2 FTD (ISO Fine Test Dust).
    - ❖ ISO 12103-A3 MTD (ISO Medium Test Dust).
    - ❖ ISO 12103-A4 CTD (ISO Course Test Dust).

91

91

The **ISO 11943:1999** calibration standard covers the calibration of automatic online particle counters for fluids using MTD 12103 Dust

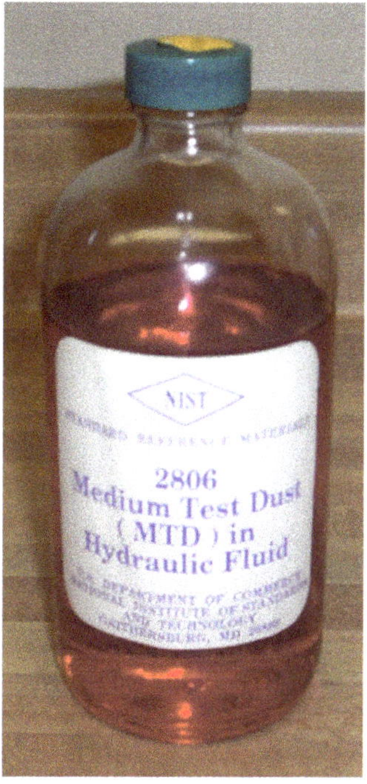

Fig. 8.55- Fluid Sample for Particle Counter Calibration

Fig. 8.56- Amount of Dirt to Create 10 Gallons of ISO 20/18/13 (Courtesy of MSOE)

92

92

8.6- Interpretation of Fluid Analysis Report

- Replacing oil based on time or operation hours is expensive.
- Replacing oil based on oil condition is the best.
- As a rule of thumb, the additive level in used oil has to be at least 70% of the additive level of new oil (Ref. Noria Corporation).
- Reading a fluid analysis report can be an overwhelming and sometimes seemingly impossible task without an understanding of the fluid properties and the various types of contamination.

93

93

A good oil analysis report will answer the following key questions:

- Is the fluid suitable for further use?
- What level of contaminants are evident?
- Are the base fluid properties and additives still intact?
- Has a critical wear situation developed?
- Are seals, breathers and filters operating effectively?
- Is fluid degradation speeding up?
- Could a severe varnish problem occur soon?

At a minimum, an oil analysis should include:

- Viscosity.
- Particle counts and ISO Code 4406.
- Moisture/water content in ppm.
- Acidity level.
- Element analysis (wear and additives level).

94

94

SPECTRUM ANALYSIS		
Wear Metals And Additives	**ppm By Weight**	**Status**
Iron	120.0	H
Copper	510.0	H
Chromium	< 1.0	N
Lead	< 1.0	N
Aluminum	1.0	N
Tin	< 1.0	N
Silicon	< 1.0	N
Zinc	423.0	N
Magnesium	< 1.0	N
Calcium	540.0	H
Phosphorus	10.0	L
Barium	1.0	N
Boron	< 1.0	N
Sodium	< 1.0	N
Molybdenum	< 1.0	N
Silver	< 1.0	N
Nickel	< 1.0	N
Titanium	< 1.0	N
Maganese	< 1.0	N
Antimony	< 1.0	N
L = LOW N= NORMAL H = HIGH		

Viscosity Analysis ASTM D445	
SSU @ 100° F: 100.0	cst 40° C: 21.6

Water Analysis ASTM D1744
Water Content (ppm): 101.0

Neutralization Analysis - ASTM D974
TAN: 0.1

Remarks
1. Please check spectro-metric analysis abnormal conditions

Table 8.16- Fluid Analysis Report, Example 1 (Excerpted from Lightening Reference Handbook)

95

95

In testing a hydraulic fluid sample, the amount of Silicon found was 30 ppm. This is considered?

A. Acceptable level.
B. Cautiously acceptable.
C. Critical condition and action need to be taken.
D. We should not care about the silicon content in hydraulic fluids.

96

96

Oil analysis log book

Parameter	Baseline	Caution	Critical
Particle count ISO 4406	15/13/10 (pre-filtered)	17/15/12	19/17/15
Viscosity (cSt)	32	low 29 high 35	low 25 high 38
Acid number (AN, mg KOH/g)	0.5	1.0 - 1.5	above 1.5
Moisture (KF in ppm)	100	200 - 300	above 300
Elements (in ppm) Fe	7	10 - 15	above 15
Al	2	20 - 30	above 30
Si	5	10 - 15	above 15
Cu	5	30 - 40	above 40
P	300	220	150 and less
Zn	200	150	100 and less
Oxidation (FTIR)	1	5	above 10
Ferrous Density (PQ, WPC, DR)	-	15	above 20

Table 8.17- Example of Analysis Log Book (Courtesy of C.C. Jensen Inc.)

97

97

- Some of the fluid analysis reports use color-coded sliding scale.
- It tells at a glance whether the analysis results are in the normal range or the overall degree to which problems have been detected.

1. → At least one or more items developing minor problem.
2. → Trend is developing.
3. → Simple maintenance and/or diagnostics are recommended.
4. → Failure is likely going to occur if maintenance is not performed.

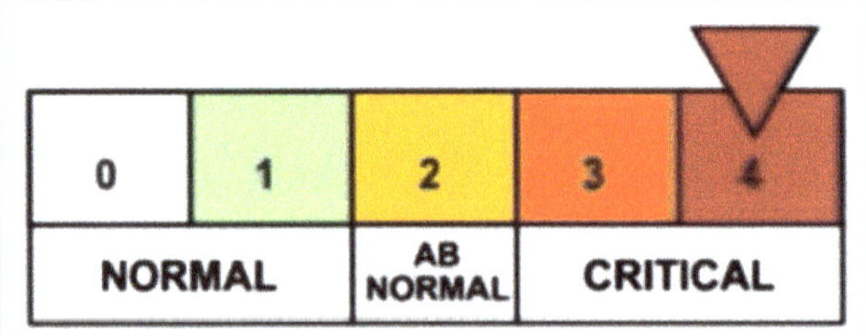

Fig. 8.57- Color-Coded Sliding Scale (Courtesy of Donaldson)

Video 415 (3 min)

98

98

The most accurate and quantitative particle analysis is.

A. Automatic Particle Counting.

B. Microscopic (Visual) Particle Counting.

C. Gravimetric Analysis.

D. Patch Test.

99

99

Method	Units	Advantages	Limitations
Patch Test	visual comparison	Fast and qualitative	Not quantitative
Gravimetric Analysis	mg/Liter	Identifies the total amount of contamination	Can't identify the particle size
Microscopic Particle Counting	number/ml	Provides accurate size and number	Sample preparation and time
Electronic Particle Counting	number/ml	Fast and repeatable results	Counts water as particles

Table 8.18- Features of Contamination Tests

100

100

Chapter 8 Reviews

1. Recommended Sampling Interval for hydraulic equipment in aerospace industry?
   A. 100 hours.
   B. Quarterly.
   C. Semi-annually.
   D. Annually.

2. Among the places shown below, the correct sampling location is?
   A. After the return filter.
   B. After the heat exchanger.
   C. At the inlet of the pump.
   D. On the return line at a point of a turbulent flow.

3. The following counts of particles of sizes were found in a 1 milliliter of a hydraulic sample (2500 particles of size > 4 microns, 640 particles of size > 6 microns, and 80 particles of size > 14 microns. What is the corresponding cleanliness level?
   A. 17/15/12.
   B. 18/16/13.
   C. 19/17/14.
   D. 20/18/15.

4. As per the ISO standard ISO 4406-1999, the following particles concentration were found (1000 particles of size > 4 microns, 200 particles of size > 6 microns, and 30 particles of size > 14 microns. What is the corresponding cleanliness level?
   A. 17/15/12 per ISO 4406/1999.
   B. Class 6 per NAS 1638.
   C. Class 7 per SAE AS 4059.
   D. All of the above.

5. Sample of fluid is filtered through a paper filter. Color and shade determine the contamination level. This test is known as:
   A. Silt Index.
   B. Microscopic (Visual) Particle Counting.
   C. Gravimetric Analysis.
   D. Patch Test.

6. Sample of fluid is pushed through a porous filter under constant pressure. The difference in the pressure during passing the 1st and 2nd half of the fluid sample determines the contamination level. This test is known as:
   A. Silt Index.
   B. Microscopic (Visual) Particle Counting.
   C. Gravimetric Analysis.
   D. Patch Test.

7. After a sample of fluid is passed through a paper filter of known weight, the paper is dried. Difference in the weight of the paper before and after passing the sample determines the contamination level. This test is known as:
   A. Silt Index.
   B. Microscopic (Visual) Particle Counting.
   C. Gravimetric Analysis.
   D. Patch Test.

8. Sample of fluid is passed through a special filter membrane and then seen under an optical sensor. The counts and size of the particles determines the contamination level. This test is known as:
   A. Silt Index.
   B. Microscopic (Visual) Particle Counting.
   C. Gravimetric Analysis.
   D. Patch Test.

9. In testing a hydraulic fluid sample, the amount of Silicon found was 30 ppm. This is considered:
   A. Acceptable level.
   B. Cautiously acceptable.
   C. Critical condition and action need to be taken.
   D. We should not care about the silicon content in hydraulic fluids.

10. The most accurate and quantitative particle analysis is:
   A. Automatic Particle Counting.
   B. Microscopic (Visual) Particle Counting.
   C. Gravimetric Analysis.
   D. Patch Test.

Chapter 8 Assignment

Student Name: -- Student ID: ------------------

Date: -- Score: ------------------------

A: find the cleanliness level (as per ISO 4406-1999) for a sample that has the following counts of particles:

- 1000 particle of size 5 microns
- 500 particle of size 8 microns
- 100 particle of size 15 microns

Chapter 9
Hydraulic Filters Performance Ratings

Objectives:

This chapters discusses the standard methods for evaluating the performance of a hydraulic filter. The purpose is to make the reader aware of the factors based on which type of filter may be more suitable for a specific application.

0

0

Brief Contents:

1

1

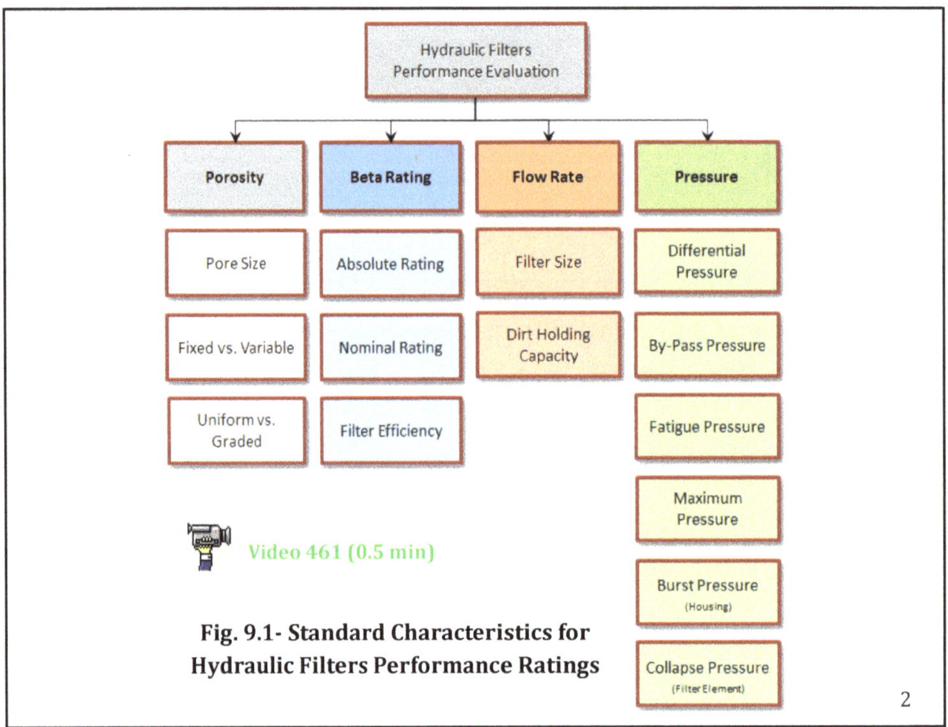

Fig. 9.1- Standard Characteristics for Hydraulic Filters Performance Ratings

2

2

9.1- Porosity

Filtration is defined as the physical mechanical process of retention or "capture" of particles in a fluid by passing the fluid through a porous filter medium.

9.1.1- Filter Porosity

- Is defined as how many pores per unit area of the filter medium.
- Also known as Mesh Size.

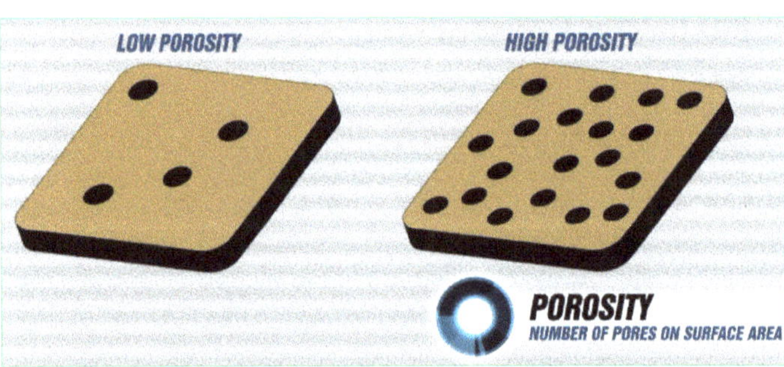

Fig. 9.2- Filter Porosity (Courtesy of Noria Corporation)

3

3

9.1.2- Pore Size

- *Pore Size* of a filter is the actual size of the openings in the surface of the filter medium. Pore size is measured in microns.
- Also Known as Micron Size.

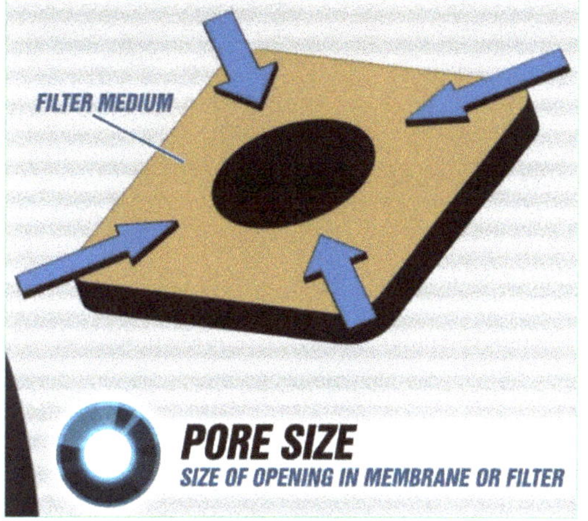

Fig. 9.3- Pore Size of a Filter Medium (Courtesy of Noria Corporation)

4

4

9.1.2.1- Thin vs. Thick Fibers of a Filter Medium

Based on the fiber size, constructing the filter media from thin fibers allows:

- More pores.
- High dirt holding capacity.
- Less differential pressure.

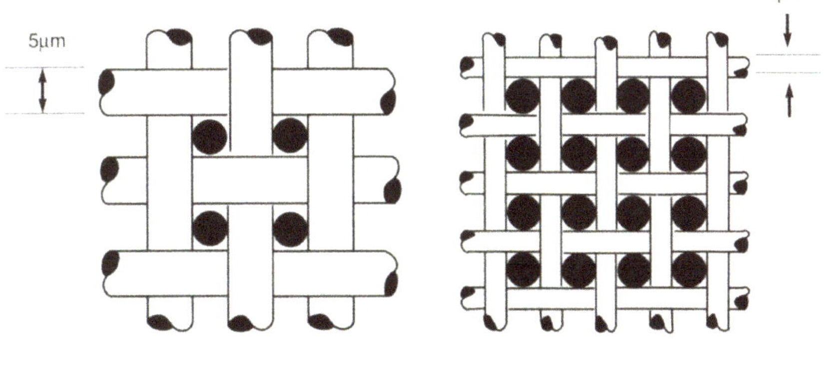

Fig. 9.4- Thin vs. Thick Fibers of a Filter Medium (Courtesy of Pall)

5

5

Glass Fibers that have the smallest pore sizes.

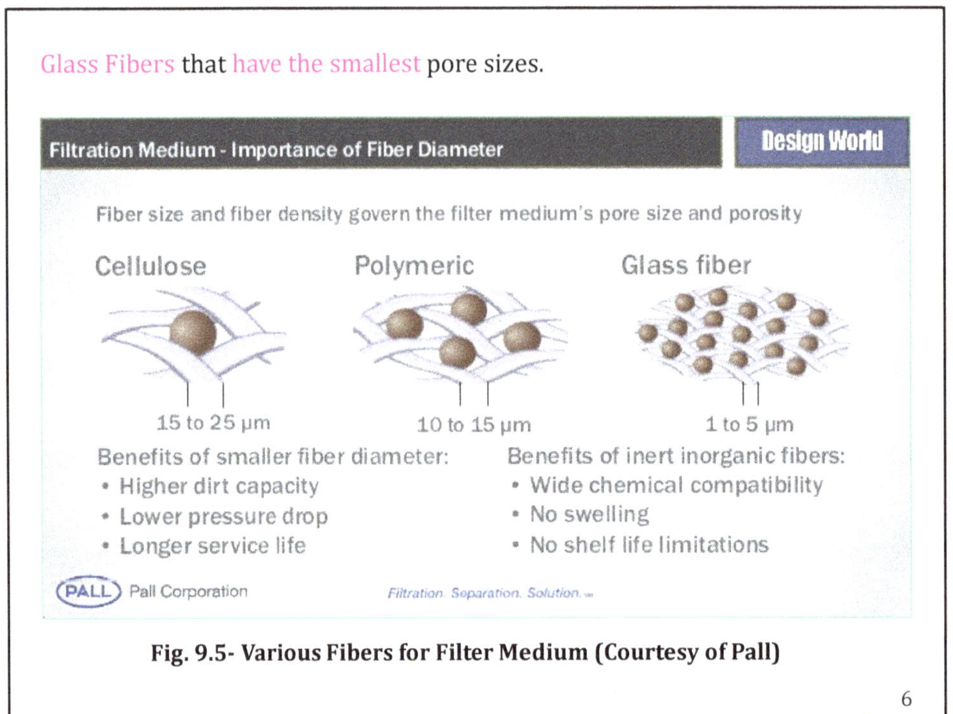

Fig. 9.5- Various Fibers for Filter Medium (Courtesy of Pall)

6

6

9.1.2.2- Fixed vs. Non-Fixed Pore Size.

- Based on the method of bonding the fibers together on each layer:
- Fixed Pore Media:
  - Fibers are bonded with specifically formulated resin.
  - Resist deterioration from pressure and flow fluctuations, temperature and aging conditions.

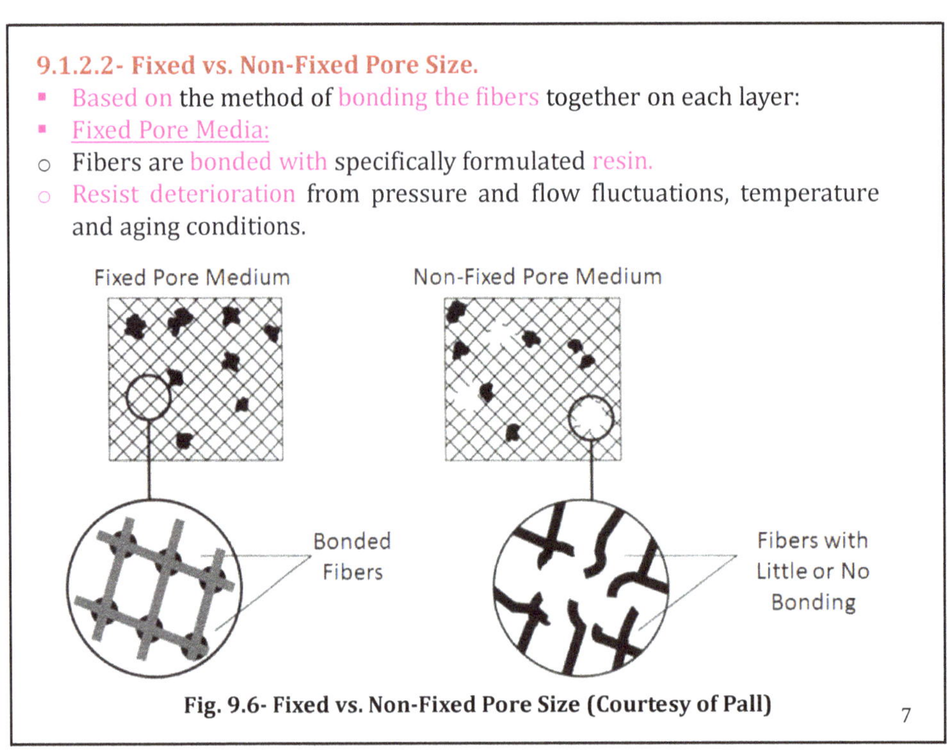

Fig. 9.6- Fixed vs. Non-Fixed Pore Size (Courtesy of Pall)

7

7

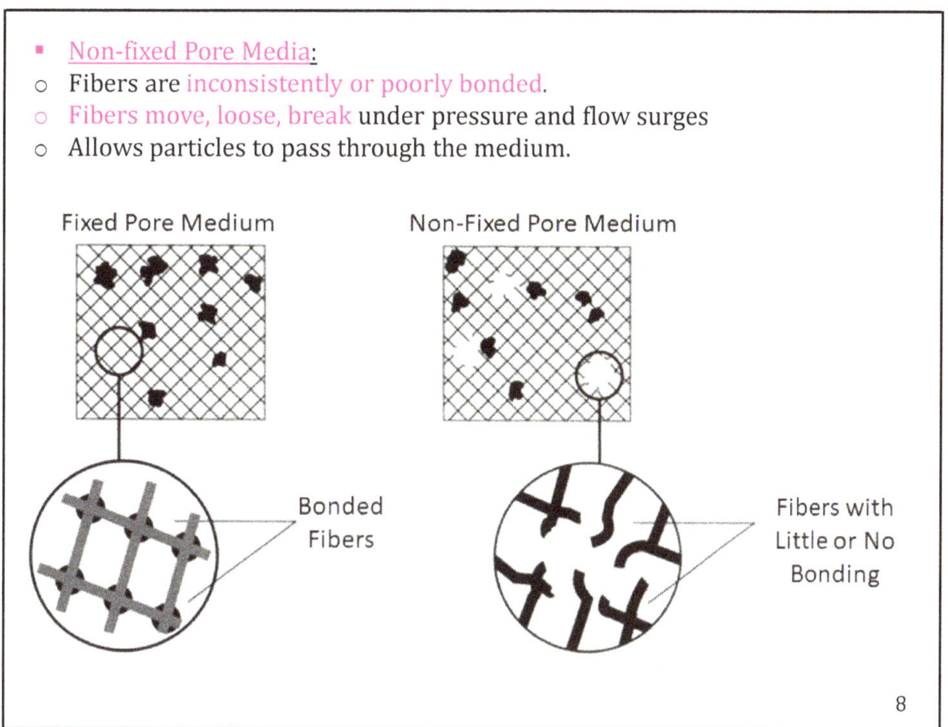

▪ Non-fixed Pore Media:
○ Fibers are inconsistently or poorly bonded.
○ Fibers move, loose, break under pressure and flow surges
○ Allows particles to pass through the medium.
Fixed Pore Medium
Non-Fixed Pore Medium
Bonded Fibers
Fibers with Little or No Bonding
8

8

9.1.2.3- Uniform vs. Graded Pore Size.

- Based on the pore size along the depth of the filter media:
  - Uniform Pore Size or Graded Pore Size:
  - Graded Pore Size (larger size on the surface).
  - Graded Pore Size allows holding more dirt, but it causes higher pressure-drop across the filter media.

Flow

Flow

Uniform Pore Medium

Graded Pore Medium

Fig. 9.7- Uniform vs. Graded Pore Size (Courtesy of Pall) 9

9

9.2- Beta Rating

Definition: the ability of a filter to separate particulate contaminants larger than certain size from a fluid.

Video 251 (0.5 min)

9.2.1- Multipass Test Performance Test (ISO 16889)

- Performed under controlled laboratory conditions:
  - Hydraulic fluid (Mil-H-5606)
  - Uniform amount of contaminant (such as ISO 12103-A3 MTD)
  - Controlled working temperature.

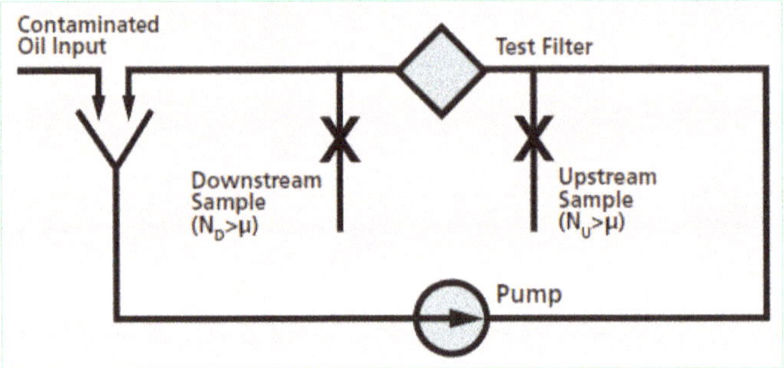

Contaminated Oil Input
Test Filter
Downstream Sample ($N_D > \mu$)
Upstream Sample ($N_U > \mu$)
Pump

Fig. 9.8- Multipass Test Performance Test (ISO 16889) 10

10

9.2.2- Beta Ratio Calculation

$$\beta_x = \frac{N_U}{N_D} \qquad 9.1$$

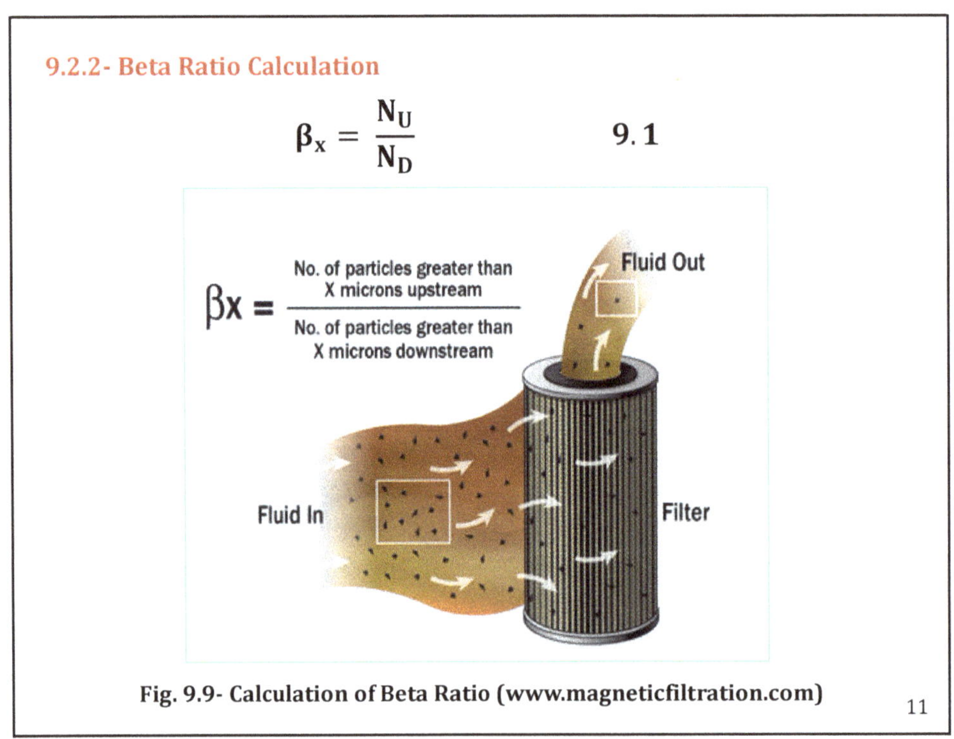

Fig. 9.9- Calculation of Beta Ratio (www.magneticfiltration.com) 11

11

Video 478 (2 min)

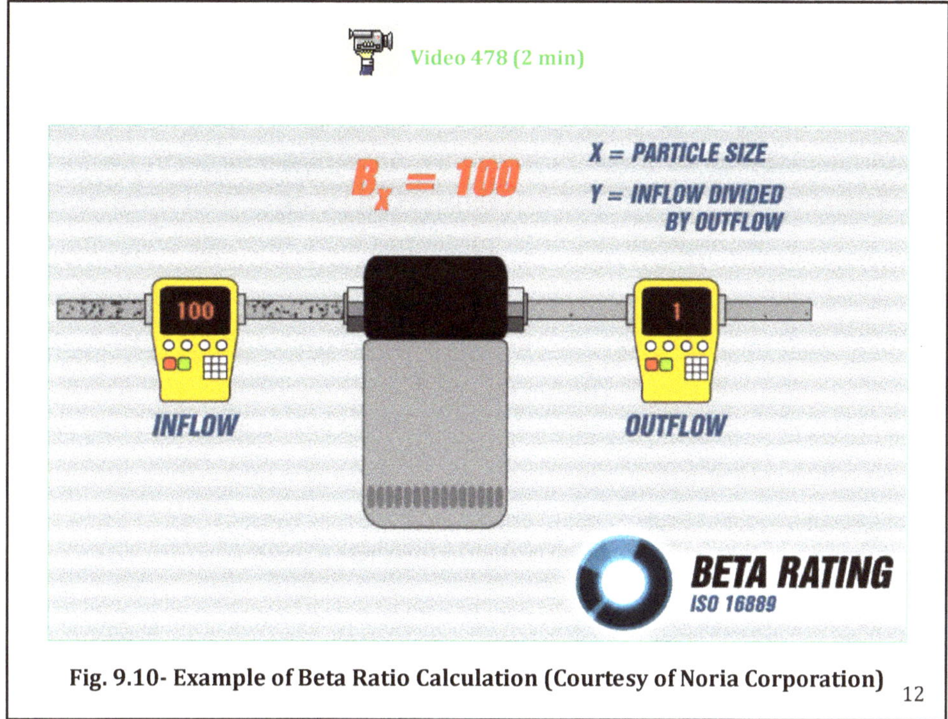

Fig. 9.10- Example of Beta Ratio Calculation (Courtesy of Noria Corporation) 12

12

In a Multipass Test, a filter was found to retain 800 out of 1000 particles larger than 4 microns. What is the beta ratio?

A. $\beta_{1.25=4}$
B. $\beta_{4=1.25}$
C. $\beta_{5=4}$
D. $\beta_{4=5}$

13

13

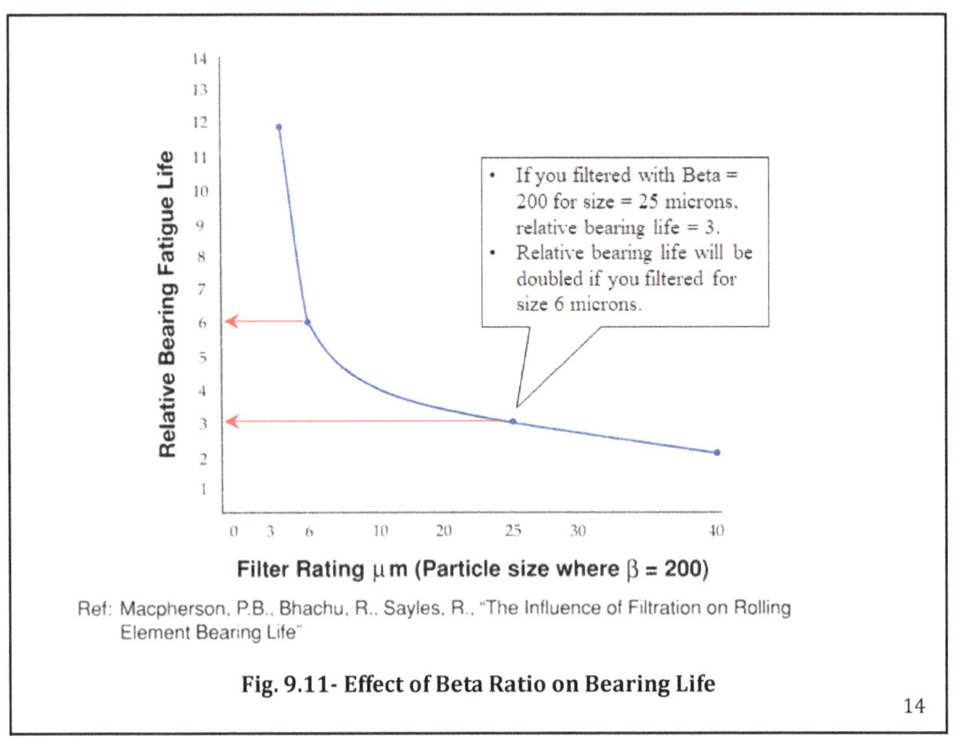

Fig. 9.11- Effect of Beta Ratio on Bearing Life

14

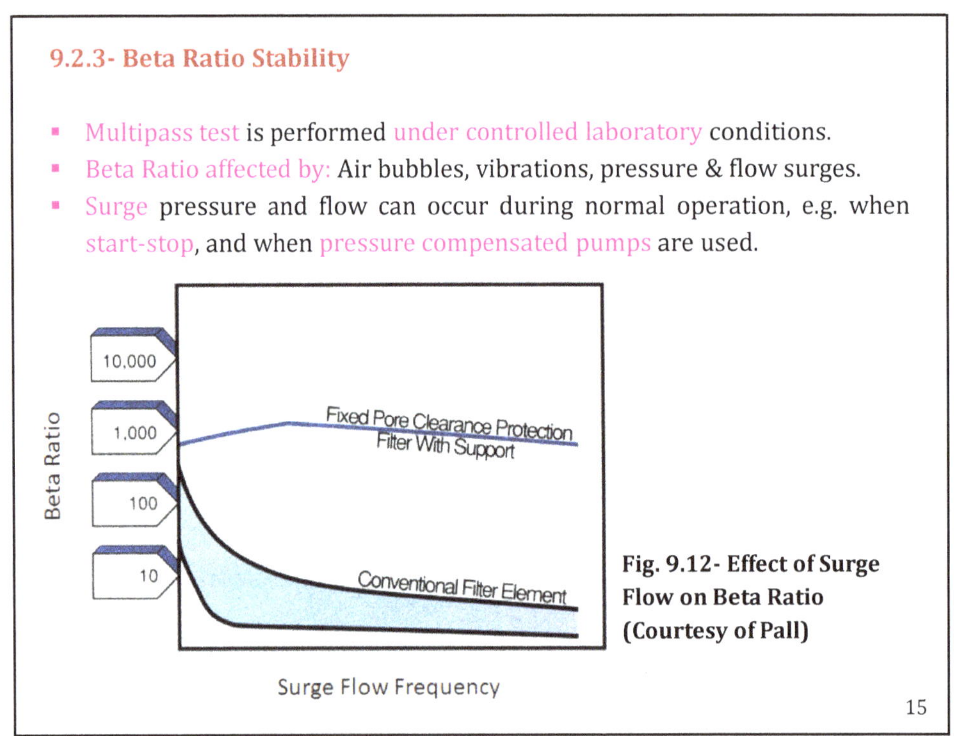

9.2.3- Beta Ratio Stability
▪ Multipass test is performed under controlled laboratory conditions.
▪ Beta Ratio affected by: Air bubbles, vibrations, pressure & flow surges.
▪ Surge pressure and flow can occur during normal operation, e.g. when start-stop, and when pressure compensated pumps are used.
Beta Ratio
10,000
1,000
100
10
Fixed Pore Clearance Protection Filter With Support
Conventional Filter Element
Surge Flow Frequency
15

Fig. 9.12- Effect of Surge Flow on Beta Ratio (Courtesy of Pall)

15

9.3- Filter Efficiency

Video 464 (1 min)

$$E_x = \left[1 - \frac{N_D}{N_U}\right] \times 100 = \left[1 - \frac{1}{\beta_x}\right] \times 100 = \left[\frac{\beta_x - 1}{\beta_x}\right] \times 100 \qquad 9.2$$

- Note: beta value [200 – 1000] ~ [efficiency change is very small (0.4%).]
- Therefore, selecting filters based on β > 200 is somewhat deceiving.

Beta Ratio

Upstream Particles: 100,000 > (x) microns

Downstream Particles			Beta Ratio (x)	Efficiency (x)
50,000	100,000 / 50,000	=	2	50.0%
5,000	100,000 / 5,000	=	20	95.0%
1,333	100,000 / 1,333	=	75	98.7%
1,000	100,000 / 1,000	=	100	99.0%
500	100,000 / 500	=	200	99.5%
100	100,000 / 100	=	1000	99.9%

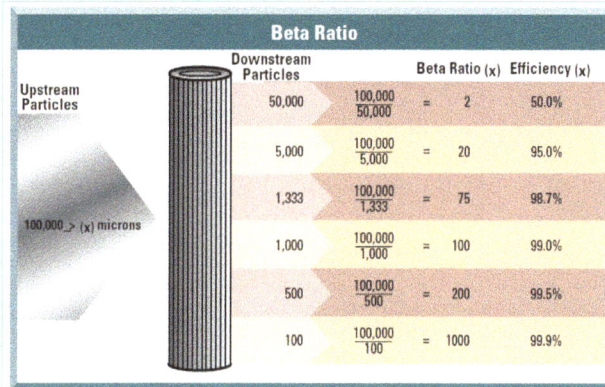

Fig. 9.13- Filter Efficiency (Courtesy of Parker) 16

16

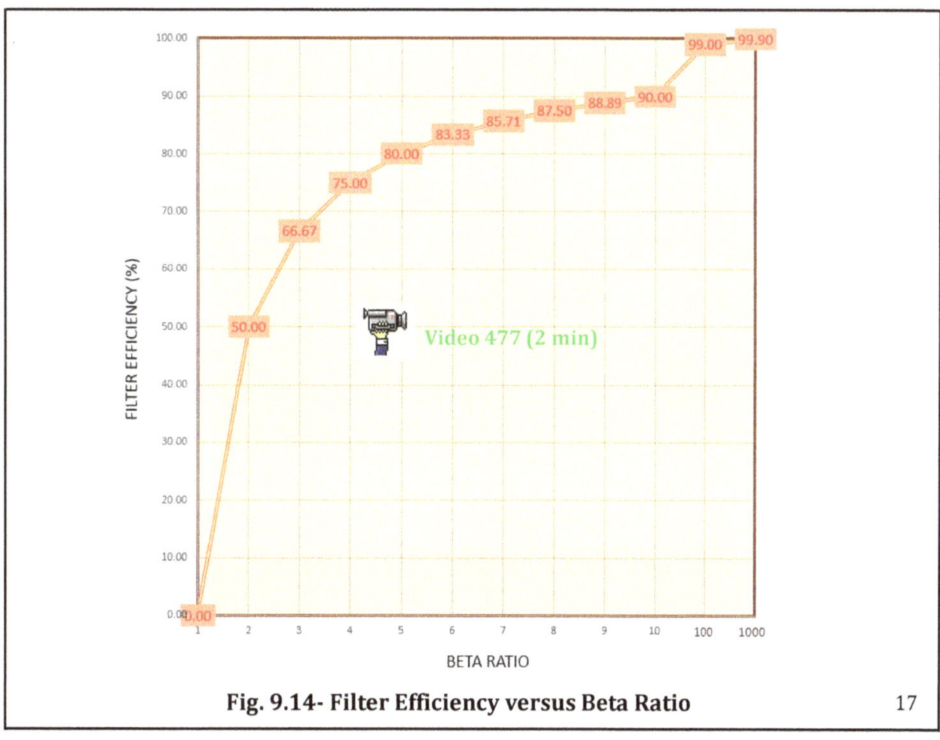

Fig. 9.14- Filter Efficiency versus Beta Ratio 17

17

In a Multipass Test, a filter was found to retain 800 out of 1000 particles larger than 4 microns. What is the filter efficiency?

A. 80%

B. 60%

C. 40%

D. 20%

18

18

9.4- Nominal and Absolute Rating

- Filters can be rated for various particle sizes.
- Example: B3/6/15 = 2, 10, 75. This means:
  - Filter is nominal at 3 microns.
  - Filter is 90% efficient at 6 microns.
  - Filter is absolute at 15 microns.

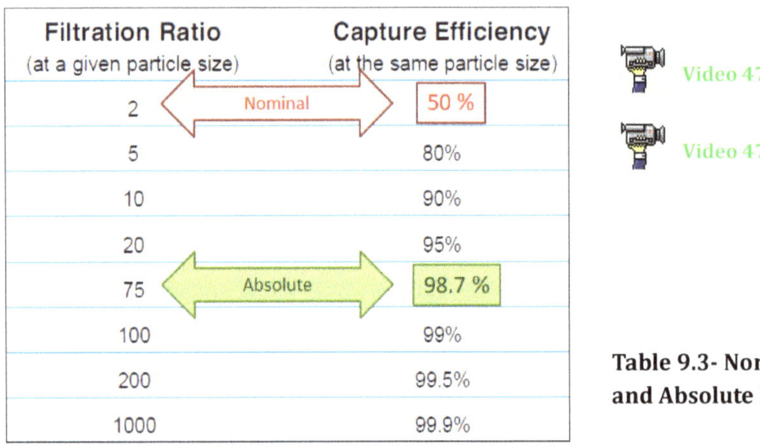

Filtration Ratio (at a given particle size)		Capture Efficiency (at the same particle size)
2	Nominal	50 %
5		80%
10		90%
20		95%
75	Absolute	98.7 %
100		99%
200		99.5%
1000		99.9%

Video 475 (1 min)

Video 476 (0.5 min)

Table 9.3- Nominal and Absolute Ratings

19

19

In a Multipass Test, the particle count greater than 5 microns upstream of a filter was 7500. The particle count greater than 5 microns downstream was 100. Which of the following filter ratings is correct?

A. The filter has $\beta_{5=75}$

B. The filter has absolute β_5 rating.

C. The filter is 98.7% efficient for particles larger than 5 microns.

D. All of the above is correct.

20

20

9.5- Filter Dirt Holding Capacity

Video 460 (0.5 min)

Definition: DHC is the weight of dirt that a filter element can hold before the pressure drop (*Terminal Pressure*) across the filter element reaches a predetermined (saturation) limit.

Standard Test: During Multipass Pass Test (ISO 16889)

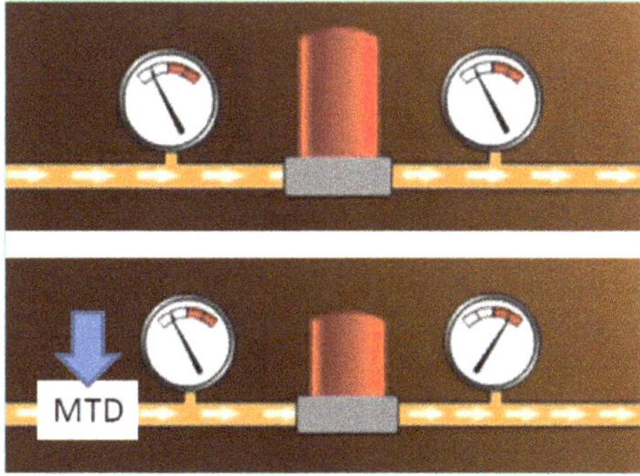

Fig. 9.15- Dirt Holding Capacity Test (Courtesy of Parker)

21

21

Cost of Removing 1 kg or lb of Dirt =

$$\frac{\textbf{Cost of Filter Element (Installation \& Disposal)}}{\textbf{Dirt Holding Capacity in kg or lb}} \qquad 9.3$$

	Example 1	Example 2
Filter type	Glass fiber based pressure filter insert	Cellulose based offline filter insert
Cost of element/insert	€ 35 / $ 50	€ 200 / $ 300
Dirt holding capacity	0.085 kg / 0.18 lbs	4 kg / 8 lbs
Cost per kg/lb removed dirt	**€ 412 / $ 278**	**€ 50 / $ 40**

Table 9.4- Cost of Removing Dirt (Courtesy of C.C. Jensen Inc.)

22

22

Which of the following filters is the most economical?

A. Cost of filter element = $50 and Dirt holding capacity = 0.1kg.

B. Cost of filter element = $100 and Dirt holding capacity = 0.4kg.

C. Cost of filter element = $200 and Dirt holding capacity = 0.4kg.

D. Cost of filter element = $300 and Dirt holding capacity = 0.2kg.

23

23

Typical Example from Industry:

- Disc filter element (3 μm nominal and 8 μm absolute).
- The filter can hold from 1.5-8 kg of dirt depends on the filter size.
- Have high efficiency and DHC but their flow is very low.
- That is why they commonly used for offline filtration.

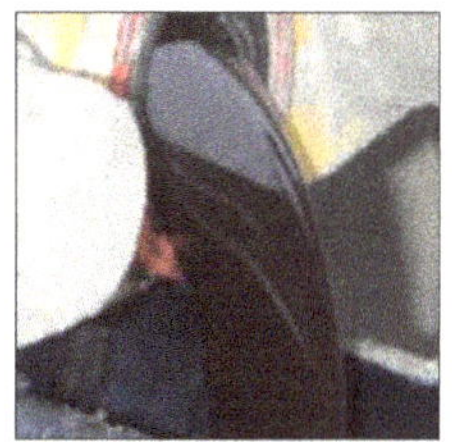
Before

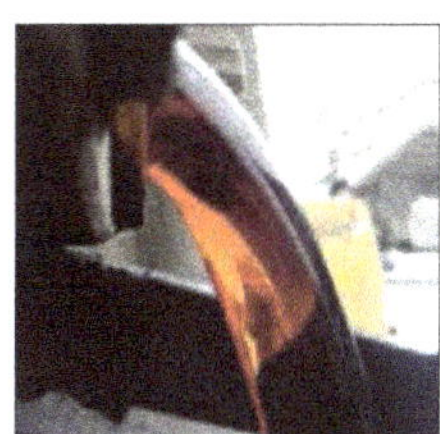
After

Fig. 9.16- Example of Nominal and Absolute Rating and DHC of a Filter (Courtesy of C.C. Jensen Inc.)

24

24

9.6- Filter Size

- A system with a higher flow rate will need larger filters.
- Too small filter → excessive pressure drop → opening the bypass valve.
- Filter Selection: Δp recommended by filter manufacturer is not exceeded at the intended flow rate and maximum fluid viscosity.
- Note: Pump flow may be not the maximum flow in the system because of the use of (differential cylinders or flow surges).

Video 459 (0.5 min)

25

25

9.7- Filter Capacity versus Efficiency

- Regular filters in the system are not designed to deal with large quantities of dirt that occur in connection with component machining, system assembly, system filling, system commissioning, or repair work.
- High restrictive media → efficiency ↑→ blocked by small amount of dirt.
- Less restrictive media → efficiency ↓→ retain more dirt before blocked.
- Balance between the DHC and efficiency of a filter must be considered.

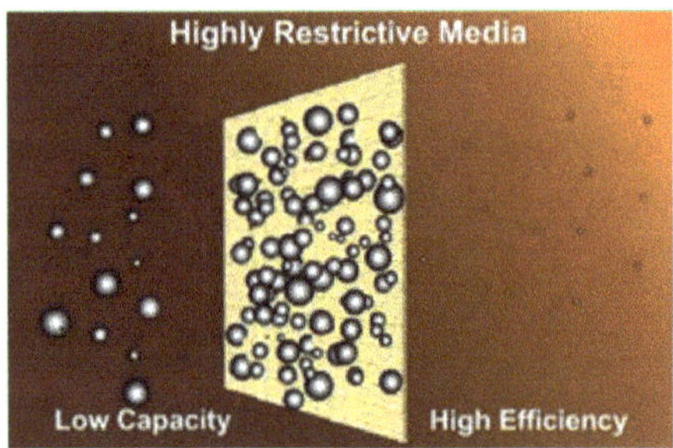

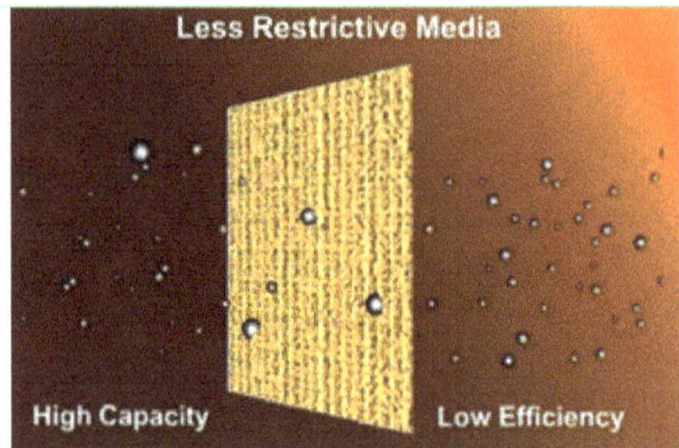

Fig. 9.17- Filter Efficiency vs. DHC (Courtesy of Parker)

26

26

Assuming the filters are the same physical size and flow rating, which of the following filters do you think has the highest dirt holding capacity?

A. $\beta_{5=2}$

B. $\beta_{5=10}$

C. $\beta_{5=20}$

D. $\beta_{5=40}$

27

27

9.8- Filter Pressure

9.8.1- Rated Burst Pressure (RBP) of a Filter Housing

- **Definition:** Static pressure at which **filter housing** structural failure occurs.
- **Standard Test:** NFPA Standard (T-2.6.1).

9.8.2- Rated Fatigue Pressure (RFP) of a Filter Housing

- **Definition:** Maximum allowable pressure for a **filter housing**.
- Safety factor is typically 4 - 6.

$$\mathbf{RFP} = \frac{\mathbf{RBP}}{\mathbf{Safety\ Fator}} \qquad 9.4$$

28

28

9.8.3- Cyclic Test Pressure (CTP) of a Filter Housing

- There are many hydraulic systems perform repetitive functions.
- **Definition:** Maximum pressure applied for certain number of cycles (typically 1 million cycles) to verify the RFP.
- K factor is tabulated (in the above standard) based on confidence and assurance levels.

$$\mathbf{CTP} = \mathbf{RFP} \times \mathbf{K} \qquad 9.5$$

Example:

- RBP = 20,000 psi.
- Safety Factor = 4.
- RFP = 20,000/4= 5,000 psi.
- K = 1.5
- RCP = 5,000 X 1.5 = 7,500 PSI

29

29

9.8.4- Filter Differential Pressure (ISO 3968)

- **Definition:** Difference between the pressure at the upstream and the downstream side of the filter.
- **Standard Test:** ΔP-Q characteristics are measured according to ISO 3968.
- Filter differential pressure **depends on:**
  - Construction of the filter housing.
  - Construction and type of filter element.
  - Filter size and flow rate through the filter.
  - Fluid Properties (Viscosity and specific gravity).

Video 474 (0.5 min)

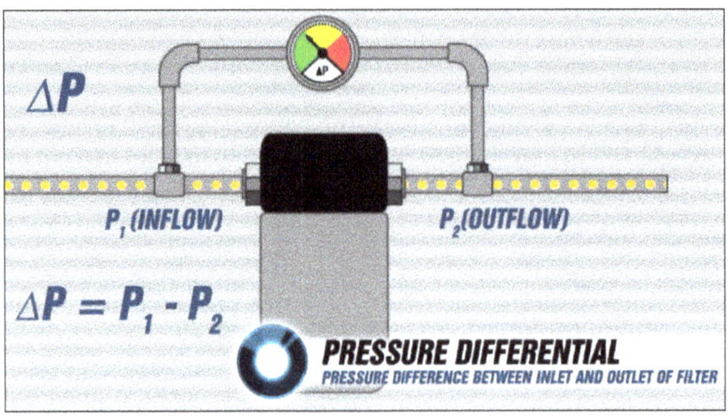

Fig. 9.18- Filter Differential Filter (Courtesy of Noria Corporation)

30

30

- Catalog data: Clean filter elements + a given viscosity.
- Corrections should be made to the actual fluid being used.
- There is no one equation that is applicable for all brands of filters.
- However, at least, filters manufacturers provide instructions.
- Filter housing Δp is corrected based on SG while
- Filter element Δp is corrected based on SG & fluid viscosity.

$$\Delta p_{total} = (\Delta p_H + \Delta p_E) \qquad 9.6$$

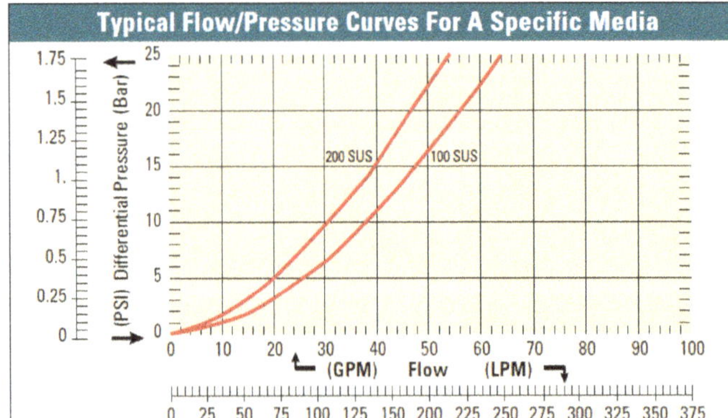

Fig. 9.19- Typical Flow-Pressure Curve for a Specific Filter (Courtesy of Parker) 31

31

Example 1 (Ref. Donaldson):

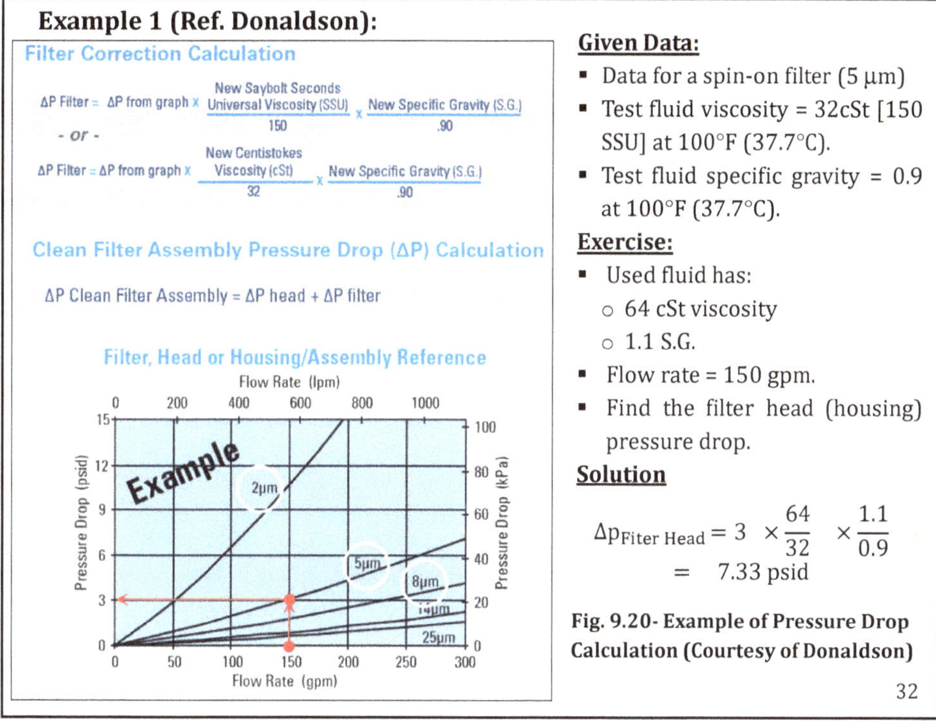

Given Data:

- Data for a spin-on filter (5 µm)
- Test fluid viscosity = 32cSt [150 SSU] at 100°F (37.7°C).
- Test fluid specific gravity = 0.9 at 100°F (37.7°C).

Exercise:

- Used fluid has:
  - 64 cSt viscosity
  - 1.1 S.G.
- Flow rate = 150 gpm.
- Find the filter head (housing) pressure drop.

Solution

$$\Delta p_{\text{Fiter Head}} = 3 \times \frac{64}{32} \times \frac{1.1}{0.9} = 7.33 \text{ psid}$$

Fig. 9.20- Example of Pressure Drop Calculation (Courtesy of Donaldson)

32

32

Example 2 (Ref. Schroeder):

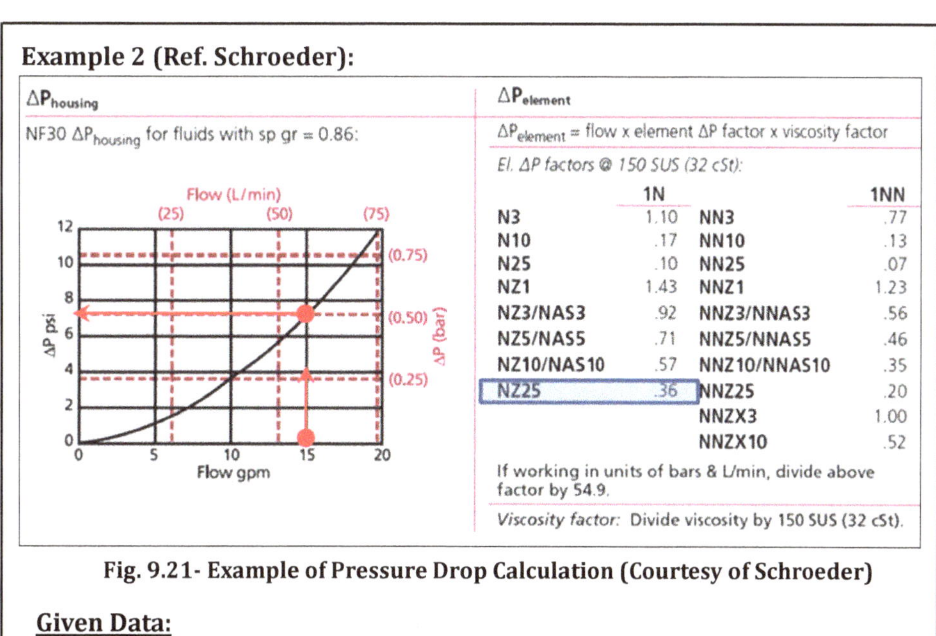

	1N		1NN
N3	1.10	NN3	.77
N10	.17	NN10	.13
N25	.10	NN25	.07
NZ1	1.43	NNZ1	1.23
NZ3/NAS3	.92	NNZ3/NNAS3	.56
NZ5/NAS5	.71	NNZ5/NNAS5	.46
NZ10/NAS10	.57	NNZ10/NNAS10	.35
NZ25	.36	NNZ25	.20
		NNZX3	1.00
		NNZX10	.52

Fig. 9.21- Example of Pressure Drop Calculation (Courtesy of Schroeder)

Given Data:

- Filter Series NZ25-1N
- Test fluid viscosity = 32cSt [150 SSU] at 100°F (37.7°C).

33

33

Exercise:

- Used fluid has:
  - 44 cSt (200 SUS) viscosity.
  - 0.86 S.G.
- Flow rate = 15 gpm.
- Find the total pressure drop.

$\Delta P_{filter} = \Delta P_{housing} + \Delta P_{element}$

Exercise:

Determine ΔP at 15 gpm (57 L/min) for NF301NZ25SMS5 using 200 SUS (44 cSt) fluid.

Solution:

$\Delta P_{housing}$ = 7.0 psi [.50 bar]

$\Delta P_{element}$ = 15 x .36 x (200÷150) = 7.2 psi
or
= [57 x (.36÷54.9) x (44÷32) = .51 bar]

ΔP_{total} = 7.0 + 7.2 = 14.2 psi
or
= [.50 + .51 = 1.01 bar]

34

34

Example 3 (Ref. Hydac):

EXAMPLE - an application with the following criteria would be sized as shown.

Conditions:
- **Fluid** – Hydraulic Oil (ISO-32)
- **Specific Gravity** – 0.86
- **Viscosity** – 141 SSU
- **Flow Rate** – 30 GPM
- **Fluid Temperature** - 104°F normal

Filter Type Selected - Pressure Filter
HYDAC Model No. DF ON 240 TE 10 D 1.0 / 12 V -B6

HOUSING

$$\Delta P\ \text{Housing} = \Delta P\ \text{Calculation}\ \textit{(From Curve in catalog)} \times \frac{\text{Actual Specific Gravity}}{0.86}$$

$$\Delta P\ \text{Housing} = 1.5\ \text{psid} \times \frac{0.86}{0.86} = 1.5\ \text{psid}$$

ELEMENT

$$\Delta P\ \text{Clean Element} = \Delta P\ \text{Calculation} \times \frac{\text{Actual Specific Gravity}}{0.86} \times \frac{\text{Actual Viscosity}}{141\ \text{SSU}}$$

$$\Delta P\ \text{Clean Element} = 30\ \text{GPM} \times 0.175 \times \frac{0.86}{0.86} \times \frac{141\ \text{SSU}}{141\ \text{SSU}}$$

$$\Delta P\ \text{Clean Element} = 5.25 \times 1 \times 1 = 5.25\ \text{psid}$$

FILTER ASSEMBLY

ΔP Filter Assembly = ΔP Housing + ΔP Clean Element
1.5 psid + 5.25 psid = 6.75 psid

Fig. 9.22- Example of Pressure Drop Calculation (Courtesy of Hydac)

35

35

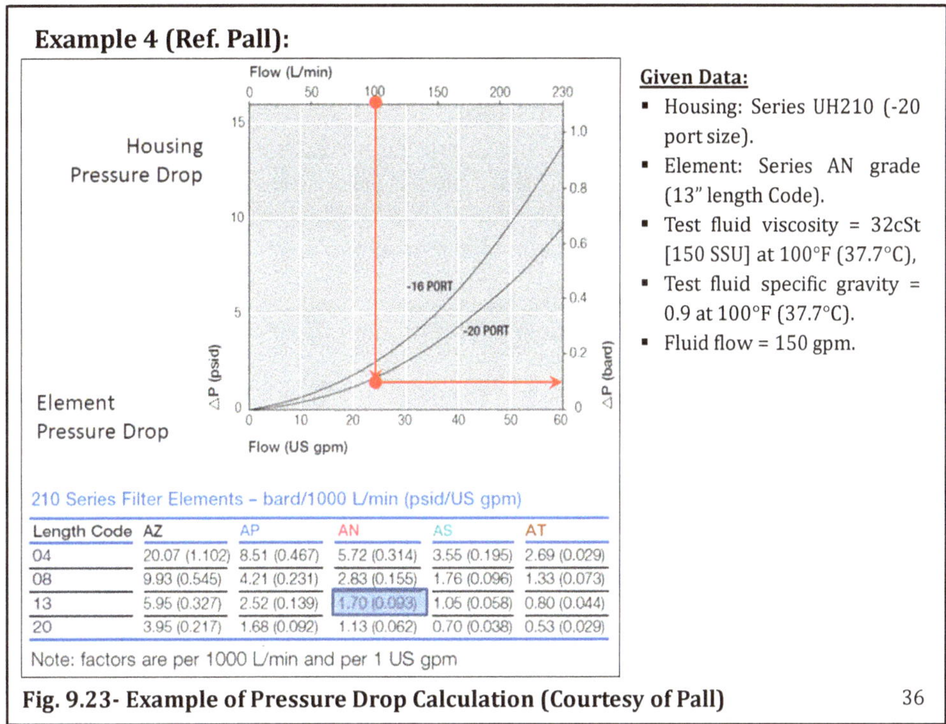

210 Series Filter Elements – bard/1000 L/min (psid/US gpm)

Length Code	AZ	AP	AN	AS	AT
04	20.07 (1.102)	8.51 (0.467)	5.72 (0.314)	3.55 (0.195)	2.69 (0.029)
08	9.93 (0.545)	4.21 (0.231)	2.83 (0.155)	1.76 (0.096)	1.33 (0.073)
13	5.95 (0.327)	2.52 (0.139)	1.70 (0.093)	1.05 (0.058)	0.80 (0.044)
20	3.95 (0.217)	1.68 (0.092)	1.13 (0.062)	0.70 (0.038)	0.53 (0.029)

Note: factors are per 1000 L/min and per 1 US gpm

Fig. 9.23- Example of Pressure Drop Calculation (Courtesy of Pall) 36

36

Exercise:

- Used fluid has:
  - 50 cSt viscosity.
  - 1.2 S.G.
- Flow rate = 100 gpm.
- Find the total (assembly) pressure drop.

Solution: **Total Filter ΔP**

= ΔP housing + ΔP element

= (0.13 x 1.2/0.9) bard (housing)

+ ((100 x 1.70/1000) x 50/32 x 1.2/0.9) bard (element)

= 0.17 (housing) + 0.35 bard (element)

= 0.52 bard (7.6 psid)

37

37

9.8.5- Filter Bypass Pressure

- **Definition:** Δp = p at inlet (dirty side) – p at the outlet (clean side).
- Δp → indicator for the state of the filter.
- Δp ↑ → filter element collapse + back pressure + unfiltered oil pass through.
- Therefore, filters are equipped with a filter bypass valve.
- **Design Value:** Ideally, a filter element should be sized so that:
- Δp (clean element + housing) < half the bypass valve setting.

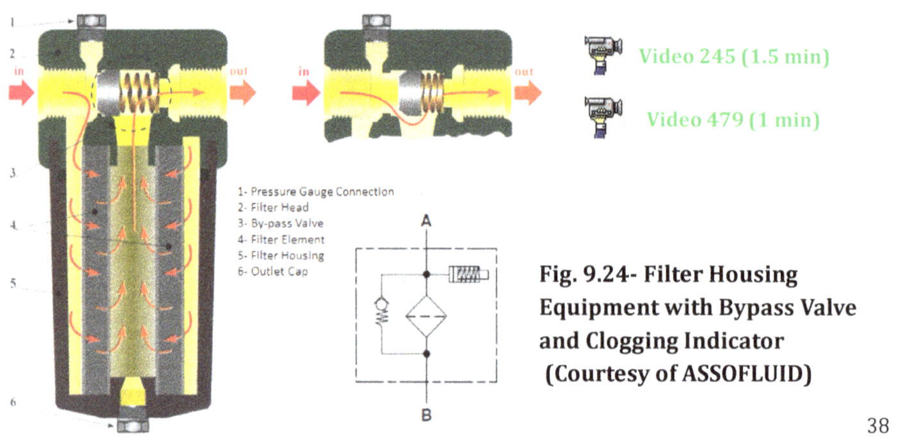

Fig. 9.24- Filter Housing Equipment with Bypass Valve and Clogging Indicator (Courtesy of ASSOFLUID)

38

38

For a filter that has an initial differential pressure of 20 psi across a clean filter element, what bypass pressure is recommended?

A. 10 psi.

B. 20 psi.

C. 30 psi.

D. 80 psi.

39

39

9.8.6- Collapse Pressure of a Filter Element

- **Definition:** It is the Δp at which a structural failure of the filter element and/or center tube occurs.
- **Design Value:**
- Collapse pressure of a filter element > 2 x bypass valve pressure drop.
- Pressure filters with no bypass are recommended with servo valves.
- Collapse pressure of a filter element (with no bypass valve) = as a minimum the same as the setting of the system PRV.
- **Standard:** Collapse Pressure is determined by (ISO 2941/ANSI B93.25):

40

40

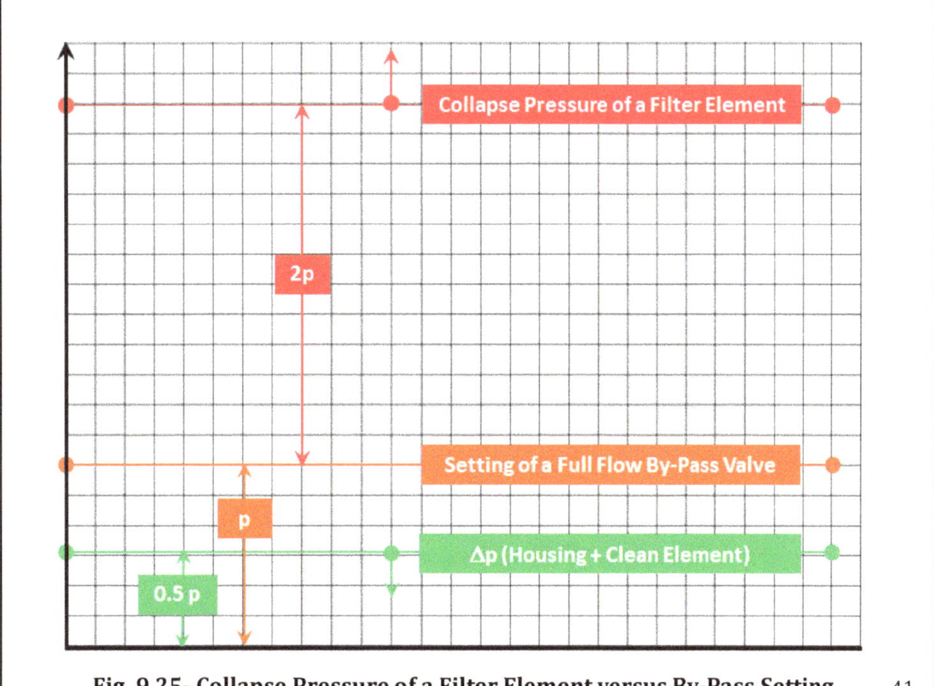

Fig. 9.25- Collapse Pressure of a Filter Element versus By-Pass Setting

41

41

For a filter that has an initial differential pressure of 20 psi across a clean filter element, what filter element collapse pressure is recommended?

A. Greater than 20 psi.
B. Greater than 40 psi.
C. Greater than 60 psi.
D. Greater than 80 psi.

42

42

9.8.7- Flow Fatigue of a Filter Element

- **Definition:** *Flow Fatigue* is the ability of a filter element to withstand structural failure of the filter medium due to flexing of the pleats caused by cyclic differential pressure.

- **Flow Fatigue Test for Filter Element (ISO 3724 OR ISO 23181):** Flow fatigue tests are run, usually based on (10-200)k cycles, to evaluate the structural strength of filter elements according to ISO 3724 or ISO 23181 Standard.

- **Fatigue Stability:** High fatigue stability is achieved by better filter element design including:
  - Supporting both sides of the element and high
  - Inherent stability of the filter materials.

43

43

Which of the following filters is most adequate for a hydraulic system that drives a digging hammer?

A. A filter that has high efficiency.

B. A filter that has high flow fatigue stability.

C. A filter that has high dirt holding capacity.

D. A filter that has large size.

44

44

Which of the following filters is most adequate for a hydraulic system that contains large pump?

A. A filter that has high efficiency.

B. A filter that has high flow fatigue stability.

C. A filter that has high dirt holding capacity.

D. A filter that has large size.

45

45

Which of the following filters is most adequate for a hydraulic system that work in a highly contaminated environment?

A. A filter that has high efficiency.

B. A filter that has high flow fatigue stability.

C. A filter that has high dirt holding capacity.

D. A filter that has large size.

46

46

Chapter 9 Reviews

1. In a Multipass Test, a filter was found to retain 800 out of 1000 particles larger than 4 microns. Which is the beta ratio?
   A. $\beta_{1.25 = 4}$
   B. $\beta_{4 = 1.25}$
   C. $\beta_{5 = 4}$
   D. $\beta_{4 = 5}$

2. In a Multipass Test, a filter was found to retain 800 out of 1000 particles larger than 4 microns. What is the filter efficiency?
   A. 80%.
   B. 60%.
   C. 40%.
   D. 20%.

3. In a Multipass Test, the particle count greater than 5 microns upstream of a filter was 7500. The particle count greater than 5 microns downstream was 100. Which of the following filter rating is correct?
   A. The filter has $\beta_{5=75}$
   B. The filter has absolute β_5 rating.
   C. The filter is 98.7% efficient for particles larger than 5 microns.
   D. All of the above is correct.

4. Which of the following filters is the most economical?
   A. Cost of filter element = $50 and Dirt holding capacity = 0.1kg.
   B. Cost of filter element = $100 and Dirt holding capacity = 0.4kg.
   C. Cost of filter element = $200 and Dirt holding capacity = 0.4kg.
   D. Cost of filter element = $300 and Dirt holding capacity = 0.2kg.

5. Assuming the filters are the same physical size and flow rating, which of the following filters do you think has the highest dirt holding capacity?
   A. $\beta_{5 = 2}$
   B. $\beta_{5 = 10}$
   C. $\beta_{5 = 20}$
   D. $\beta_{5 = 40}$

6. For a filter that has an initial differential pressure of 20 psi across a clean filter element, what bypass pressure is recommended?
   A. 10 psi.
   B. 20 psi.
   C. 30 psi.
   D. 100 psi.

7. For a filter that has an initial differential pressure of 20 psi across a clean filter element, what filter element collapse pressure is recommended?
   A. Greater than 20 psi.
   B. Greater than 40 psi.
   C. Greater than 60 psi.
   D. Greater than 80 psi.

8. Which of the following filters is most adequate for a hydraulic system that drives a digging hammer?
   A. A filter that has high efficiency.
   B. A filter that has high flow fatigue stability.
   C. A filter that has high dirt holding capacity.
   D. A filter that has high large size.

9. **Which of the following filters is most adequate for a hydraulic system that contains servo valve?**
   A. A filter that has high efficiency.
   B. A filter that has high flow fatigue stability.
   C. A filter that has high dirt holding capacity.
   D. A filter that has high large size.

10. **Which of the following filters is most adequate for a hydraulic system that contains large pump?**
   E. A filter that has high efficiency.
   F. A filter that has high flow fatigue stability.
   G. A filter that has high dirt holding capacity.
   H. A filter that has high large size.

Chapter 9 Assignment

Student Name: -- Student ID: ------------------

Date: --- Score: ------------------------

Use the data sheet shown below to find the overall assembly (housing + element) differential pressure, given that:

- Housing: Series UH210 (-16 Port Size).
- Element: Series AZ grade (20" length Code).
- Test fluid viscosity = 32cSt [150 SSU] at 100°F (37.7°C),
- Test fluid specific gravity = 0.9 at 100°F (37.7°C).
- Fluid flow = 200 l/min.
- Actual application fluid viscosity = 60 cSt and 1.5 SG

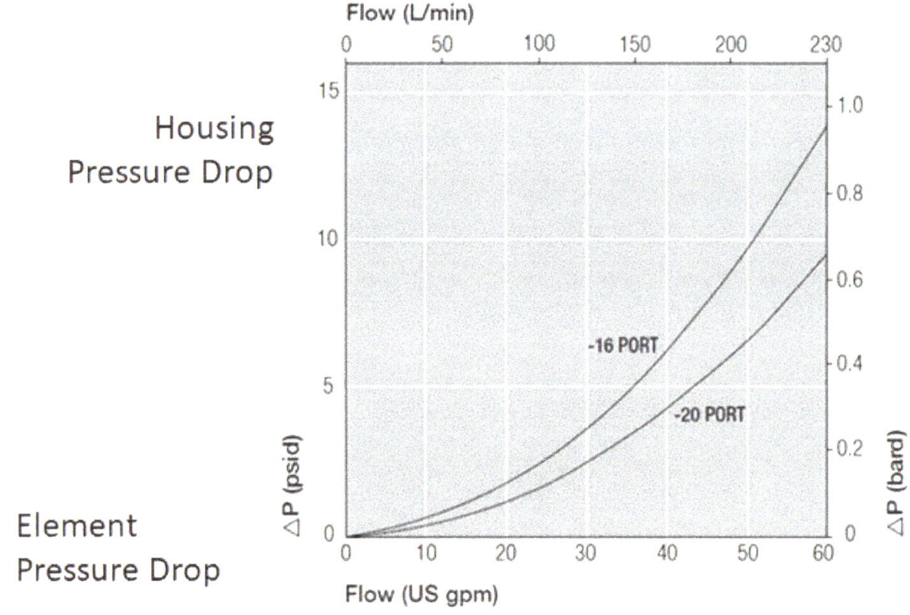

210 Series Filter Elements – bard/1000 L/min (psid/US gpm)

Length Code	AZ	AP	AN	AS	AT
04	20.07 (1.102)	8.51 (0.467)	5.72 (0.314)	3.55 (0.195)	2.69 (0.029)
08	9.93 (0.545)	4.21 (0.231)	2.83 (0.155)	1.76 (0.096)	1.33 (0.073)
13	5.95 (0.327)	2.52 (0.139)	1.70 (0.093)	1.05 (0.058)	0.80 (0.044)
20	3.95 (0.217)	1.68 (0.092)	1.13 (0.062)	0.70 (0.038)	0.53 (0.029)

Note: factors are per 1000 L/min and per 1 US gpm

Chapter 10 Contamination Control in Hydraulic Transmission Lines

Objectives:

This chapter discusses best practices for controlling contamination in hydraulic transmission lines including projectile cleaning and hydraulic system flushing.

0

0

Brief Contents:

1

1

10.1- Contamination in Hydraulic Transmission Lines

- The initial cleanliness level of a hydraulic system can affect its service life.
- **ISO 1643** describes a clean-up procedure after system final assembly.

10.1.1- Sources of Contamination in Hydraulic Transmission Lines

- As it has been previously discussed, particulate contamination in hydraulic transmission lines is due to:
- Built-in: during manufacturing and/or assembly (cutting, crimping, bending, and flaring)
- Introduced: Ingested from surrounding (long storage)
- Introduced: Induced during system repair (untightening & retightening)
- Generated: settled inside transmission lines during system operation.

2

2

10.1.2- Methods for Cleaning Hydraulic Transmission Lines

- The common basic methods for cleanliness of transmission lines are:
  - Projectile Cleaning.
  - Pickling using Chemicals.
  - Hydraulic System Flushing.

- In order to minimize generated contaminants :
  - Use Clean Air: Firing a projectile
  - Clean Work Area
  - Cut the Line to the Correct Length:
    - This will eliminate additional processes for adjusting the line length that produce more contaminants.

  - Use the Right Saw Blade:
    - Abrasive wheeled chop saws are the worst offenders.
    - Metal blades reduce the amount of debris generated by the cutting.
    - Use a "Clean Cut" saw or blade that is specifically designed to minimize debris ingression.

3

3

Video 651 (0.5 min)

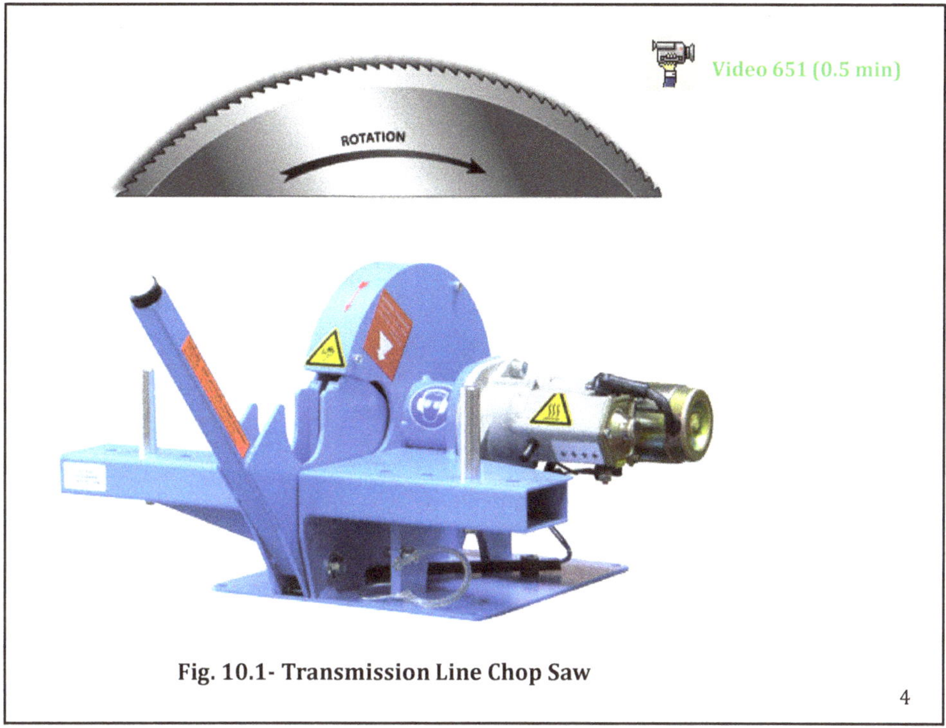

Fig. 10.1- Transmission Line Chop Saw

4

4

10.2- Projectile Cleaning

10.2.1- Projectile Cleaning Overview

- **Distributors** of transmission lines are under more pressure from their customer base and OEM's to meet or maintain cleanliness levels.
- **Manufacturers** may not be as concerned as distributors, but their actions can still affect a distributors' cleanliness results.
- **End users** should never assume that transmission lines are clean.
- **Projectile Cleaning** is a sponge-like projectile which is shot through hose and tubing assemblies by a blast of compressed air.
- **FPI-MSOE Research** indicated that projectile cleaning can reduce the contamination level by up to four ISO Codes

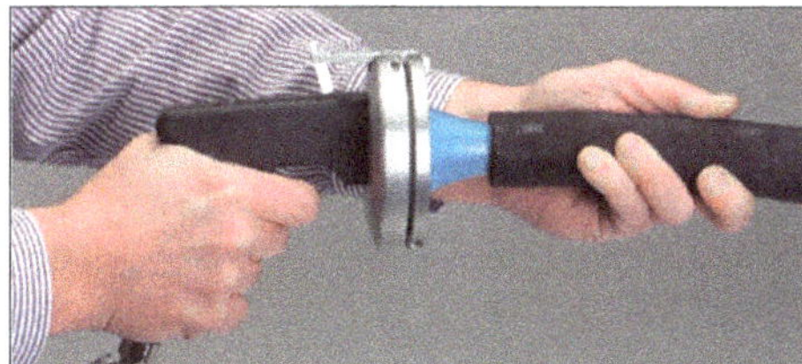

Fig. 10.2- Transmission Lines Projectile Cleaning
(Courtesy of Ultra Clean Technologies)

5

5

10.2.2- Projectile Cleaning Equipment

❑ **Projectile Cleaning Equipment:**

- Much more affordable than that for high-velocity flushing.
- It takes much less time than high-velocity flushing.
- Cleans hydraulic lines much more effectively than compressed air alone.

❑ **Compressed Air:**

- Clean and moisture-free compressed air.
- 80 PSI (5.5 Bar) minimum to 110 PSI (7.5 Bar) maximum.
- 1/2" ID air hose to ensure 110 SCFM (3.1 m^3/min) air flow.
- 5-micron filter and regulator with gauge are strongly suggested!

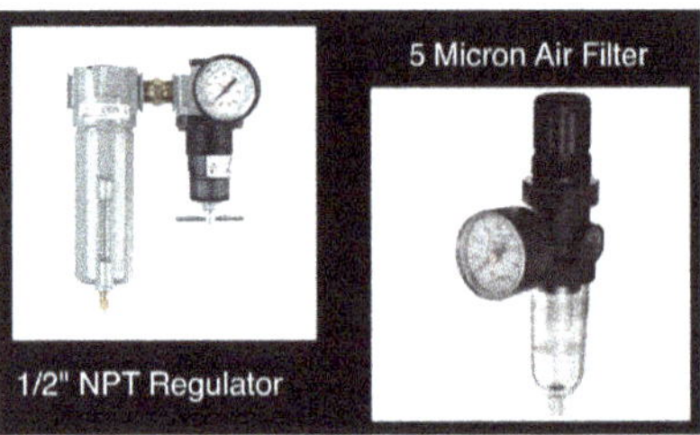

Fig. 10.3- Compressed Air Requirements (Courtesy of Ultra Clean Technologies)

6

6

❑ **Launcher:**

- Various sizes and styles are available.
- A one launcher can be used for various types of nozzles.

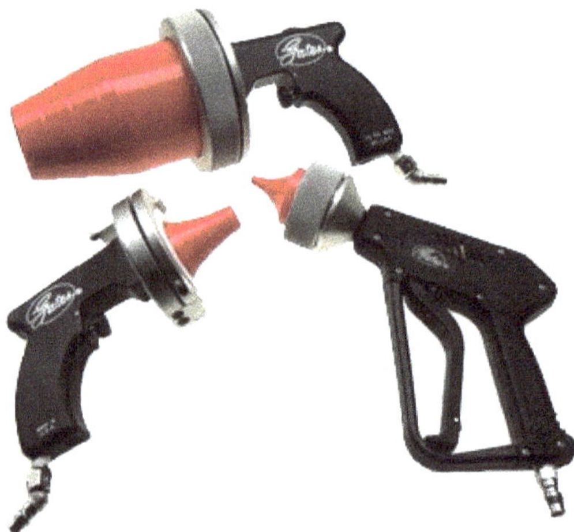

Fig. 10.4- Launchers (Courtesy of Gates)

7

7

❑ Projectiles:

- Various sizes and Types are available.
- Usually, projectiles are sized 20% to 30% > I.D. of the cleaned line.
- It is to be noted that projectiles are non-reusable.
- Regular Projectiles (1): used for all lines, particularly hoses.
- Abrasive projectiles (2): used for tubes or pipes that contain rust, weld slag or other corrosion particles on the inside surface,
- Grinding Projectiles (3): recommended for all types of carbon steel pipes.

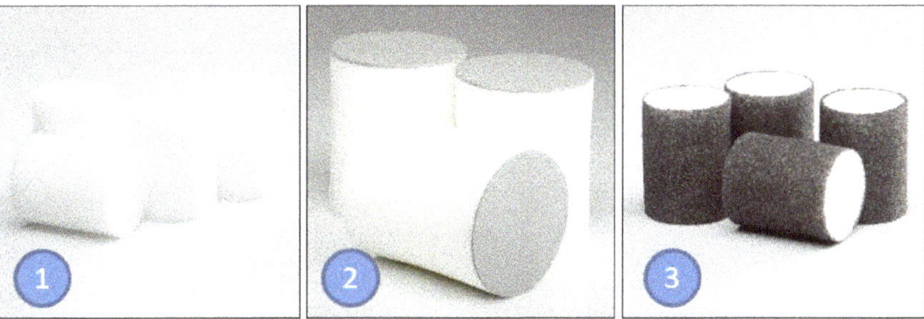

Fig. 10.5- Projectiles (Courtesy of Ultra Clean Technologies)

8

8

Cap Seals: Protective caps or plugs are used at both ends after cleaning.

A Complete Kit: a complete kit is available that includes a launcher, set of nozzles, set of projectiles, and a bucket for used projectiles.

Fig. 10.6- Cap Seals (Courtesy of Ultra Clean Technologies)

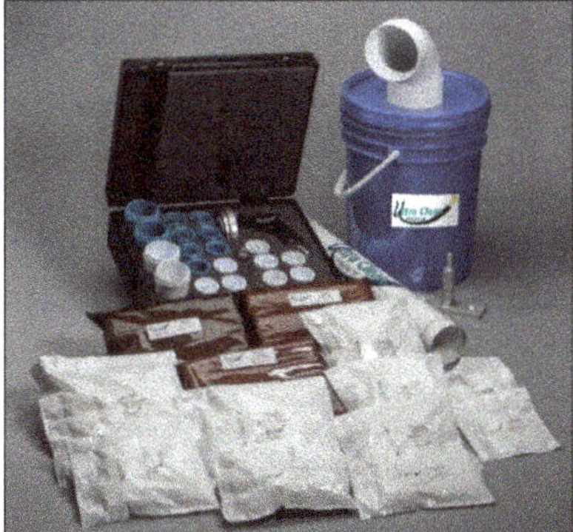

Fig. 10.7- Complete Kit (Courtesy of Ultra Clean Technologies)

9

9

10.2.3- Hydraulic Hose Projectile Cleaning

- Looking inside a hose to check if it is clean or not is not a good practice.
- Naked eyes can't see particles below 40 µm.

Fig. 10.8- Hydraulic Hose Cleanliness (Courtesy of Ultra Clean Technologies) 10

10

- A hydraulic hose is highly recommended to be cleaned right after cutting for the following reasons:
  - Heat from the cutting process will cause the rubber & metal dust to stick or adhere to the hose tube as it cools.
  - Contamination from freshly cut hose is much easier to remove.
  - Hose ends are difficult to insert over the contamination.
  - Contamination that is trapped between the hose ends and rubber tube could become an eventual leak path for the hydraulic fluid.

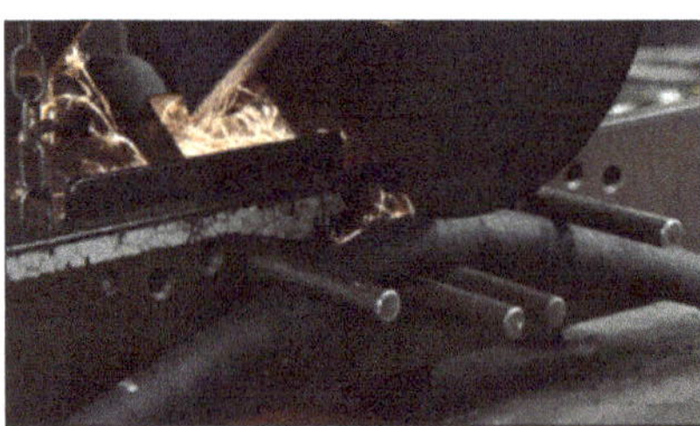

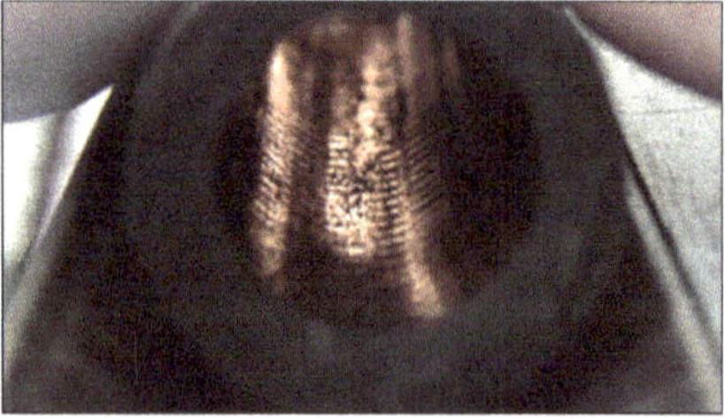

Fig. 10.9- Hydraulic Hose Cutting Causes Contamination (Courtesy of Ultra Clean Technologies) 11

11

Hose Cleaning Procedure:

1. Connect your launcher to a dry filtered and regulated air source.
2. Load the recommended nozzle and projectile into the launcher.

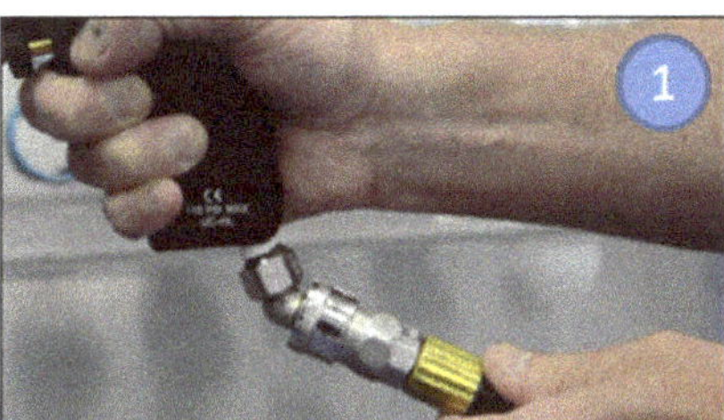

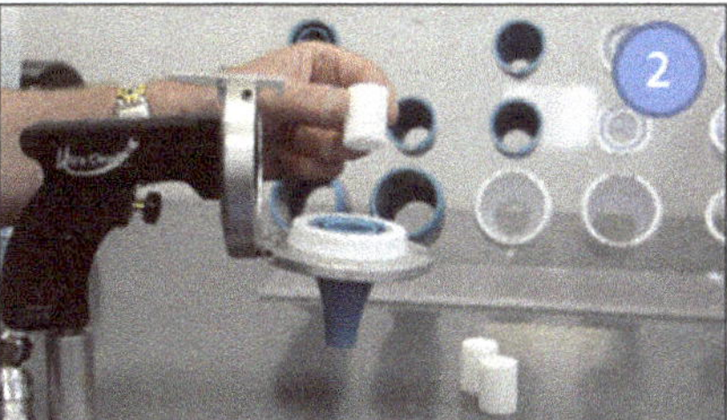

Fig. 10.10- Hydraulic Hose Cleaning Procedure (Courtesy of Ultra Clean Technologies)

12

12

3. Now close the face plate of the launcher. The safety release bar will lock it into position.
4. Insert the nozzle into the hose. Secure the other end of the hose into the containment barrel or catcher bucket.

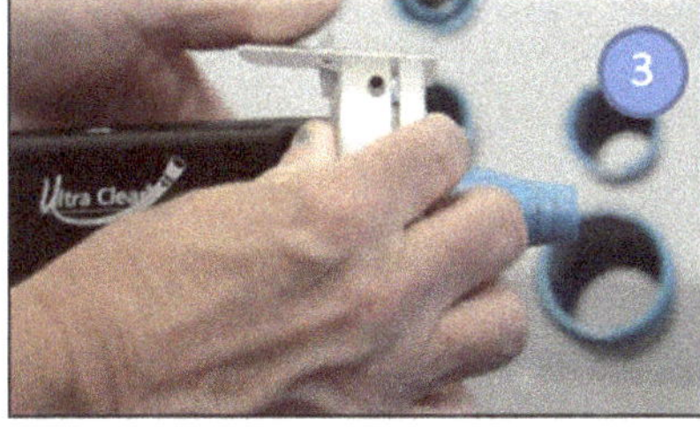

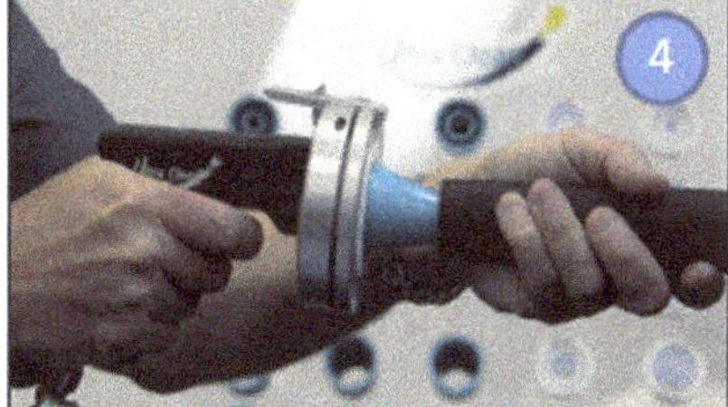

13

13

5. Depress the trigger until the projectile has exited the opposite end of the hose. The projectile strips out the internal contamination as it travels through the hose and around bends forcing the contamination out in front of it.
6. Wipe the end of the nozzle clean and repeat this process now through the other end of the hose.
7. Continue launching projectiles until they exit the transmission line visibly clean. At this point, launching additional projectiles typically will not remove any more contaminants.

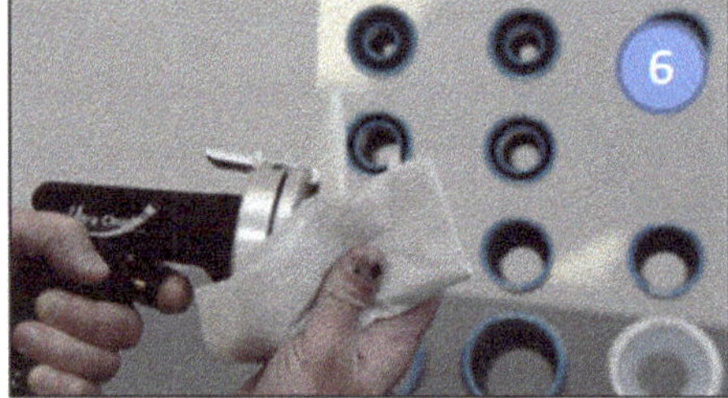

Video 445 (2 min) 14

14

Hose Cleaning after Hose Ends Attached:

Video 446 (1 min)

- The attachment process causes stem deformation to achieve the proper coupling retention.
- Internal metal flash contamination will occur, and it must be removed.
- Projectile must continue to be fired through the entire assembly until they exit visibly clean.
- Then hose ends are covered by cap seals.

Fig. 10.11- Hydraulic Hose Cleaning after Attaching Hose Ends (Courtesy of Ultra Clean Technologies) 15

15

10.2.4- Hydraulic Tubes and Pipes Projectile Cleaning

- Microscopic particles settle inside pipes/tubes due to (tube cutting, bending, flaring, and de-burring).
- Such contaminants can effectively destroy precision of hydraulic components in the system.

Video 448 (1 min)

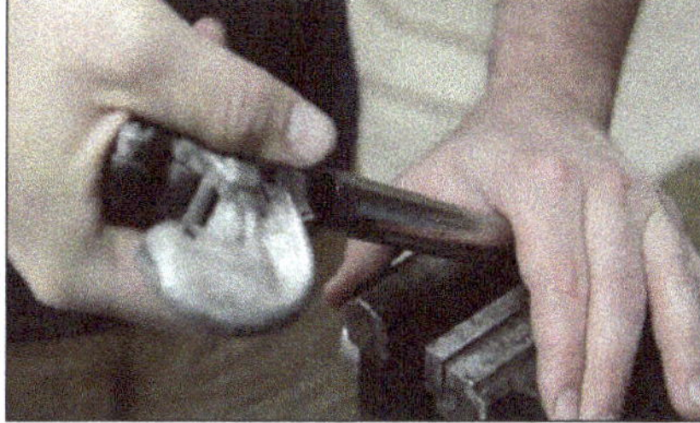

Fig. 10.12- Tube Bending and Flaring (Courtesy of Ultra Clean Technologies)

16

16

Tube or Pipe Cleaning after Processing:

- **First**, check if the inside surface contains rust, weld slag or other corrosion particles, then abrasive or grinding projectiles are used untill these contaminants removed.
- **Second**, regular projectiles may then be implemented to clean the tube from both sides.
- **Third**, when the projectiles exit the tube/pipe visibly clean, tube or pipe ends are covered by cap seals.

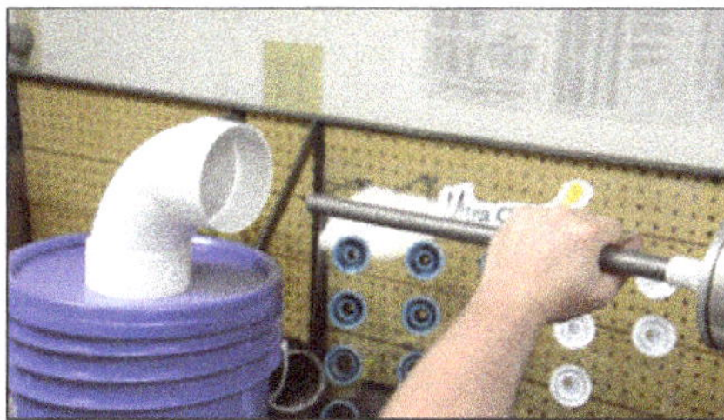
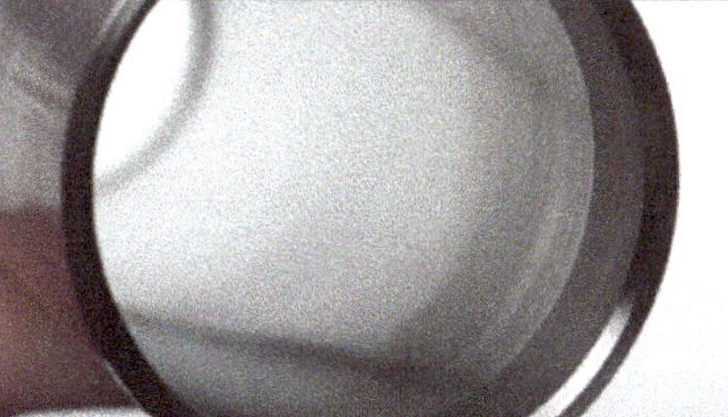

Fig. 10.13- Tube Projectile Cleaning (Courtesy of Ultra Clean Technologies) 17

17

10.2.5- Clean Seal Capsules

- Traditional caps and plugs are forced onto assemblies:
  - → Plastic particles shear off into the hose or tube
  - → Re-contamination occurs .
- Advanced capping method uses *Clean Seal Capsules* to:
  - → Raise the cleanliness level of transmission lines.
  - → Eliminates possible re-contamination.
- Clean seal capsules are available in 16 sizes to fit most hose and tube assemblies.

Fig. 10.14- Pre-Stacked Clean Seal Capsules (Courtesy of Ultra Clean Technologies) 18

18

- **Heat Source for Shrinking the Clean Seal Capsule is:**
  - Time is controlled by a timer that can be set up to 60 minutes.
  - The temperature is set to approximately 165 ^{0}C(325 ^{0}F).
  - The machine needs few minutes to reach the correct temperature from a cold start.

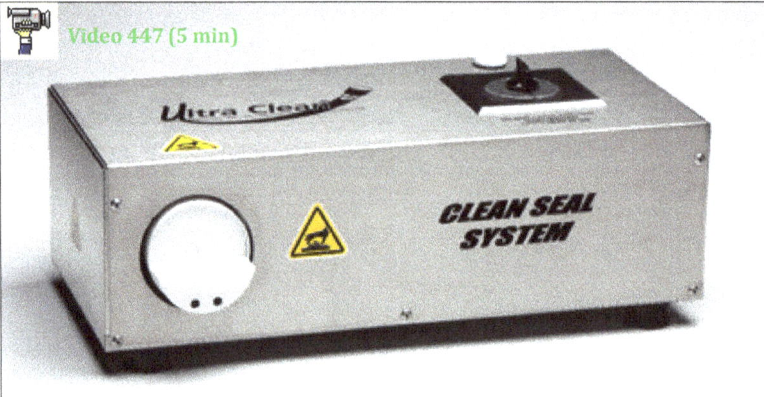

Fig. 10.15- Heat Source for Shrinking the Clean Seal Capsules (Courtesy of Ultra Clean Technologies)

19

19

Process of Applying Clean Seal Capsules:

1. Choose the closest fit capsule and place it over the coupling. Slide the correct size clean seal capsule over the end fitting.
2. Place the clean seal capsule against the white plunger and push in. A complete seal takes place in less than two seconds.

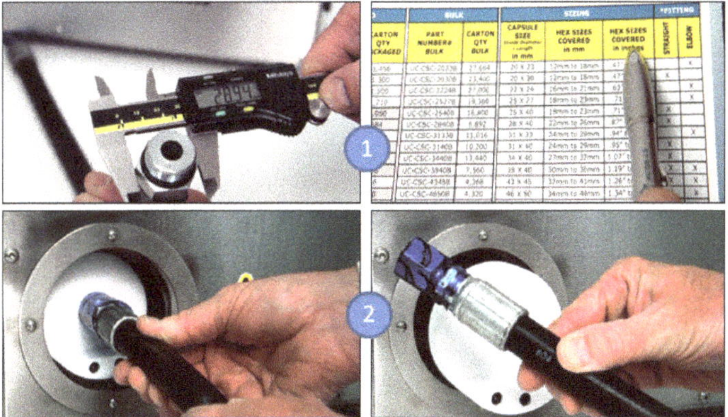

Fig. 10.16- Process of Applying Clean Seal Capsules (Courtesy of Ultra Clean Technologies)

20

20

At the time of using the transmission line, follow these three easy steps:

1. **Grip** the black pull tab.
2. **Rip** the pull tab upwards.
3. **Slip** the Clean Seal Capsule off of the assembly.

Fig. 10.17- Process of Removing Clean Seal Capsules (Courtesy of Ultra Clean Technologies)

21

21

10.2.6- Clean Seal Flange

- Is a tool that easily attaches to SAE flanges, particularly in the field.
- Used to prevent dirt from entering hydraulic transmission lines.
- Used to prevent oil from spilling out of hydraulic transmission lines.
- No tools are needed to connect or remove the Clean Seal Flange!

Fig. 10.18- Clean Seal Flanges (Courtesy of Ultra Clean Technologies)

22

22

Recommended projectiles size is?

A. (1-2) % > line ID
B. (5-10) % > line ID
C. (10-20) % > line ID
D. (20-30) % > line ID

23

23

Pickling process means?

A. Cleaning by oil.
B. Cleaning by acids.
C. Cleaning by compressed air.
D. Cleaning by wipes.

24

24

Flushing process means?

A. Cleaning by oil.
B. Cleaning by acids.
C. Cleaning by compressed air.
D. Cleaning by wipes.

25

25

10.3- Pickling of Hydraulic Transmission Lines

10.3.1- What is Pickling?

- **Cleaning by acids** to remove:
  - Contamination such as mandrel lubricants and other greasy products.
  - Contamination such as rust or oxide produced from long storage.
- These contaminants are difficult to be removed after running the machine.
- Therefore, these products must be removed before the first use of a line.

10.3.2- Pickling Process

For small individual lines:

- Spray an alcohol-based solvent inside the line.
- The solvent should stay inside the line for certain time to break up the greasy products residue.
- The line is then cleaned using dry projectiles.

26

26

For multiple long hydraulic pipes and tubes:

- A more in-depth process is required.
- Normally this process is done by a specialized vendor.
- If no instructions are found, the following sequence can be used as a guide:
- **Degreasing:**
  - Lines are closed from both ends and filled with the degreasing liquid.
  - Fill lines degreasing liquid (typically 5% soda).
  - Pressure slightly higher than atmospheric pressure.
  - Time (typically 1 hour).
  - Rinsing by soft water for 15-20 minutes.
- **De-rusting:**
  - Dip the line in Hydrochloric acid (18-20% concentration) tank.
  - Time (20-30) minutes.
  - Rinsing by soft water for 20-30 minutes.
- **Drying:**
  - Dry by compressed air or nitrogen.
  - Apply Projectile cleaning.
- **Flushing:**
  - Line flushing with compatible hydraulic fluid, for making a layer of oil.

27

27

10.4- Flushing of Hydraulic Transmission Lines

10.4.1- What is Flushing?

- Cleaning by oil (Kidney Wash or Offline Filtration).
- Performed under certain conditions following a certain procedure.

10.4.2- Reasons to Flush a Hydraulic System

- Newly Built Hydraulic Systems.
- Majorly Repaired Hydraulic Systems.
- Hydraulic Fluid Degradation.
- Contamination Class was Exceeded.
- Failure of an Oil Water-Cooler.
- Mixing of Incompatible Hydraulic Fluids.
- Bacteriological Contamination.
- After Pickling Process.

28

28

10.4.3- Flushing System Requirements (ISO 23309)

❑ **Standard:** The following sections present the flushing system requirement according to ISO 23309 and ISO 16431.

❑ **Flushing Flow:**

- Turbulent through out the whole system.
- The *Reynolds Number* must be higher than 3,000 in all parts of flushing circuits.

❑ **Flushing Fluid:**

- Fluid with low viscosity (ISO 15 is recommended) to
- Be able to easily reach into the sharp corners and
- Be able to pass through flushing filters with reduced differential pressure.
- Shall be specified by the equipment manufacturer.
- Must be compatible with components and seals in the flushed system.

29

29

❑ **Flushing Filters:**

- Flushing Filters Efficiency:
  - Targeted cleanliness level ↑→ flushing filter efficiency ↑

- Flushing Filters DHC:
  - Typically higher than normal filters.
  - Targeted cleanliness level ↑→ flushing filter DHC ↑
  - Initial cleanliness level ↓ → flushing filter DHC ↑
  - Flushing oil volume ↑→ flushing filter DHC ↑

- Flushing Filter Medium:
  - Depends on the reason for flushing.
  - Selected to remove (water contents, varnish, chemicals, fluid degradation products, are combination of them.

- Flushing Filters Size:
  - Large surface area → high flushing flow rate at an acceptable Δp
  - Several filters can be arranged in parallel to increase the surface area.

30

30

❑ **Flushing Temperature:**

- Typically higher than normal operating temperature.
- Typically not less than 50 ^{0}C (122 ^{0}F).
- Temperature control system maintains constant flushing temperature.

❑ **Flushing Oil Volume:**

- Fills the volume of the hydraulic lines + volume in the reservoir to maintain satisfactory and safe suction conditions for the flushing pumps.

❑ **Flushing Duration:**

- Is based on # of fluid circulations (typically 200 times).
- Flushing may continue for more 30 minutes to assure well cleaning.

$$\text{Minimum Flushing Time (minutes)} = \frac{200 \text{ x Flushing Fluid Volume (liters)}}{\text{Flow Rate } \left(\frac{\text{liters}}{\text{min}}\right)} \quad 10.1$$

Example:

- Flushing fluid volume =100-liter,
- Flushing pump flow rate of 10 liter/min,
- Minimum flushing time = (200 x 100)/ 10 = 2000 min

31

31

- **Flushing Evaluation:**
  - Intermittent evaluation by fluid sampling.
  - Continuous evaluation by installing online particle counter.
  - 3 consecutive sustained test results → system is clean.
  - The process should continue until cleanliness level is one code below the system's target cleanliness level.
  - For example, if the target is ISO 15/13/11, continue to flush the system until ISO 14/12/10 is reached.

- **Flushing Power Units:**
- Commercial units available. They can be customized also.

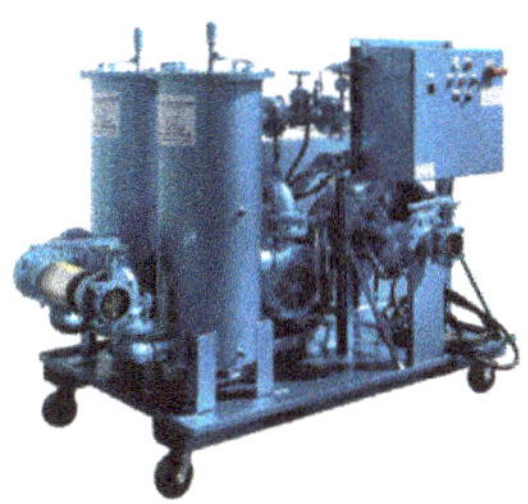

Fig. 10.19- Examples of Hydraulic System Flushing Power Units

32

32

10.4.4- Flushing Process

If no instructions are provided, the following (in sequence) provides a guideline for system flushing:

1. Operate the system for 15 min. after reaching the regular operating Temperature.
2. Completely drain the system including all components & reservoir.
3. With a lint-free rag, clean the reservoir of all sludge and deposits.
4. Prepare the system for flushing with the following considerations.
   - Avoid dead ends.
   - Circuits being flushed shall be connected in series (not in parallel).
   - Replace critical components by "Jumpers".
5. Flushing unit shall be located as close as possible to the flushed system to minimize pressure losses.

33

33

6. Hook the flushing unit to the system.
7. Make every effort to avoid spilling oil.
8. Run the flushing unit to force the flushing fluid through the system.
9. Measure the flushing Temp and flow near the return line.
10. Stroke DCVs to reverse the direction of flushing.
11. Continue until reaching one cleanliness level below the target.
12. Drain the flushing fluid when hot and as quickly as possible.
13. Inspect and clean the reservoir again. Chemical cleaning techniques will not only clean the hydraulic oil reservoir but also provide a protective oxide layer that will further inhibit build up in the future.

34

34

14. Replace the system filters.
15. Fill the system with the fluid to be used.
16. Return the system to its original circuit design.
17. Bleed and vent the pump.
18. Run the system's pump at no load for about 10 minutes.
19. Cycle the actuators to return oil to the reservoir and bleed air from the system.
20. Keep an eye on the fluid level in the reservoir, and refill if needed.
21. Run the system for 30 minutes to bring it to normal operating temperature.

35

35

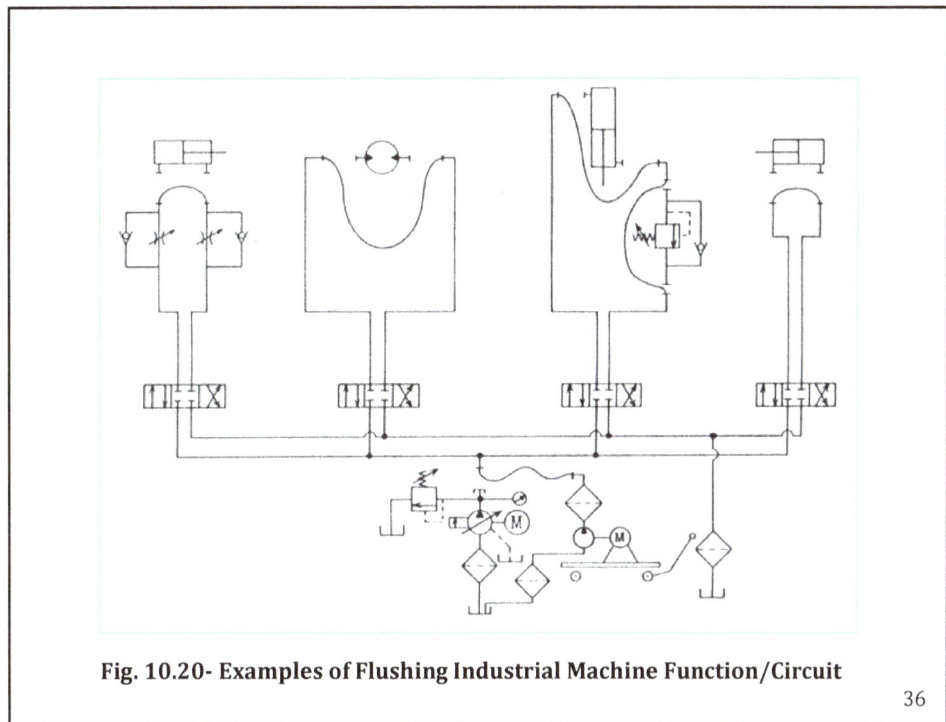

Fig. 10.20- Examples of Flushing Industrial Machine Function/Circuit

36

36

Reynolds Number for flushing flow in all parts of flushing circuit?

A. Greater than 1000.
B. Greater than 2000.
C. Greater than 3000.
D. Greater than 4000.

37

37

A flushing oil volume of 300 liter should circulate 300 times in the flushed system. If a 90 liter/min flushing pump is used, what is the flushing time?

A. 10 min.

B. 100 min.

C. 1000 min.

D. 10000 min.

38

38

Chapter 10 Reviews

1. Recommended projectile size is?
   A. (1-2) % > line ID
   B. (5-10) % > line ID
   C. (10-20) % > line ID
   D. (20-30) % > line ID

2. Pickling process means?
   A. Cleaning by oil.
   B. Cleaning by acids.
   C. Cleaning by compressed air.
   D. Cleaning by wipes.

3. Flushing process means?
   A. Cleaning by oil.
   B. Cleaning by acids.
   C. Cleaning by compressed air.
   D. Cleaning by wipes.

4. Reynolds Number for flushing flow in all parts of flushing circuit?
   A. Greater than 1000.
   B. Greater than 2000.
   C. Greater than 3000.
   D. Greater than 4000.

5. A flushing oil volume of 300 liter should circulate 300 times in the flushed system. If a 90 liter/min flushing pump is used, what is the flushing time:
   A. 10 min.
   B. 100 min.
   C. 1000 min.
   D. 10000 min.

Chapter 10 Assignment

Student Name: -- Student ID: -------------------

Date: -- Score: ------------------------

A: Describe how to select the Dirt Holding Capacity of a flushing filter.
B: List reasons for flushing a hydraulic system.

Answers to Chapters Reviews

Chapter 1:

1	2	3	4						
D	A	A	D						

Chapter 2:

1	2	3	4	5	6	7	8	9	10
D	B	D	C	B	A	B	C	D	C

11	12	13	14	15	16	17	18	19	20
B	D	A	D	C	A	B	C	D	D

Chapter 3:

1	2	3	4	5					
A	D	C	B	D					

Chapter 4:

1	2	3	4	5					
A	D	D	C	C					

Chapter 5:

1	2	3	4	5					
C	D	D	B	D					

Chapter 6:

1	2	3	4	5					
D	C	A	B	B					

Chapter 7:

1	2	3	4	5					
D	A	B	C	D					

Chapter 8:

1	2	3	4	5	6	7	8	9	10
A	D	B	D	D	A	C	B	C	A

Chapter 9:

1	2	3	4	5	6	7	8	9	10
D	A	D	B	A	C	D	B	A	D

Chapter 10:

1	2	3	4	5					
D	B	A	C	C					

www.ingramcontent.com/pod-product-compliance
Lightning Source LLC
Chambersburg PA
CBHW052340210326
41597CB00037B/6208

* 9 7 8 0 9 9 7 7 8 1 6 4 9 *